AF588140

International Association of Geodesy Symposia

Jeffrey T. Freymueller, Series Editor
Laura Sánchez, Series Assistant Editor

Series Editor
Jeffrey T. Freymueller
Endowed Chair for Geology of the Solid Earth
Department of Earth and Environmental Sciences
Michigan State University
East Lansing, MI, USA

Assistant Editor
Laura Sánchez
Deutsches Geodätisches Forschungsinstitut
TUM School of Engineering and Design
Technische Universität München
Munich, Germany

International Association of Geodesy Symposia

Jeffrey T. Freymueller, Series Editor
Laura Sánchez, Series Assistant Editor

Symposium 116: Global Gravity Field and Its Temporal Variations
Symposium 117: Gravity, Geoid and Marine Geodesy
Symposium 118: Advances in Positioning and Reference Frames
Symposium 119: Geodesy on the Move
Symposium 120: Towards an Integrated Global Geodetic Observation System (IGGOS)
Symposium 121: Geodesy Beyond 2000: The Challenges of the First Decade
Symposium 122: IV Hotine-Marussi Symposium on Mathematical Geodesy
Symposium 123: Gravity, Geoid and Geodynamics 2000
Symposium 124: Vertical Reference Systems
Symposium 125: Vistas for Geodesy in the New Millennium
Symposium 126: Satellite Altimetry for Geodesy, Geophysics and Oceanography
Symposium 127: V Hotine-Marussi Symposium on Mathematical Geodesy
Symposium 128: A Window on the Future of Geodesy
Symposium 129: Gravity, Geoid and Space Missions
Symposium 130: Dynamic Planet - Monitoring and Understanding ...
Symposium 131: Geodetic Deformation Monitoring: From Geophysical to Engineering Roles
Symposium 132: VI Hotine-Marussi Symposium on Theoretical and Computational Geodesy
Symposium 133: Observing our Changing Earth
Symposium 134: Geodetic Reference Frames
Symposium 135: Gravity, Geoid and Earth Observation
Symposium 136: Geodesy for Planet Earth
Symposium 137: VII Hotine-Marussi Symposium on Mathematical Geodesy
Symposium138: Reference Frames for Applications in Geosciences
Symposium 139: Earth on the Edge: Science for a Sustainable Planet
Symposium 140: The 1st International Workshop on the Quality of Geodetic Observation and Monitoring Systems (QuGOMS'11)
Symposium 141: Gravity, Geoid and Height Systems (GGHS2012)
Symposium 142: VIII Hotine-Marussi Symposium on Mathematical Geodesy
Symposium 143: Scientific Assembly of the International Association of Geodesy, 150 Years
Symposium 144: 3rd International Gravity Field Service (IGFS)
Symposium 145: International Symposium on Geodesy for Earthquake and Natural Hazards (GENAH)
Symposium 146: Reference Frames for Applications in Geosciences (REFAG2014)
Symposium 147: Earth and Environmental Sciences for Future Generations
Symposium 148: Gravity, Geoid and Height Systems 2016 (GGHS2016)
Symposium 149: Advancing Geodesy in a Changing World
Symposium 150: Fiducial Reference Measurements for Altimetry
Symposium 151: IX Hotine-Marussi Symposium on Mathematical Geodesy
Symposium 152: Beyond 100: The Next Century in Geodesy
Symposium 153: Terrestrial Gravimetry: Static and Mobile Measurements (TG-SMM 2019)
Symposium 154: Geodesy for a Sustainable Earth
Symposium 155: X Hotine-Marussi Symposium on Mathematical Geodesy
Symposium 156: Gravity, Positioning and Reference Frames
Symposium 157: Together Again for Geodesy
Symposium 158: International Symposium on Gravity, Geoid and Height Systems 2024 (GGHS2024)

International Symposium on Gravity, Geoid and Height Systems 2024 (GGHS2024)

Proceedings of the Joint IAG Commission 2, IGFS and GGOS Symposium on Gravity, Geoid, and Height Systems 2024, Thessaloniki, Greece, 4 – 6 September 2024

Edited by

Jeffrey T. Freymueller, Laura Sánchez

ISSN 0939-9585 ISSN 2197-9359 (electronic)
International Association of Geodesy Symposia
ISBN 978-3-032-22864-2 ISBN 978-3-032-22865-9 (eBook)
https://doi.org/10.1007/978-3-032-22865-9

This work was supported by International Association of Geodesy.

This Springer imprint is published by the registered company Springer Nature Switzerland AG
The registered company address is: Gewerbestrasse 11, 6330 Cham, Switzerland

Preface

The International Association of Geodesy (IAG) symposium "Gravity, Geoid and Height Systems 2024" (GGHS2024) continues the long, 25 years, history of the Joint IAG Commission 2 "Gravity Field" and International Gravity Field Service (IGFS) symposia. GGHS2024 was a joint international symposium organized by IAG Commission 2, the IGFS and the Global Geodetic Observing System (GGOS). It took place in Thessaloniki, Greece, in September 4–6, 2024, at the premises of the Aristotle University of Thessaloniki Research Dissemination Center (KEDEA). It focused on methods for observing, estimating and interpreting the Earth's gravity field and the essential role of gravity field modelling in measuring, understanding and predicting changes in the Earth system. GGHS2024 featured six sessions spanning the entire 3 days of the program, with oral, short-oral and poster sessions, and the new short-oral format received a very positive response from the majority of participants. For GGHS2024, 125 abstracts were received, out of which 58 were scheduled as oral presentations and 67 as posters. One hundred and thirty participants from 26 countries participated in the conference. It should be particularly emphasized that this symposium was also able to attract the young generation of scientists, since 48 participants (~38% of the total number) denoted that they are Early Career Scientists (ECS). The scientific program of GGHS2024 was of outstanding quality and showed significant scientific advancements in several fields of gravity field research, which are briefly summarized below:

Participants of the GGHS2024 Joint Commission 2, IGFS and GGOS Symposium, Thessaloniki, Greece, September 4–6, 2024

ECS participants during the GGHS2024 Joint Commission 2, IGFS and GGOS Symposium, Thessaloniki, Greece, September 4–6, 2024

- Session 1: Reference systems and frames in physical geodesy
  (Chairs: Riccardo Barzaghi, Laura Sánchez, Hartmut Wziontek)
- Session 2 (Co-organized with the IAG QuGe Project): Novel technologies in terrestrial, airborne and satellite gravity field determination
  (Chairs: Jürgen Müller, Derek van Westrum, Sylvain Bonvalot)
- Session 3: Static and time-variable global gravity field modelling
  (Chairs: Srinivas Bettadpur, Roland Pail, Adrian Jäggi)
- Session 4: Regional gravity field modelling and geophysical interpretation
  (Chairs: Pavel Novák, Mirko Reguzzoni, Ismael Foroughi)
- Session 5: Gravity for climate, atmosphere, ocean and natural hazard research
  (Chairs: Annette Eicker, Carmen Blackwood, Rebecca McGirr)
- Session 6: Data management, dissemination of results and networking of stakeholders
  (Chairs: Sinem Ince, Daniela Carrión, Martin Sehnal)

This proceedings volume includes a selection of 23 papers presented at GGHS2024. All papers have been peer-reviewed, and we would like to acknowledge the contribution of the associate editors and reviewers. Their support is greatly appreciated.

Thessaloniki, Greece Georgios S. Vergos

Contents

Part III Gravity for Climate and Geohazards

Part IV Methods

Part I

Novel Technologies in Terrestrial, Airborne and Satellite Gravity Field Determination

Future Satellite Gravimetry with a Network of Miniaturized Satellites

N. Pfaffenzeller, R. Pail, J. Jensen, and K. Schilling

Abstract

This feasibility study evaluates the potential of a CubeSat-based gravimetry mission for monitoring Earth's gravity field, focusing on scientific benefits and technical implementation. Scientific requirements are derived from user needs, and suitable miniaturized instruments for accelerometry, GNSS, and inter-satellite ranging are assessed. Various CubeSat constellations and formations are explored to mitigate temporal aliasing, the primary error in current gravity missions, and in this context, adapted gravity field processing schemes are also applied. Gravity field performance is analyzed based on satellite configuration, instrumental noise, and orbital parameters. We identify the configuration consisting of four polar and four inclined satellite pairs distributed across several orbital planes as the most promising candidate. The technical feasibility prioritizes a platform that balances scientific precision with CubeSat constraints and cost limitations. Key challenges include the miniaturization of scientific payloads – requiring a 35% reduction in instrument size – and securing suitable launch opportunities. Despite these challenges, results indicate that a CubeSat constellation could improve temporal gravity field retrieval and enhance climate process monitoring. Further research is recommended to refine the mission concept.

Keywords

Future satellite mission · Gravity field · Satellite constellations · Technological demonstrator

1 Introduction

Satellite gravimetry provides a unique tool to monitor global mass change processes in the Earth system in a limited time span. Currently, data products are provided on a monthly basis by the processing centers of the GRACE (Tapley et al. 2004) or the GRACE-FO (Kornfeld et al. 2019; Landerer et al. 2020) missions. Based on the successful heritage of the previously mentioned satellite pair missions in in-line formation, GRACE and GRACE-FO, and the single-satellite missions of CHAMP (Reigber et al. 1999) and GOCE (Drinkwater et al. 2003), new concepts are elaborated. This necessity arises because current satellite gravity missions face intrinsic limitations in spatiotemporal resolution. They hinder the accurate retrieval of high-frequency signals originating from oceanic and atmospheric mass variations, as well as ocean tides, which tend to alias into the gravity field solutions (Flechtner et al. 2016). Consequently, new mission architectures and observation strategies are required to overcome these challenges and improve the resolution and accuracy of temporal gravity field measurements. The double satellite pair concept (Bender et al. 2008) consists of one in-line polar and one in-line inclined pair and is currently investigated within the "Mass Change and Geosciences International Constellation" (MAGIC) study (Haag-

N. Pfaffenzeller (✉) · R. Pail
Chair of Astronomical and Physical Geodesy, TUM School of Engineering and Design, Technical University of Munich, Munich, Germany
e-mail: nikolas.pfaffenzeller@tum.de

J. Jensen · K. Schilling
Zentrum für Telematik e.V., Würzburg, Germany

J. T. Freymueller, L. Sànchez (eds.), *International Symposium on Gravity, Geoid and Height Systems 2024 (GGHS2024)*,
International Association of Geodesy Symposia 158, https://doi.org/10.1007/1345_2025_288

Table 1 Summary of simulation aspects. The section number is indicated in brackets

Orbit	Satellite Network Design (2.3), Inter-Satellite Distance and Altitude Variations (4.1.1), Orbit Design and Mission Lifetime Analysis Based on Solar Activity Variations (4.2.1)
Sensors	Accelerometer (2.2.1), Inter-satellite Ranging (2.2.2), GNSS Receiver (2.2.3)
Parametrization approach	Nominal (4.1), Co-parametrization of Short-Term Gravity Solutions and of Ocean Tide Constituents
Satellite system	System Overview and Budgets (4.2.2), Preliminary Satellite Design (4.2.3), Design Constraints (5)

mans and Tsaoussi 2020). Other mission concepts covering various satellite formations (e.g., Pendulum, Cartwheel) were investigated by Elsaka et al. (2014), but are limited in their technical feasibility. The realization of the double-pair mission is very cost intensive with large conventional satellites, as used in the GRACE/GRACE-FO mission. With the completed 3-year DFG (German Research Foundation)-funded project "Cube-satellite networks for geodetic Earth observation on the example of gravity field retrieval" (Cube-Grav), we investigated the potential of a network of small satellites for gravity field retrieval. In this manuscript, the science and mission requirements and the related miniaturized payload are addressed in Sect. 2. In Sect. 3, various numerical simulation setups are presented, and results are shown in Sect. 4, covering the scientific outcomes in Sect. 4.1 and the technical findings in Sect. 4.2. In Sect. 5, these main results are discussed, and in Sect. 6, conclusion and recommendations are provided. Table 1 summarizes the parameters investigated within this study.

2 Requirements

2.1 User Needs

In the context of the International Union of Geodesy and Geophysics (IUGG) in 2015, the user requirements for mass change monitoring from space were consolidated within an international initiative (Pail et al. 2015). A similar approach was adopted for the ongoing MAGIC study (Haagmans and Tsaoussi 2020) and its inclined pair NGGM (Daras 2023). The Earth observation user requirements for a future CubeSat mission constellation are expected to align closely with those established by the IUGG and NGGM/MAGIC, as they were derived independently of specific satellite mission architectures and available instrumentation. As a baseline, the CubeSat constellation should, at minimum, satisfy the threshold requirements defined in the IUGG study and the NGGM/MAGIC MRDs.

2.2 Instruments and Candidate Sensors

As the basis for a future network of small satellites, we rely on the established observation concepts in the form of high-low satellite-to-satellite tracking (hl-sst) and low-low satellite-to-satellite tracking (ll-sst) adopted by GRACE/GRACE-FO. Thus, as key payload, we assume accelerometers mounted at the center of mass of all participating satellites, a ranging device to measure the inter-satellite distance between two satellites forming a satellite pair (ll-sst observations), and a GNSS receiver for hl-sst observations. We do not consider any information on the attitude system in our simulations, which is a necessary task for future work, in particular in regard to inter-satellite ranging measurements. We consider multiple satellites in different configurations, but always in terms of satellite pairs. The selected instrument candidates and their specifications are evaluated in terms of size (1 U = 10 cm^3), weight, and power (SWaP) and technological readiness level (TRL).

2.2.1 Accelerometer

Three accelerometer candidates have been assessed for a miniaturized satellite gravimetry mission. The first, CubSTAR, is a miniaturized electrostatic accelerometer developed by Onera (Liorzou et al. 2023), with noise performances varying from 50 to 300 μm, depending on the gap between the proof mass and the electrodes. CubSTAR has a size of 4 U, weighs 5 kg, has a power demand of 4.5 W, and has a TRL of 6 (Rodrigues et al. 2022). The second, the Simplified Gravitational Reference Sensor (SGRS), builds on the heritage of the Gravitational Reference Sensor from the LISA Pathfinder mission and is being developed under the leadership of the University of Florida (Dávila Álvarez et al. 2022). However, with a weight of 13 kg, a power demand of 20 W, a size of 10^4 cm^3, and a TRL of 5, SGRS is a rather large solution for a miniaturized mission. The third candidate, the optomechanical accelerometer, incorporates a mechanical resonator with a laser interferometric readout and is the most compact accelerometer with 2.18 U in size, a weight of 2 kg, a power demand of 10 W (Hines et al. 2022), and a TRL of 5 (Nelson et al. 2024). All three accelerometers demonstrate promising capabilities. However, based on all the key parameters from the previously mentioned instruments, the optomechanical accelerometer emerges as the most suitable candidate.

2.2.2 Inter-Satellite Ranging Instrument

Three potential developments for an inter-satellite ranging instrument are identified and evaluated. The CubeSat Laser

Infrared CrosslinK (CLICK) mission aims to investigate inter-satellite communication links for CubeSats and serves as a technological demonstrator (Tomio et al. 2022). However, its expected accuracy of approximately 50 cm is not suitable for precise ranging applications and is therefore excluded from further consideration. Another candidate is the Micro Non-Planar Ring Oscillator (μNPRO), proposed for miniaturized ranging in satellite gravimetry (Wiese et al. 2022). This sensor, developed by NASA/JPL and based on the LISA Pathfinder mission, remains largely undocumented in publicly available sources. Due to the lack of accessible technical specifications, this option is also dismissed. The most promising candidate is the Dynamic Optical Ranging and Timing System (DORT), a short-pulse laser system designed for absolute ranging measurements in satellite gravimetry and other applications (Paul et al. 2025). Developed by MUnique Technology, DORT offers an expected accuracy of at least 1 μm and has a size of 3 U, a power demand of 10 W, and a weight of 1.5 kg. However, as noise specifications for the DORT system were not available during the CubeGrav project, we assume that the noise performance of the GRACE-FO microwave ranging instrument (MWI) is representative of a future ranging instrument such as DORT.

Therefore, a baseline accuracy of 1 μm is considered the minimum requirement for inter-satellite ranging. A substudy performed within the CubeGrav project investigated various instruments for accelerometry and ranging, revealing that if the MWI performance were degraded to 10 or 100 μm, the resulting gravity field solutions would be dominated by increased ranging noise rather than temporal aliasing (Pfaffenzeller and Pail 2021).

2.2.3 GNSS Instrument

A suitable GNSS sensor is the ZED-F9P developed by the company u-blox. It is a dual-frequency multi-GNSS receiver with an expected accuracy in the decimeter range (Moeller et al. 2024). The ZED-F9P has a size of 10^2 cm^3, weighs 0.065 kg, has a power demand of 0.5 W, and has a TRL of 9.

2.2.4 Instrument Performance Evaluation

The noise performance of various candidate accelerometer and inter-satellite ranging instruments is presented in terms of amplitude spectral density (ASD) in Fig. 1, with the corresponding analytical descriptions provided in Eqs. (1)–(7). Numerical specifications are available for the accelerometers used in MAGIC (a_{MAGIC}) (Lenoir et al. 2011), and the simplified gravitational reference sensor (a_{SGRS}) (SGRS) (Dávila Álvarez et al. 2022), as well as for the microwave ranging instrument (MWI) (r_{MWI}) (Kornfeld et al. 2019) and the laser ranging interferometer (LRI) (r_{LRI}) (Nicklaus et al. 2020). The numerical approximations for the CubSTAR accelerometer (50 μm: a_{C50} and 300 μm: a_{C300}) are derived from Liorzou et al. (2023), while those for the optomechanical (optom) accelerometer (a_{optom}) are based on the work of Hines et al. (2020). Additionally, we assume different orbit position errors achievable by a future GNSS sensor, covering three cases: 1, 30, and 100 cm. In the following, the noise specifications of the abovementioned instruments are defined:

$$a_{\mathrm{MAGIC}} = 1\cdot 10^{-11} \sqrt{\frac{\left(\frac{10^{-3}\,\mathrm{Hz}}{f}\right)^2}{\left[\left(\frac{10^{-5}\,\mathrm{Hz}}{f}\right)^2 + 1\right]} + 1 + \left(\frac{f}{10^{-1}\,\mathrm{Hz}}\right)^4}\ \frac{\mathrm{m}}{\mathrm{s}^2\sqrt{\mathrm{Hz}}} \tag{1}$$

$$a_{\mathrm{C50}} = 1.58\cdot 10^{-11}\cdot \sqrt{\left(\frac{2.2\cdot 10^{-2}\mathrm{Hz}}{0.9\cdot f}\right)^{\frac{5}{4}} + 695 + \left(\frac{4.8\cdot f}{1.5\,\mathrm{Hz}}\right)^{4.3}}\ \frac{\mathrm{m}}{\mathrm{s}^2\sqrt{\mathrm{Hz}}} \tag{2}$$

$$a_{\mathrm{C300}} = 1.58\cdot 10^{-11}\cdot \sqrt{\left(\frac{2.2\cdot 10^{-2}\mathrm{Hz}}{0.9\cdot f}\right)^{1.2} + 1 + \left(\frac{4.5\cdot f}{9.9\cdot 10^{-1}\mathrm{Hz}}\right)^{4.42}}\ \frac{\mathrm{m}}{\mathrm{s}^2\sqrt{\mathrm{Hz}}} \tag{3}$$

$$a_{\mathrm{SGRS}} = 5\cdot 10^{-13}\cdot \sqrt{1 + \left(\frac{1}{f}\right)^{\frac{2}{3}}}\ \frac{\mathrm{m}}{\mathrm{s}^2\sqrt{\mathrm{Hz}}} \tag{4}$$

$$a_{\mathrm{optom}} = 3\cdot 10^{-11}\cdot \sqrt{\frac{0.1\,\mathrm{Hz}}{f}}\ \frac{\mathrm{m}}{\mathrm{s}^2\sqrt{\mathrm{Hz}}} \tag{5}$$

$$r_{\mathrm{MWI}} = 2.62\cdot (2\pi f)^2\cdot 10^{-6}\cdot \sqrt{1 + \left(\frac{0.003\,\mathrm{Hz}}{f}\right)^2}\ \frac{\mathrm{m}}{\mathrm{s}^2\sqrt{\mathrm{Hz}}} \tag{6}$$

$$r_{\mathrm{LRI}} = 2\cdot (2\pi f)^2\cdot 10^{-8}\cdot \sqrt{1 + \left(\frac{10^{-2}\,\mathrm{Hz}}{f}\right)^2}\ \frac{\mathrm{m}}{\mathrm{s}^2\sqrt{\mathrm{Hz}}} \tag{7}$$

2.3 Satellite Network Design

In the following, we focus on the satellite network design. We distinguish between satellite constellations and formations. Formation flying involves satellites that are dynamically linked through a shared control law, which governs at least one state variable of each satellite in relation to the others (Bandyopadhyay et al. 2016). Consequently, any movement of one satellite within the formation directly affects the others. In contrast, if this dynamic coupling is absent, the system is classified as a constellation. Notably, a satellite constellation can include multiple satellite formations within its structure. As a foundation for our study, we adopt the in-line polar and in-line inclined satellite pair (70° inclination),

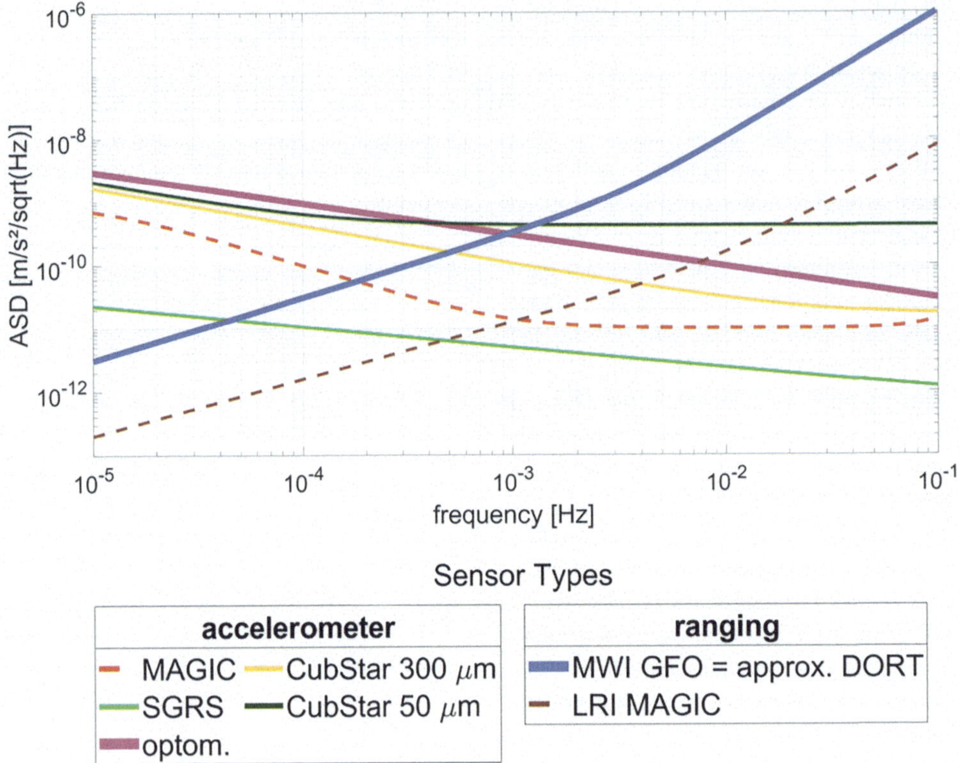

Fig. 1 Amplitude spectral density (ASD) of the accelerometer and the ranging candidates from Sects. 2.2.1 and 2.2.2. In addition, the performance of the accelerometer from MAGIC and the performance of the laser ranging interferometer (LRI) are visualized in dashed lines

as outlined in the MAGIC study, considering it as a state-of-the-art constellation. Specifically, we focus on these two orbital planes (polar: 468 km altitude, inclined: 432 km altitude) selected as one of seven orbit sets, which were investigated in detail within the MAGIC study (Table 3 in Massotti et al. 2021). Extended satellite constellations are achieved by adding satellites on additional planes with the same altitude and inclination but rotated by their right ascension of the ascending node (RAAN) to enhance spatial resolution and adjusting the mean anomaly to improve temporal sampling. The resulting orbital configurations are presented in the work of Pfaffenzeller and Pail (2023), which include four constellations with polar-only satellite pairs and three constellations combining polar and inclined satellite pairs. Exemplarily, we select the GRACE-like formation of 1 polar pair (1p) and the constellations of a MAGIC-like double-pair 1 polar + 1 inclined pair (1p1i), 18 satellite pairs distributed on 9 orbital planes (18pRAAN), and 4 polar pairs + 4 inclined pairs, each distributed on 2 orbital planes (4p4iRAAN). Furthermore, orbital planes with varying inclinations are considered in the next design. Specifically, we rely on a 6-satellite-pair constellation called 1p5i, extending the double constellation, with 4 satellite pairs on the orbital planes with an altitude of 432 km each and inclination of 80°, 60°, 48°, and 33°. Additionally, an alternative constellation design, deviating completely from the MAGIC configuration, is based on the orbital parameters of the International Space Station (ISS), with an inclination of 51.6° and an altitude of 385 km, and a sun-synchronous orbit (SSO) with an inclination of 97° and altitude of 480 km (see Sect. 4.2.1). We consider four satellite pairs on each of the previously mentioned orbital planes (4SSO4ISS).

The satellite formation is considered not only in an in-line configuration, where a single link exists between two adjacent satellites forming a pair, but also in a pearl-string formation. In this formation, each satellite has an inter-satellite link with all other satellites in the same orbital plane, resulting in multiple inter-satellite links. A schematic visualization of this configuration is provided in the work of Pfaffenzeller and Pail (2023). Additionally, the study investigates the cross-track formation concept. The inter-satellite ranging is done in a cross-track direction, perpendicular to the satellite's movement, similar to the formation described by Zingerle et al. (2024). The combination of multiple satellite pairs in in-line formation with some in cross-track formation is referred to as the mixed formation setup. The inter-satellite distance between adjacent satellites forming a satellite pair is set to 220 km.

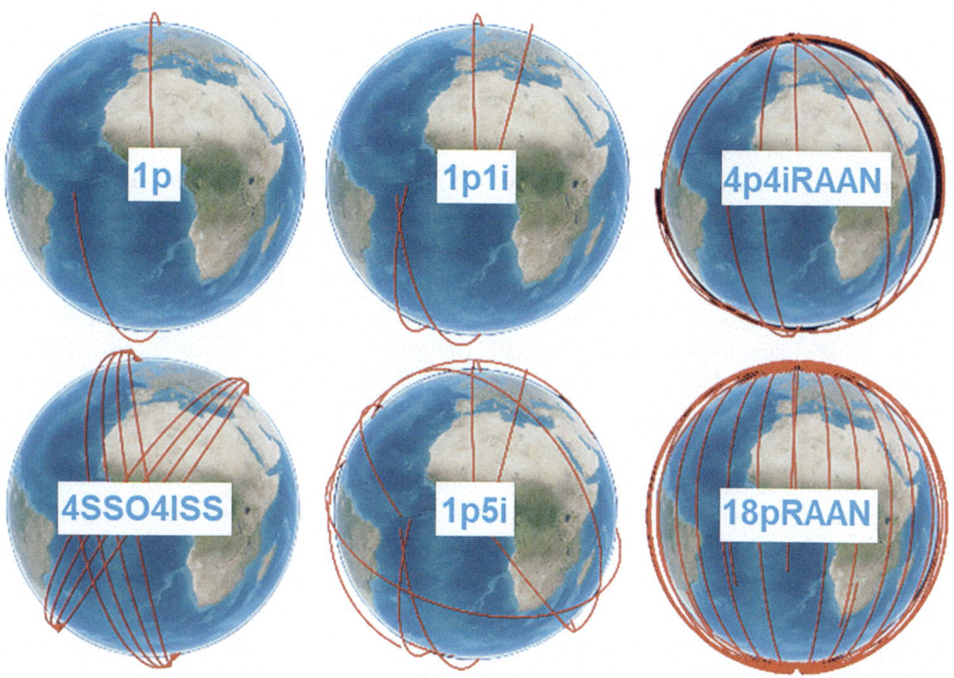

Fig. 2 Visualization of the first revolution of the following satellite configurations: 1 polar pair (1p), 1 polar pair + 1 inclined pair (1p1i), 4 polar + 4 inclined pairs each distributed on 2 planes (4p4iRAAN), 4 pairs on 1 orbital plane of SSO + 4 inclined pairs on 1 orbital plane of ISS (4SSO4ISS), 1 polar + 5 different inclined pairs (1p5i), 18 polar pairs distributed on 9 orbital planes (18pRAAN)

In Fig. 2, the first revolution of the selected satellite configurations is visualized, which is analyzed regarding its gravity field retrieval performance in Sect. 4.1.

3 Simulation Environment

For numerical performance evaluation, we use two simulation tools: a reduced-scale and a full-scale simulator, both available at IAPG, TUM. The reduced-scale simulator, computationally efficient due to its linear observation equation, models satellite orbits as circular using Keplerian elements and processes only ll-sst observations. It derives acceleration differences along the line of sight from satellite positions (Murböck 2015; Murböck et al. 2015). The full-scale simulator, in contrast, employs fully numerically integrated orbits and processes both hl-sst and ll-sst observations. Gravity field processing follows a modified integral equation approach with short-arc orbit representation (Pfaffenzeller and Pail 2023). Simulated gravity field observations are computed based on the static gravity field model GOCO05S (Mayer-Gürr et al. 2015). Additionally, contributions from temporal gravity field variations are considered using the Earth system model (ESM) HIS (Dobslaw et al. 2014), which accounts for signals from hydrology, ice, and solid Earth, as well as an error product for high-frequency signals from the atmosphere and ocean (AO error) (Dobslaw et al. 2016). Uncertainties in ocean tide models (OT error) are simulated by calculating the difference between two ocean tide models: the Empirical Ocean Tide (EOT) model 11a (Savcenko and Bosch 2012) and the Goddard Ocean Tide (GOT) model 4.7 (Ray 1999). The lifetime and control analysis were conducted using ESA's Debris Risk Assessment and Mitigation Analysis (DRAMA) tool (Braun et al. 2020) and Ansys Systems Tool Kit (Release 12.7).

4 Results

4.1 Scientific Evaluation

The simulation results are expressed in degree amplitudes of equivalent water height (EWH; Wahr et al. 1998):

$$\sigma_n\,(\text{EWH}) = \frac{a\rho_\text{E}}{3\rho_\text{W}}\frac{2n+1}{1+k_n}\sqrt{\sum_{m=0}^{n}\left(\overline{C}^2_{nm}+\overline{S}^2_{nm}\right)} \quad (8)$$

In this equation, C_{nm} and S_{nm} are the estimated spherical harmonic coefficients of degree n and order m (d/o). The parameter a represents the Earth's semimajor axis, while ρ_E and ρ_W denote the average densities of the Earth and water, respectively. Additionally, k_n corresponds to the Load Love number for degree n. Subsequent figures illustrate the temporal gravity field signal (HIS) to be recovered, along with the gravity field retrieval error (i.e., noise) of various constellations. The performance of the satellite formations and constellations, described in Sect. 2.3, is evaluated in terms of monthly gravity field retrieval with nominal processing (see Sect. 4.1.3). Selected results are also summarized by Pfaffenzeller and Pail (2023). The pearl-string formation does not provide the expected improvement in spatiotemporal performance. While multi-inter-satellite links increase redundancy, the global and temporal coverage remains largely unaffected. Increasing the number of satellite pairs in the state-of-the-art double-pair constellations yields significant benefits for the resulting gravity field: in the low spherical harmonic (SH) spectrum, by distributing satellites across multiple orbital planes (via shifts in RAAN), and in the high SH spectrum, by distributing satellites within a single orbital plane (via shifts in mean anomaly). This distribution has minimal impact on monthly gravity field estimation, but is critical for short-term solutions and ocean tide co-estimation (see Sect. 4.1.3). For the selected configurations shown in Fig. 2, the retrieved monthly gravity fields are illustrated in Fig. 3, with corresponding global RMS values provided. The 1p formation exhibits the typical vertical striping pattern seen in GRACE-like missions (Fig. 3a). This pattern is significantly reduced with the MAGIC-like double-pair constellation (1p1i) shown in Fig. 3b. The 4SSO4ISS constellation (d), with pairs on an ISS-like orbit at a low altitude, results in low residuals. However, due to the lack of pole coverage from satellites in SSO orbit, large residuals are observed in high-latitude regions. As a result, this constellation does not support the extension of the time series over polar ice sheets but may be preferable from a technical perspective (cf. Sect. 4.2.1).

In the 4p4iRAAN configuration (c), the residuals show a less distinct pattern. As expected, larger residuals are observed over midlatitude regions, due to the higher altitude of the inclined pairs and the RAAN shift compared to the 4ISS4SSO configuration (d). However, this constellation improves the global RMS by a factor of four compared to the previous one.

The 1p5i constellation (Fig. 3e) provides a broad variety of multidirectional observations with five different inclined pairs. However, this configuration has a disadvantage for global gravity field retrieval: with lower inclination, fewer parts of the Earth are covered, resulting in fewer observations and larger residuals, which increase the RMS value. Finally, the 18pRAAN constellation is presented with three formation concepts: in-line (Fig. 3f), cross-track (Fig. 3g), and mixed (Fig. 3h). In the in-line and cross-track formations, observations are restricted to North-South and East-West directions, respectively, leading to residual patterns aligned with the vertical/horizontal directions. The cross-track formation performs significantly worse than the in-line formation, as indicated by a much more distinct residual pattern and a sevenfold increase in RMS. Combining both formations into a mixed configuration combines the benefits of both setups. With multidirectional observations, temporal aliasing is significantly reduced, leading to the lowest RMS.

4.1.1 Inter-Satellite Distance and Altitude Variations

In this section, we examine the impact of variations in altitude and inter-satellite distances based on the satellite configurations selected in Sect. 2.3. Altitude variation is closely related to mission lifetime (cf. Sect. 4.2.1). At lower altitudes, higher atmospheric drag results in reduced mission lifetime. Conversely, increasing the altitude leads to greater attenuation of the gravity field signal, particularly affecting the higher SH spectrum. Although changes in the signal-to-noise ratio (SNR) with altitude are observed, altitude is not considered a primary performance driver for the mission, because temporal aliasing remains the main error contributor.

The inter-satellite distance is directly linked to the signal-to-noise ratio of the ranging instrument and its ability to measure long distances accurately. We assess the impact of inter-satellite distances ranging from 1 to 300 km on gravity field solutions derived from in-line satellite pair formations. The results indicate that, in general, longer inter-satellite distances are preferable, as shorter gravity field wavelengths cannot be accurately detected with smaller separations. In multi-satellite configurations, the ability to mitigate temporal aliasing effects significantly influences the relevance of inter-satellite distance. For constellations capable of reducing temporal aliasing, the impact of shorter inter-satellite distances is less pronounced. Specifically, for constellations composed solely of polar satellites, shorter distances lead to more significant degradation of the gravity field solution. In contrast, constellations combining polar and inclined satellite pairs exhibit less sensitivity to shorter distances. After considering all factors, we identify 25 km as the minimum required inter-satellite distance for CubeSat missions. This distance causes only a minor reduction in the overall performance while remaining achievable with miniaturized ranging technologies.

4.1.2 Instrument Performance

The performance of potential miniaturized instrument candidates is assessed with nominal processing for monthly gravity field retrieval. To this end, various instrument combinations are generated, consisting of different accelerometer sensors and an assumed future ranging instrument modeled on the basis of the noise behavior of the MWI from the GRACE-FO mission (cf. Sect. 2.2). To quantify the pure impact of the accelerometer and ranging instruments, a GNSS geolocation error of 1 cm is assumed for the hl-sst observations across all scenarios. The evaluation of instrument performance is conducted using the 4p4iRAAN satellite constellation. The results are presented in terms of equivalent water height (EWH) degree amplitudes in

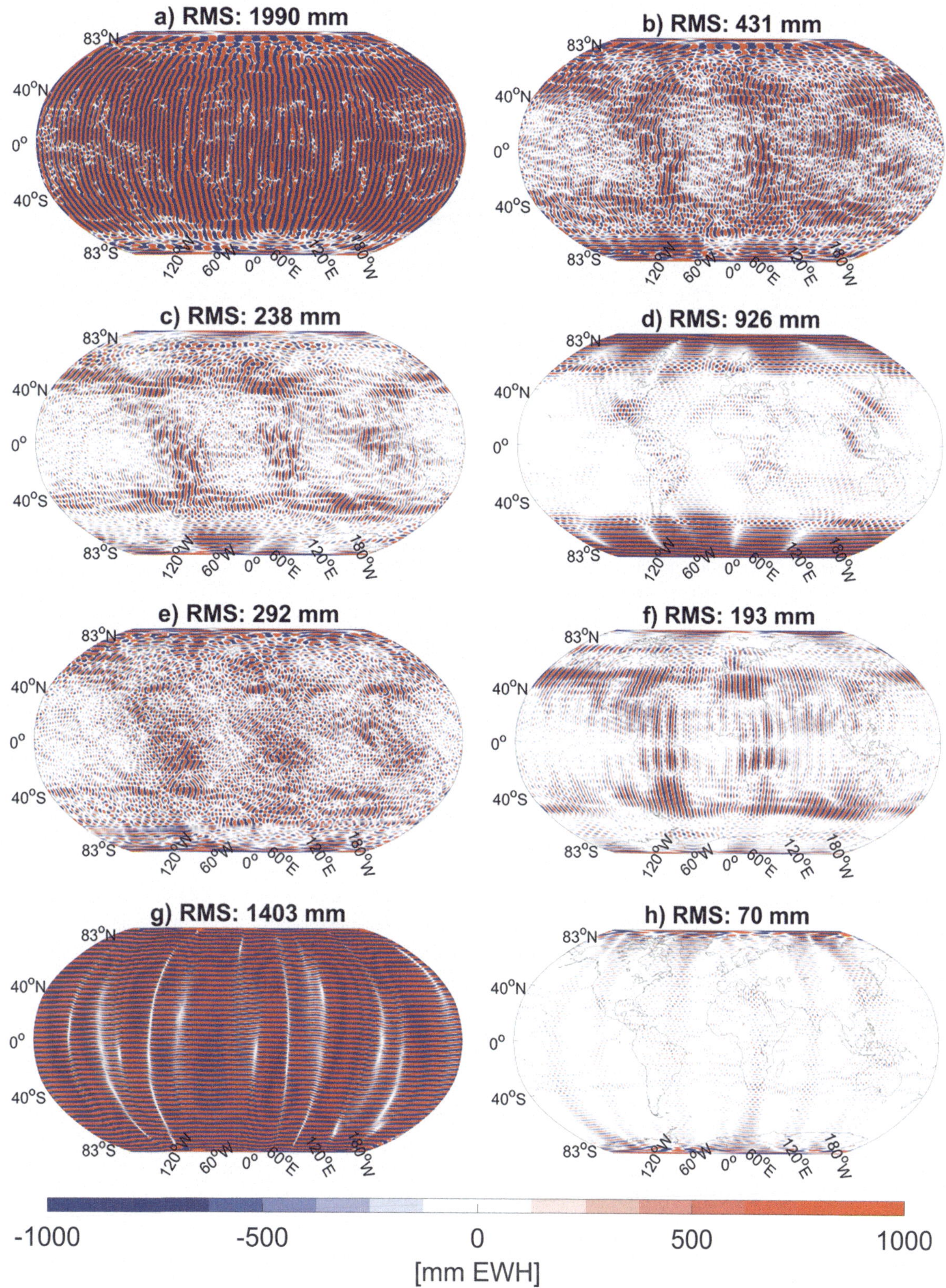

Fig. 3 Global monthly EWH gravity field residuals from simulations given in mm from d/o 3 to d/o 100. The residuals represent the difference between the retrieved gravity fields and the time-variable signal from HIS. The results are generated with the following constellations and formations (cf. Fig. 2): (**a**) 1p, (**b**) 1p1i, (**c**) 4p4iRAAN, (**d**) 4SSO4ISS, (**e**) 1p5i, (**f**) 18pRAAN in-line formation, (**g**) 18pRAAN cross-track formation, and (**h**) 18pRAAN mixed formation. The global RMS values based on EWH grids (weighted with cosine latitude) estimated from d/o 3 to 100 are indicated in the upper right corner of each configuration

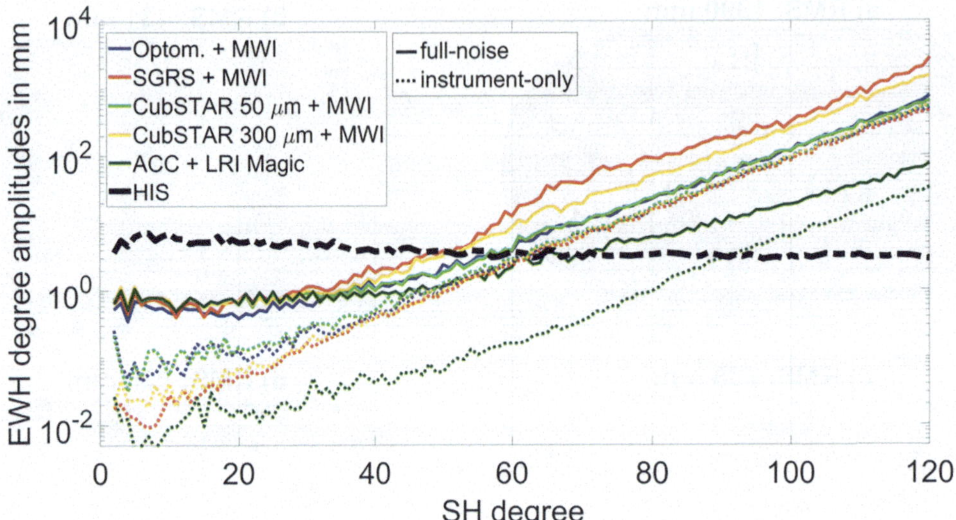

Fig. 4 EWH degree amplitudes on monthly averaged gravity field retrieval in mm. The gravity field retrieval (=noise) is visualized together with the time-variable signal HIS from January 2002 to be retrieved. The results are generated with the satellite constellation 4p4iRAAN and with the different instrument noise assumptions (cf. Sect. 2.2). The dotted lines represent the instrument-only solutions, considering only instrumental noise contributions within the simulation setup. The solid lines represent the full-noise solutions, taking into account instrumental noise and time-variable gravity field contributions (HIS + AO error + OT error)

millimeters in Fig. 4. The dotted lines in Fig. 4 represent the "instrument-only" solutions, which consider only instrumental noise contributions, while the solid lines represent the "full-noise" solutions, accounting for both instrumental noise and time-variable gravity field contributions from nontidal and tidal signals. For quantification purposes, the mean time-varying gravity signal (HIS) over 1 month is shown as a black dashed line. Results generated with the MAGIC-like accelerometer and laser ranging interferometer (LRI) are provided for comparison, depicted in dark green. The enhanced performance and lower noise levels of the MAGIC-like instruments lead to a higher SNR, demonstrated by the crossover point of the HIS signal at higher d/o compared to the miniaturized instrument candidates, for both instrument-only and full-noise scenarios. This behavior is anticipated, as the accelerometer and LRI exhibit lower noise across the entire frequency spectrum, as shown in Fig. 1.

For the instrument-only scenarios of the miniaturized instrument candidates, this finding is confirmed. Generally, instrument noise is dominated by accelerometer noise at low frequencies, impacting the low SH spectrum, and by ranging noise at higher frequencies, affecting the high SH spectrum. Changes in instrument performance, such as a comparison between the lower noise of the SGRS accelerometer and the higher noise of the optomechanical accelerometer, lead to corresponding variations in behavior. However, different behavior is observed in the full-noise scenarios for the miniaturized instrument combinations. Despite the SGRS and CubSTAR with a 300 µm gap exhibiting better noise performance than the optomechanical accelerometer and the CubSTAR with a 50 µm gap, the former candidates yield a lower SNR. This discrepancy is attributed to the weighting scheme applied within the least squares adjustment, which also accounts for temporal gravity field contributions. Notably, instrument noise is not the dominant factor; temporal aliasing exceeds the influence of instrument noise, as clearly indicated in Fig. 4. The ability to mitigate temporal aliasing depends on the selected satellite constellation and formation, as discussed in Sect. 4.1 and by Pfaffenzeller and Pail (2023).

So far, we have assumed a GNSS geolocation error of 1 cm. Increasing this error to 30 cm results in the degradation of the coefficients in the low-degree SH spectrum, while the remaining SH coefficients remain largely unaffected. However, when considering a severe GNSS geolocation error of 100 cm, the impact extends across the entire SH spectrum, leading to a lower SNR. This deterioration arises due to the increased weighting of hl-SST observations relative to ll-SST observations. Given that the GNSS candidate sensor discussed in Sect. 2.2.3 provides accuracy within the decimeter range, its influence on gravity field solutions is less significant compared to that of the accelerometer and ranging instruments. Consequently, GNSS geolocation plays a subordinate role for temporal gravity field retrieval.

4.1.3 Parametrization Scheme

In this section, we review the parametrization scheme and its impact on gravity field solutions. In the case of nominal processing, we retrieve the SH coefficients of gravity fields on a monthly basis. In our analysis, we apply an extended parametrization scheme to model high-frequency gravity signals leading in parallel to mitigate temporal aliasing effects arising from nontidal and tidal contributions (cf. Sect. 1).

Co-parametrization of Short-Term Solutions

The selected parameterization approach enables the simultaneous estimation of both short-term and long-term gravity field solutions, achieving varying spatial resolutions. Lower SH coefficients (e.g., up to d/o 30) are estimated for short-term intervals (e.g., 24, 12, or 6 h), while higher SH coefficients are estimated on a monthly basis. This methodology, which divides the results into short- and long-term components, is based on the work of Wiese et al. (2011) and has been applied to multi-satellite configurations in the study of Pfaffenzeller and Pail (2023). The findings from the latter study demonstrate that this parameterization approach allows for the retrieval of standalone gravity fields with temporal resolutions of 24, 12, and 6 h, depending on the selected satellite constellation. A key requirement for the satellite configuration is that it must provide adequate ground track coverage within the chosen short-term time frame. As shown by Pfaffenzeller and Pail (2023), this requirement can be met through global distribution of satellite pairs, such as the 18 satellite pairs distributed across 9 polar orbital planes (18pRAAN) or the 4 polar and 4 inclined satellite pairs distributed across 2 orbital planes (4p4iRAAN).

Co-parametrization of Ocean Tide Constituents

The temporal aliasing of tidal signals can be mitigated using an alternative parametrization approach involving a two-step procedure. In the first step, signals from major ocean tide constituents are co-estimated alongside the regular gravity field solutions over various timescales, such as periods spanning several months to years. The resulting enhanced ocean tide de-aliasing product is then applied to the nominal gravity field processing. As demonstrated by Hauk and Pail (2018), this approach improves the final gravity field's SH coefficients, particularly for double-pair constellations. Existing ocean tide products suffer from insufficient observations and data quality, particularly in high-latitude regions. Multi-satellite constellations, including multiple pairs on polar orbits, provide valuable observations that help reduce these aliasing effects. The key to improving gravity field solutions lies not only in the selection of the satellite constellation but also in the duration of the co-estimation period for ocean tide signals. The added benefit of a multi-satellite network is that it enables the use of shorter time periods (e.g., several months) to produce comparable or even superior results to the state-of-the-art double-pair constellation, which typically requires years of data. A summary of these findings is presented by Pfaffenzeller and Pail (2024).

Based on the results presented above, we identify the constellation 4p4iRAAN as the most promising candidate in terms of scientific performance and select this satellite constellation for further analysis regarding its technical feasibility in Sect. 4.2.

4.2 Technical Evaluation

4.2.1 Orbit Design

CubeSats are commonly deployed as secondary payloads or via rideshare programs, which offer a cost-effective means of launching small satellites. These opportunities are available through various missions, including SSO missions in LEO, resupply missions to the ISS, and SpaceX Starlink launches. These missions typically operate at an inclination of approximately 45°. However, CubeSats often lack access to polar orbits or orbits with inclinations around 70°, necessitating dedicated rocket launches for these specialized orbital paths. Launching small payloads on dedicated rockets typically incurs costs starting at approximately 4 million Euros, whereas rideshare missions are less expensive, with costs ranging between 250,000 and 400,000 Euros per satellite. For the 4p4iRAAN satellite constellation, achieving the required orbital configurations with four dedicated rocket launches would incur a minimum cost of 16 million Euros. In contrast, launching 16 satellites via rideshare in suboptimal orbits would cost approximately 6.4 million Euros. However, modifying the orbital planes postlaunch is unfeasible due to the significant additional energy demands (delta-v), which exceed the capabilities of small satellites, even those equipped with electric propulsion. The delta v budget quantifies the total change in velocity required for a space mission, serving as a critical indicator of the propellant needed for various maneuvers and thus reflecting the mission's overall energy requirements (Walter 2008). This limitation results in two primary options: utilizing an additional orbital transfer vehicle to facilitate orbital adjustments, which would introduce substantial additional costs, or adopting a constellation to the SSO/ISS orbital planes (cf. 4SSO4ISS). Unfortunately, the SSO/ISS configurations fail to provide adequate coverage of polar regions (cf. Fig. 3d), which are critical for monitoring glaciers and the cryosphere being key components of climate observation. Moreover, the ISS orbital inclination falls outside the optimal range of 65–75°, as indicated in studies such as those of Wiese et al. (2012) and Heller-Kaikov et al. (2023). As a result, extensive coverage of the polar regions can only be achieved through satellite pairs operating in polar orbits. Despite these limitations, the SSO/ISS constellation may still offer a cost-effective solution for technology demonstration missions. For the CubeGrav mission, we assume an operational mission lifetime of 5 years that we orientate on the nominal lifetime of the existing satellite mission GRACE/-FO. One significant factor influencing its longevity is drag disturbance

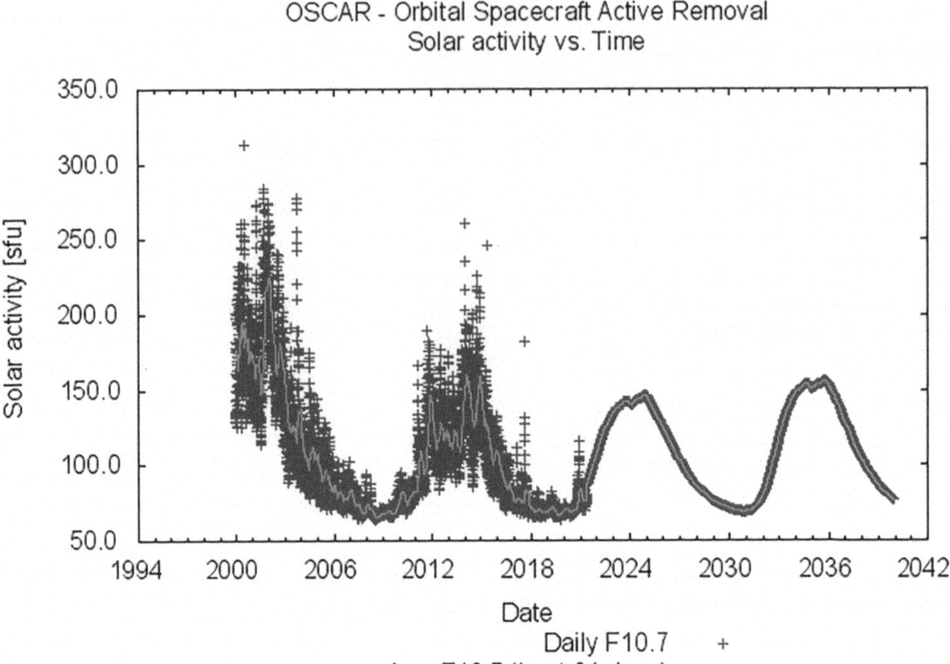

Fig. 5 Solar activity values depending on the years. The values are provided from the year 2000 to the year 2040, including measurements from 2000 to 2022 and estimations from 2022 to 2040

acting on the satellite. Consequently, analyzing solar activity is critical for determining the mission's lifetime. Figure 5 illustrates the solar magnetic activity cycle from the year 2000, with predictions extending to the year 2040.

The orbital lifetime is analyzed depending on the launch of the satellites for two solar activity cases (Fig. 6). For the mission start in the year 2000 during periods of high solar activity, a polar orbit with the selected initial altitude of 468 km of the 4p4iRAAN constellation lasts approximately 2 years. In contrast, during low solar activity, such as in 2006, the mission lifetime of a polar orbit extends to nearly 8 years. Thus, to minimize the demand on the delta-v budget for altitude correction maneuvers, it is advantageous to launch during a phase of low solar activity. Altitude corrections are necessary to maintain a 5-year lifetime during periods of high solar activity. Specifically, the satellite must remain within a 15 km margin for the first 3 years, requiring eight station-keeping maneuvers with an average impulse of 91 Ns each, amounting to a total of 735 Ns. The timing of the launch significantly impacts the orbital lifetime. To avoid the peak solar activity phase, the launch should occur either before mid-2030 or after 2035. For instance, a launch in June 2030 allows for a 5-year mission without the need for altitude correction maneuvers, whether in a polar or inclined orbit.

The initial satellite configuration is assumed to be established at launch, with the entire delta-v budget allocated to maintain the relative distances between satellite pairs. Two practical orbital maneuvers for a single satellite in a pair are simulated. The position in the local-vertical-local-horizontal coordinate frame for both scenarios reflects the relative deviation between the target position in the two-pair constellation and the satellite's actual location, representing the error that the satellite must correct. Maneuver costs are computed using a model predictive control (MPC) method, which employs a cost function designed to minimize thrust consumption while modeling the thrust vector for a chemical propulsion system. In the first scenario, the cross-track separation is 46.40 m, and the error is corrected within approximately two orbits. The total impulse required for this maneuver is 0.72 Ns. In the second scenario, a corrective maneuver addresses the along-track drift resulting from orbital altitude differences. The initial offsets to be corrected are 24.95 m in the radial direction and 999.42 m in the along-track direction. Using the MPC approach, these offsets are rectified over the course of three orbits, with a total impulse expenditure of 0.83 Ns. For a thruster with a total impulse capacity of 400 Ns, up to 480 in-plane maneuvers or 550 cross-track maneuvers could be performed. This would allow for approximately one maneuver per satellite every 5 days over a 5-year mission duration.

4.2.2 System Overview and Budgets

An overview of the satellite system and the budgets is provided in this section. The mass and volume specifications

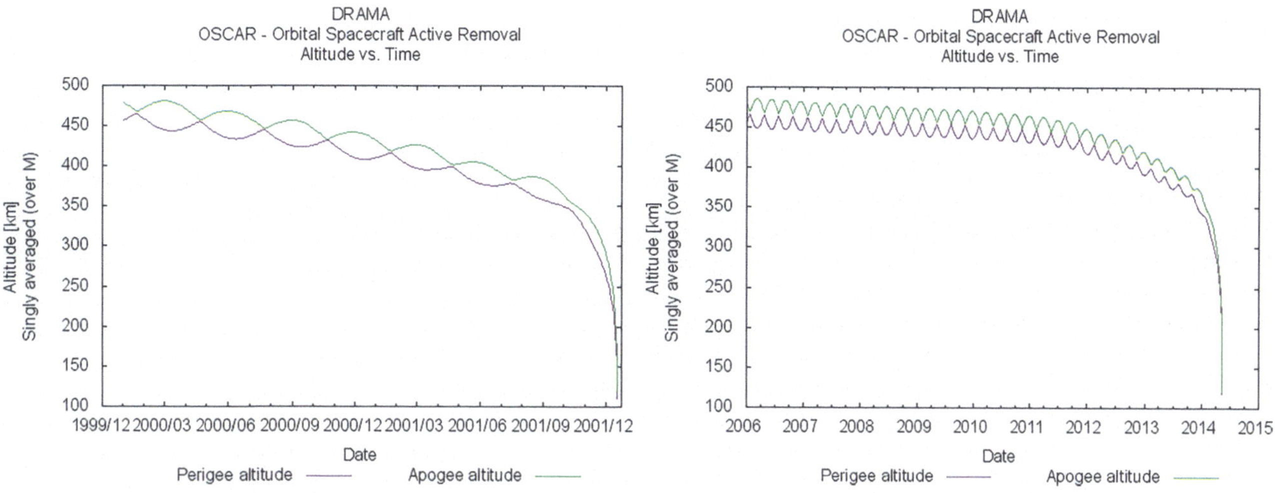

Fig. 6 Orbital lifetime depending on the launch date in high solar activity (left) and low solar activity (right)

Table 2 Mass and volume specifications of the CubeSat systems

	Mass (kg)	Volume (U)
Total structural mass	0.4	–
Satellite bus	1.85	1.8
Propulsion system	1.25	0.8
Satellite w/o payload	3.5	2.6
Scientific payload current	3.5	5.18
Total current	6.92	7.98
Total miniaturized	7.42	6.0

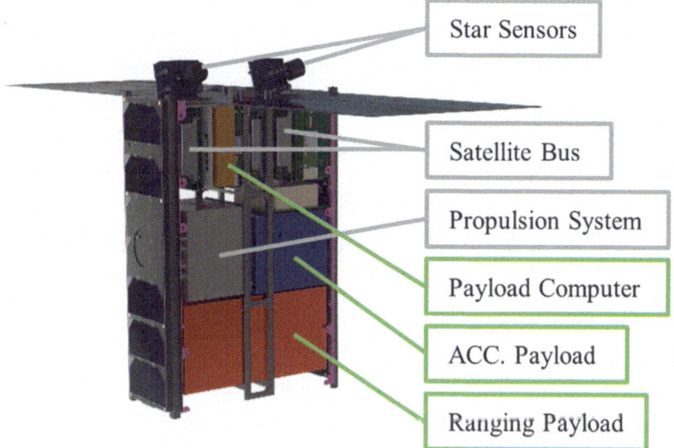

Fig. 7 Preliminary satellite design

of the different CubeSat systems are listed in Table 2. The selected satellite configuration (satellite without payload), consisting of the satellite bus and the chemical propulsion system, covers 2.6 U and 3.5 kg. According to the scientific payload specifications, the optomechanical accelerometer and the inter-satellite ranging instrument cover in sum a volume of 5.18 U and 3.5 kg. As it becomes obvious, the satellite configuration including the scientific payload exceeds the selected framework of 6 U. Since the provided numbers of the scientific payload are at the maximum level and are expected to reduce in the future, miniaturization is expected. A minimum of 2 U must be reserved for the bus and subsystems to ensure the satellite's proper functionality.

Consequently, the payload volume needs to be reduced by at least 35% to fit within the 6 U constraints. Ideally, this reduction should exceed 40% to provide flexibility in positioning the accelerometer and thruster. Other factors, such as the energy budget and communication budget, have also been taken into account and have influenced the initial system design.

4.2.3 Preliminary Satellite Design

A visualization of the preliminary system design is provided in Fig. 7. The setup includes the flight-proven TRL 9 components star sensor, chemical propulsion systems, and satellite bus. The satellite bus visualized as the grey boxes represents all necessary components for a successful mission including the On-Board Computer (OBC), the Attitude and Orbit Control System (AOCS) including actuators and sensors, the Electrical Power System (EPS), the Power Distribution Unit (PDU), and the communication modules. The scientific payload covering predominantly the accelerometer and the inter-satellite ranging instrument is indicated with the green boxes. The latter ones should represent available space that is adjustable for payload requirements.

5 Design Constraints and Requirements

In the following, the key challenges associated with the design and performance of the CubeSat mission are presented, along with their origins, proposed solutions, and impacts on the overall system. One of the primary challenges is the accommodation of the scientific payload within the CubeSat's limited space of 6 U. To address this challenge, it is necessary to miniaturize both the payload and platform packaging; however, achieving the required tight tolerances requires considerable effort to meet these space optimization requirements. Another significant challenge is the precise placement of the optomechanical accelerometer at the CubeSat's Center of Mass (COM), which is crucial for ensuring measurement accuracy and mission success. Precise mass distribution methods and advanced modeling techniques are necessary to minimize COM deviations. Achieving this requires careful iterative design analysis. Regarding propulsion, while solid electric propulsion (SEP) is favored due to its simpler and faster development compared to liquid alternatives, it presents a trade-off. While SEP aligns well with the payload constraints, it may limit postlaunch orbital adjustments, which could affect mission flexibility. Additionally, CubeSat size constraints in the selected 6 U design limit power generation capacity. To overcome this, deployable solar arrays are proposed to maximize power output. However, the deployment mechanism introduces added complexity, and the balance between power needs and observation opportunities requires further optimization. In addition, the deployment of solar arrays induces vibrations that necessitate dampening mechanisms and vibration isolation techniques at the payload mounting points. These solutions add complexity and mass to the CubeSat but are crucial to maintaining power generation efficiency while mitigating vibrations. These findings underscore the need for careful design trade-offs, emphasizing that while solutions to each challenge are feasible, their integration into the CubeSat system requires careful consideration of mission constraints and iterative refinement. Further studies are necessary to fully optimize these solutions.

6 Conclusion and Recommendations

The CubeGrav project has demonstrated the effectiveness of CubeSats in retrieving Earth's gravity field, offering substantial scientific benefits by enhancing spatiotemporal resolution and mitigating temporal aliasing compared to conventional single- or dual-satellite mission configurations. The constellation of 4 polar and 4 inclined satellite pairs, each distributed on 2 orbital planes (4p4iRAAN), is one of the most promising satellite constellations in terms of scientific performance, and a potential realization in orbit is possible but more cost intensive than other CubeSat constellations. Appropriate instrument options for accelerometry and GNSS sensing could be identified. However, a key challenge remains in developing a suitable inter-satellite ranging system capable of achieving micrometer-level accuracy while adhering to stringent size, weight, and power constraints. Advancements in miniaturized instrumentation are expected to address this challenge, thereby influencing the design of future satellite missions. The lifetime of the satellite mission depends on solar activity, and it is recommended that the satellites are launched close to the solar minimum to ensure a sufficient operational duration of several years. Continued investigation into both the scientific performance and the technical implementation of such systems is essential to exploit the full potential of CubeSats for gravity field monitoring.

References

Bandyopadhyay S, Foust R, Subramanian GP, Chung S-J, Hadaegh FY (2016) Review of formation flying and constellation missions using nanosatellites. J Spacecr Rocket 53(3):567–578

Bender PL, Wiese DN, Nerem RS (2008) A possible dual-grace mission with 90 degree and 63 degree inclination orbits. In: Proceedings of the 3rd international symposium on formation flying, missions and technologies. ESA/ESTEC, Noordwijk, pp 1–6

Braun V, Funke Q, Lemmens S, Sanvido S (2020) DRAMA 3.0-upgrade of ESA's debris risk assessment and mitigation analysis tool suite. J Space Saf Eng 7(3):206–212

Daras I (ed) (2023) Next Generation Gravity Mission (NGGM) mission requirements document, Issue 1.0. Earth and Mission Science Division, European Space Agency. https://doi.org/10.5270/ESA.NGGM-MRD.2023-09-v1.0

Dávila Álvarez A, Knudtson A, Patel U, Gleason J, Hollis H, Sanjuan J et al (2022) A simplified gravitational reference sensor for satellite geodesy. J Geod 96(10). https://doi.org/10.1007/s00190-022-01659-0

Dobslaw H, Bergmann I, Dill R, Forootan E, Klemann V, Kusche J, Sasgen I (2014) Updating ESA's Earth system model for gravity mission simulation studies: 2. Comparison with the original model. Scientific Technical Report; 14/08. Deutsches GeoForschungsZentrum GFZ, 57pp

Dobslaw H, Bergmann-Wolf I, Forootan E, Dahle C, Mayer-Gürr T, Kusche J, Flechtner F (2016) Modeling of present-day atmosphere and ocean non-tidal de-aliasing errors for future gravity mission simulations. J Geod 90(5):423–436

Drinkwater M, Floberghagen R, Haagmans R, Muzi D, Popescu A (2003) GOCE: ESA's first Earth explorer core mission. In: Earth gravity field from space – from sensors to earth science, Space sciences series of ISSI, vol 18. Kluwer Academic Publishers, Dordrecht, pp 419–432. ISBN: 1-4020-1408-2

Elsaka B et al (2014) Comparing seven candidate mission configurations for temporal gravity field retrieval through full-scale numerical simulation. J Geod 88(1):31–43. https://doi.org/10.1007/s00190-013-0665-9

Flechtner F, Neumayer K-H, Dahle C, Dobslaw H, Fagiolini E, Raimondo J-C, Güntner A (2016) What can be expected from the GRACE-FO laser ranging interferometer for Earth science applications? Surv Geophys 37(2):453–470. https://doi.org/10.1007/s10712-015-9338-y

Haagmans R, Tsaoussi L (eds) (2020) Next generation gravity mission as a mass-change and geosciences international constellation (MAGIC) mission requirements document. Earth and Mission Science Division, European Space Agency, NASA Earth Science Division. https://doi.org/10.5270/esa.nasa.magic-mrd.2020. Accessed 14 Jan 2025

Hauk M, Pail R (2018) Treatment of ocean tide aliasing in the context of a next generation gravity field mission. Geophys J Int 214(1):345–365. https://doi.org/10.1093/gji/ggy145

Heller-Kaikov B, Pail R, Daras I (2023) Mission design aspects for the mass change and geoscience international constellation (MAGIC). Geophys J Int 235(1):718–735. https://doi.org/10.1093/gji/ggad266

Hines A, Richardson L, Wisniewski H, Guzman F (2020) Optomechanical inertial sensors. Appl Optics 59(22): G167–G174. https://doi.org/10.1364/AO.393061

Hines A, Nelson A, Zhang Y, Valdes G, Sanjuan J, Stoddart J, Guzmán F (2022) Optomechanical accelerometers for geodesy. Remote Sens 14(17). https://doi.org/10.3390/rs14174389

Kornfeld RP, Arnold BW, Gross MA, Dahya NT, Klipstein WM, Gath PF, Bettadpur S (2019) GRACE-FO: the gravity recovery and climate experiment follow-on mission. J Spacecr Rocket 56(3):931–951. https://doi.org/10.2514/1.A34326

Landerer FW, Flechtner FM, Save H, Webb FH, Bandikova T, Bertiger WI et al (2020) Extending the global mass change data record: GRACE follow-on instrument and science data performance. Geophys Res Lett 47(12). https://doi.org/10.1029/2020GL088306

Lenoir B, Lévy A, Foulon B, Lamine B, Christophe B, Reynaud S (2011) Electrostatic accelerometer with bias rejection for gravitation and Solar System physics. Adv Space Res 48(7):1248–1257. https://doi.org/10.1016/j.asr.2011.06.005

Liorzou F, Lebat V, Christophe B, Boulanger D, Rodrigues M, Zahzam N et al (2023) ONERA accelerometers for future gravity mission. EGU General Assembly 2023, Vienna, Austria

Massotti L, Siemes C, March G, Haagmans R, Silvestrin P (2021) Next generation gravity mission elements of the mass change and geoscience international constellation: from orbit selection to instrument and mission design. Remote Sens 13(19). https://doi.org/10.3390/rs13193935

Mayer-Gürr T et al (2015) The new combined satellite only model GOCO05s. In: Presentation at EGU 2015, Vienna, Apr 2015

Moeller G, Wolf A, Sonnenberg F, Bauer G, Soja B, Rothacher M (2024) A low-cost commercial off-the-shelf GNSS receiver for space. EGU General Assembly 2024, Vienna, Austria

Murböck M (2015) Virtual constellations of next generation gravity missions. In: Reihe C: Dissertationen: Deutsche Geodätische Kommission. Verlag C.H. Beck, Bayerische Akademie der Wissenschaften, München. https://dgk.badw.de/fileadmin/user_upload/Files/DGK/docs/c-750.pdf. Accessed 14 Jan 2025

Murböck M, Pail R, Daras I, Gruber T (2015) Optimal orbits for temporal gravity recovery regarding temporal aliasing. Int J Geod 88(2):113–126. https://doi.org/10.1007/s00190-013-0671-y

Nelson A, Warrayat M, Dahn J, Harley-Trochimczyk I, Sanjuan J, Guzman F (2024) A six-axis optomechanical inertial sensing and navigation system for future mass change missions. In: AGU 2024, Washington, DC, USA

Nicklaus K, Cesare S, Massotti L, Bonino L, Mottini S, Pisani M, Silvestrin P (2020) Laser metrology concept consolidation for NGGM. CEAS Space J 12(3):313–330. https://doi.org/10.1007/s12567-020-00324-6

Pail R et al (2015) Science and user needs for observing global mass transport to understand global change and to benefit society. Surv Geophys 36(6):743–772

Paul E, Darrow M, Pfaffenzeller N, Eder B, Gruber T, Pail R (2025) New high-precise optical ranging device enables satellite gravimetry with small satellites. In: Proc. SPIE 13699, International Conference on Space Optics - ISCO 2024. https://doi.org/10.1117/12.3072774

Pfaffenzeller N, Pail R (2021) Multi-satellite formations and constellations of CubeSats and their potential in NGGMs. In: Scientific assembly of the International Association of Geodesy, 2021, Beijing, China. https://mediatum.ub.tum.de/doc/1615996/1615996.pdf. Accessed 14 Jan 2025

Pfaffenzeller N, Pail R (2023) Small satellite formations and constellations for observing sub-daily mass changes in the Earth system. Geophys J Int 234(3):1550–1567. https://doi.org/10.1093/gji/ggad132

Pfaffenzeller N, Pail R (2024) Capabilities of ocean tide aliasing reduction by co-estimation of major constituents with future satellite constellations. In: Symposium on gravity, geoid and height systems 2024, Thessaloniki

Ray R (1999) A global ocean tide model from TOPEX/POSEIDON altimetry: GOT99.2, Rep NASA/TM-1999-209478. Goddard Space Flight Center, Greenbelt, 58 pp. https://ntrs.nasa.gov/api/citations/19990089548/downloads/19990089548.pdf. Accessed 14 Jan 2025

Reigber C, Schwintzer P, Lühr H (1999) The CHAMP geopotential mission, Bollettino di Geofisica Teoretica ed Applicata, 40/3–4, September–December 1999. In: Marson I, Sünkel H (eds) Proceedings of the 2nd joint meeting of the International Gravity and the International Geoid Commission, 1998, Trieste, Italy, pp 285–289. ISSN: 0006-6729

Rodrigues M, Bergé J, Boulanger D, Christophe B et al (2022) Space accelerometers for micro and nanosatellites: Fundamental Physics and Geodesy missions from MICROSCOPE, GOCE and GFO return of experience. In: 4S symposium 2022, Vilamoura, Portugal. https://hal.science/hal-03854833/document. Accessed 16 Jul 2025

Savcenko R, Bosch W (2012) EOT11a – empirical ocean tide model from multi-mission satellite altimetry. DGFI Report No. 89. Deutsches Geodätisches Forschungsinstitut, München. https://epic.awi.de/id/eprint/36001/1/DGFI_Report_89.pdf. Accessed 14 Jan 2025

Tapley BD, Bettadpur S, Watkins M, Reigber C (2004) The gravity recovery and climate experiment: mission overview and early results. Geophys Res Lett 31(9). https://doi.org/10.1029/2004GL019920

Tomio H, Kammerer W, Grenfell P, Cierny O, Garcia M, Lindsay C et al (2022) Transmitter and fine pointing system development and testing for the CubeSat Laser Infrared CrosslinK (CLICK) B/C mission. In: International conference on space optics 2022, Dubrovnik

Wahr J, Molenaar M, Bryan F (1998) Time variability of the Earth's gravity field: hydrological and oceanic effects and their possible detection using GRACE. J Geophys Res 103(B12):30205–30229

Walter U (2008) Astronautics. Wiley-VCH, Weinheim (Physics textbook). ISBN: 978-3-527-40685-2

Wiese DN, Visser P, Nerem RS (2011) Estimating low resolution gravity fields at short time intervals to reduce temporal aliasing errors. Adv Space Res 48(6):1094–1107. https://doi.org/10.1016/j.asr.2011.05.027

Wiese DN, Nerem RS, Lemoine FG (2012) Design considerations for a dedicated gravity recovery satellite mission consisting of two pairs of satellites. J Geod 86(2):81–98. https://doi.org/10.1007/s00190-011-0493-8

Wiese DN, Bienstock B, Blackwood C, Chrone J, Loomis BD, Sauber J et al (2022) The mass change designated observable study: overview and results. Earth Space Sci 9(8):e2022EA002311. https://doi.org/10.1029/2022EA002311

Zingerle P, Gruber T, Pail R, Daras I (2024) Constellation design and performance of future quantum satellite gravity missions. Earth Planets Space 76:101. https://doi.org/10.1186/s40623-024-02034-3

Sensor Fusion Error and~Spectral Analyses for Airborne Quantum Gravimetry: The AeroQGrav Case Study

Francesco Darugna, Henning Albers, Temmo W. Wübbena, and Jannes B. Wübbena

Abstract

Earth's gravity field can be reconstructed by adopting several techniques that involve data collected from satellite-, aircraft-, and ground-based measurements. Airborne gravimetry is a powerful tool for remote regions, like mountains or deserts, and for obtaining a higher spatial resolution than satellite gravity measurements. The German Absolute Aero Quantengravimetrie (AeroQGrav) project aims to build an airborne atom interferometer to fill gaps in regional geoid modelling and achieve precisions of down to 1 $\mu m/s^2$ after 5 s of signal integration. In this manuscript, starting from the requirements, a spectral analysis of the sensor errors addresses the limiting factors of the project. In the sensor fusion process, separating the kinematic from the gravitational acceleration represents one of the main challenges of airborne gravimetry. Depending on the spatial resolution and frequency domain of interest, the successful achievement of the required measured gravity's accuracy and precision may be limited by the spectral characteristics of the adopted kinematic sensors. An optimal simulation scenario is adopted to model the sensors and evaluate their noise contribution. Finally, it is presented how to exploit the fusion of the kinematic sensor measurements to retrieve the gravity signal at multiple frequencies of interest and precisions, overcoming the single-sensor limitations.

Keywords

Airborne quantum gravimetry · GNSS · Sensor fusion · Spectral analysis

1 Introduction

Recently, quantum sensors have been utilised for absolute gravimetry. This technology is based on atom interferometry (e.g. Berman 1997; Peters et al. 1999; Geiger et al. 2020), which exploits the quantum states of particles that are highly sensitive to their environment, making it well suited for sensing external forces or fields.

One of the key challenges of gravimeters is distinguishing gravitational acceleration from kinematic acceleration. Since the advent of the global positioning system (GPS), later expanded by additional global navigation satellite systems (GNSS), the standard approach has been to use satellite measurements to determine kinematic acceleration (Schwarz et al. 1992).

Airborne absolute atom gravimeters have demonstrated accuracies of approximately 10 $\mu m/s^2$ (Bidel et al. 2023). However, their performance depends on the signal integration interval and, consequently, the frequency scale considered. The contribution of both inertial sensors (e.g. the quantum gravimeter) and kinematic sensors (e.g. GNSS) to the overall error budget varies with frequency (Schwarz et al. 1992), which may constrain geodetic applications requiring high spatial resolution, such as mineral exploration (Okada 2021).

The Absolute Aero Quantengravimetrie (AeroQGrav) project is led by a consortium of academic and industrial

F. Darugna (✉) · H. Albers · T. W. Wübbena · J. B. Wübbena
Geo++ GmbH, Garbsen, Lower Saxony, Germany
e-mail: francesco.darugna@geopp.de

J. T. Freymueller, L. Sànchez (eds.), *International Symposium on Gravity, Geoid and Height Systems 2024 (GGHS2024)*, International Association of Geodesy Symposia 158, https://doi.org/10.1007/1345_2025_290

partners, aiming to develop an airborne absolute quantum gravimeter with a performance level of 1 $\mu m/s^2$. The consortium's diverse expertise and ambitious objectives make AeroQGrav an exciting research endeavour, covering topics ranging from quantum sensors (Schubert 2023) to advancements in troposphere modelling for improved aircraft height estimation (Darugna et al. 2023).

One of the distinctive features of the AeroQGrav project is its integration of multiple technologies for kinematic measurements. In addition to the GNSS payload, the system includes a terrestrial laser scanner (TLS) and a laser Doppler velocimeter (LDV). By fusing these sensors, the measurement capabilities can be extended across a broader frequency range. Furthermore, the measurement of the aircraft roto-translational motion is aided by data acquired from an inertial measurement unit (IMU).

Here, an initial spectral assessment of the errors introduced by the AeroQGrav sensors is presented, evaluating the limitations that must be overcome to meet the project's requirements. Additionally, a sensor fusion approach is introduced to leverage the different AeroQGrav kinematic sensors to mitigate errors in gravity reconstruction.

The chapter is organised as follows: Sect. 2 introduces the motivation and accuracy requirements, while Sect. 3 presents the reconstructed gravity error description. Section 4 covers sensor modelling and spectral analysis, followed by Sect. 5, which explains the sensor fusion strategy and presents the results. Finally, conclusions and an outlook are provided.

2 Motivation and~Requirements

In airborne gravimetry, multiple sensors are required to isolate the gravity signal from the measured absolute acceleration in the non-inertial local frame, where the body accelerates. Specifically, a kinematic sensor is needed to provide the kinematic acceleration, allowing its separation from the gravity signal.

A direct measurement of kinematic acceleration can be obtained by differentiating the trajectory reconstructed by a GNSS-based positioning engine. As mentioned above, GNSS is a common choice in airborne gravimetry. In AeroQGrav, TLS and LDV are included as additional kinematic sensors to improve the estimation of kinematic accelerations.

In airborne gravimetry, typical spatial resolutions (full wavelength) of the extracted gravity signal extend from a few hundred metres to a few hundred kilometres. As shown in Table 1, the AeroQGrav project focuses on geodetic applications with spatial resolutions of approximately 600 m to 100 km, including quasi-geoid modelling and stability, regional geoid modelling, pure gravity mapping, and analysis of local gravity features.

Table 1 Spatial requirements for geodetic applications. For range intervals' multiple applications, the one with most stringent requirement is reported

Spatial resolution λ (km)	Gravity accuracy ($\mu m/s^2$)	Application
0.6	1	AeroQGrav goal
1.0	10	Regional modelling
5.0	6	Regional modelling
10.0	5	Regional modelling
20.0	3	Drift stability
50.0	2	Drift stability
100.0	1	Drift stability

Recent studies have demonstrated that quasi-geoid models can achieve accuracies better than 1 cm, with regional accuracies reaching the millimetre level (Guo et al. 2023). Accordingly, one requirement is to constrain drift stability between 1 and 100 km to within three to 5 mm to maintain this level of performance. As a rule of thumb, a gravity change dg is mapped into a geoid change dN as follows (Hofmann-Wellenhof and Moritz 2005):

$$dN \approx \frac{\lambda}{2}\frac{dg}{\gamma}, \tag{1}$$

where λ is the distance wavelength, and γ is the average gravity value on Earth's surface.

Using Eq. (1), the resulting requirement for gravity accuracy, related to geoid model drift stability, is reported in Table 1, which summarises the geodetic application requirements and can be visualised in Fig. 1.

One of the main geodetic objectives is regional geoid modelling. For this purpose, the focus is on scales between 1 and 10 km, where combined high-resolution global models, such as EGM2008 (Pavlis et al. 2012), could be improved.

Another application that could benefit from the AeroQGrav results is the pure gravity mapping of the free-air anomaly (e.g. reductions for levelling measurements), which requires an accuracy of 10 $\mu m/s^2$ across the entire spatial spectrum.

Finally, AeroQGrav addresses the challenge of achieving 1 $\mu m/s^2$ at approximately 600 m wavelength, a demanding objective that is essential for detecting highly localised gravity features caused by subsurface structures, such as salt domes. These features are well sampled by surface measurements in regions like those near Hannover, in Lower Saxony, Germany.

The spatial requirements reported in Table 1 can be translated into frequency domain by applying the aircraft velocity, v, which is expected to be around 60 m/s during the cruise phase of the AeroQGrav flights, operated by the Institute of Flight Guidance of the Technical University of Braunschweig (TUB). This transformation helps to illustrate

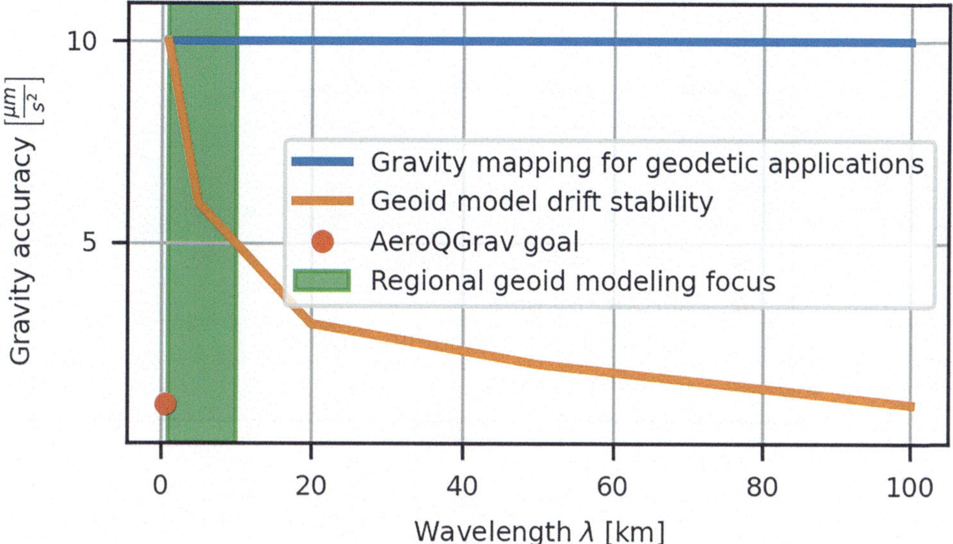

Fig. 1 Geodetic application requirements with respect to the wavelength (spatial resolution)

the need for multiple sensors, depending on their power distribution across different frequencies.

In the frequency domain, the power spectral density (PSD) is the spectral equivalent of the covariance in the temporal domain. In other words, the power spectrum decomposition of the covariance function of a signal is its PSD. Three assumptions are made (Jekeli 2003):

- The signal is a stochastic process (i.e. at any time, the process value is associated with a probability).
- The statistics of the process is stationary (i.e. the covariance depends on time differences but not on absolute time).
- The process is ergodic, i.e. the statistic over time is equivalent to the statistic over probability space.

For each point in frequency, a first-order approximation of the reconstructed gravity error PSD value is computed by using the following relationship:

$$\text{PSD} \approx \frac{\sigma^2}{f}, \tag{2}$$

where σ is the required gravitational accuracy reported in Table 1 and f is the frequency obtained as v/λ, with v being the aircraft velocity. This expression does not apply universally; it is used as an approximation for a single frequency point and is employed to generate the dashed line and the blue and red markers in Fig. 2. The PSD at that frequency defines the power threshold that an error signal must not exceed in order to meet the requirement.

To put it differently, the PSD of the reconstructed gravity signal must remain in the green-coloured acceptance area depicted in Fig. 2. Additionally, Fig. 2 highlights the frequency range relevant to the regional geoid modelling.

3 The Gravity Error Equation

The objective is to reconstruct the gravity signal along the ellipsoidal height direction. This operation is executed in a topocentric non-inertial frame, namely, the North-East-Down (NED) reference frame. It is worth mentioning that the measured gravity is not exactly in the Down direction along the ellipsoidal height, but it is along the vertical to the geoid (e.g. Seeber 2003). The angle between the two directions is called vertical deflection and can be measured indirectly or computed by adopting gravity models.

However, considering the gravity model EGM2008, for locations around Hannover (e.g. latitude = 52.5°, longitude = 9.5°), where most of the AeroQGrav test will be performed, the vertical deflection's effect on the reconstructed gravity signal amounts to roughly 7×10^{-6} rad. This means an error in the reconstructed gravity of about 2.5×10^{-10} m/s^2. Hence, for the requirements presented in Table 1, the vertical deflection's effect can be neglected. Moreover, the error decreases with the ellipsoidal height. Therefore, the gravity vector can be assumed to be aligned with the Down direction of the NED frame for the airborne gravimetry purposes of this work.

The AeroQGrav atom interferometer is mounted on a stabilised platform (designed and assembled by the consortium partner iMAR Navigation GmbH) that keeps the axis of measurement as aligned as possible to the local vertical where the gravitational acceleration acts. Differently from the strapdown approach (Becker et al. 2016), the gravimeter measurement is along this one axis only, and it will be indicated as a_{D}.

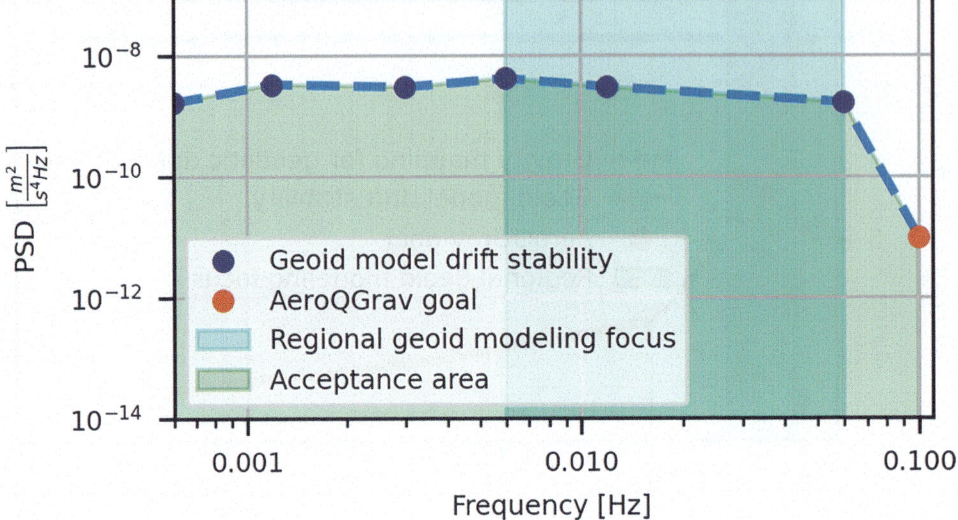

Fig. 2 PSD of the geodetic application requirements

Since the alignment with the local vertical depends on the capability of the stabilised platform, it is reasonable to introduce a tilt error δa_{tilt} (Olesen 2002). Alignment errors of the platform make the gravimeter less sensitive to vertical gravity acceleration and make it sensitive to horizontal accelerations. In other words, it shows the impact of the coupling of the orientation error (between the platform frame and the NED frame) with the acting acceleration.

Including δa_{tilt} and neglecting second-order effects, the reconstructed gravity signal is given by

$$g_{\text{D}} = \ddot{x}_{\text{D}} + a_{\text{Eöt}} - a_{\text{D}} + \delta a_{\text{tilt}}, \tag{3}$$

where $\ddot{x}_D$ is the kinematic acceleration, $a_{\text{Eöt}}$ is the so-called Eötvös effect resulting from the coupling with Earth's rotation. Indicating an error by using the letter δ, once corrected for the Eötvös effect, Eq. (3) can be rewritten as

$$\delta g_{\text{D}} = \delta \ddot{x}_{\text{D}} - \delta a_{\text{D}} + \delta a_{\text{tilt}}. \tag{4}$$

In summary, within AeroQGrav, the objective is to maintain the PSD of δg_{D} below the curve (dashed line) defined by the requirements shown in Fig. 2, i.e. remaining within the green-coloured acceptance region.

4 Sensor Spectral Analysis

As mentioned above, AeroQGrav's payload includes a multi-antenna GNSS system, TLS, and LDV. Currently, the LDV's performance is being investigated. This sensor is expected to significantly improve the precision of the kinematic acceleration reconstruction at high frequencies. While its potential is highlighted in the following sections, an in-detail analysis of the LDV fusion is left for future work.

In addition to the kinematic sensors, the quantum gravimeter and the platform accelerometer are also defined. It is important to note that the quantum gravimeter is considered as a combination of the atom interferometer and the classical accelerometer, working together to reconstruct the gravitational signal. The platform accelerometer, part of the inertial measurement unit, provides attitude stabilisation for the platform with the goal of minimising δa_{tilt} in Eqs. (3) and (4).

4.1 Sensor Modelling

The spectral limitations of the sensors are assessed by treating each sensor measurement as white noise, characterised by the sensor precision. This configuration represents an ideal scenario where no additional error contributions are considered. For instance, the evaluation of sensor sampling intervals is beyond the scope of this work. The sampling interval is assumed to be 1 s. It is important to note that this does not imply that the sensors operate at 1 Hz but that their resulting acceleration measurements are sampled at 1 Hz.

Table 2 outlines the modelling concept used in the analysis. Each sensor output is assumed to be zero-mean white noise, characterised by its standard deviation. For instance, the estimated ellipsoidal height from the GNSS-based processing is assumed to have a precision of 1 mm over twice the differentiation interval. Achieving such precision is highly challenging, but feasible (Wübbena et al. 2001; Hirt et al. 2011) over short durations (ranging from a few seconds to a few hundred seconds), provided that tropospheric and site-

Table 2 Sensor modelling: white noise standard deviation and operation to compute the acceleration for each sensor

Sensor	σ	Operation
GNSS	1 mm	Differentiate twice
TLS	0.5 mm/s	Differentiate
Quantum gravimeter	0.1 μm/s^2	Low-pass filter
Platform accelerometer	8.2 mm/s^2	Low-pass filter

dependent effects are properly accounted for. To account for these error sources, enhanced tropospheric modelling (Darugna et al. 2023) and absolute multi-frequency GNSS antenna calibration (Wübbena et al. 2019) are applied. Additionally, the multi-antenna configuration reduces the uncorrelated observation noise by the square root of the number of antennas used (up to eight, in the case of AeroQGrav).

The GNSS-estimated position is differentiated twice to obtain the kinematic acceleration by using first-order numerical differentiation. A discrete signal (e.g. the GNSS-based ellipsoidal height) is represented as a series $x[n]$ with $n \in \mathbb{N}$, corresponding to the continuous $f(t)$ with $t \in \mathbb{R}$. The differentiation interval τ_A is defined by k times the sampling interval τ_s. For example, if the sampling interval is 1 s and a differentiation interval τ_A of 5 s is required, then $k = 5$ is used. In this context, the second derivative operation is approximated by the central finite difference:

$$\frac{d^2 f(t)}{dt^2} \approx \frac{x[n+k] - 2x[n] + x[n-k]}{{\tau_A}^2}. \tag{5}$$

Conversely, for the TLS, a single-step differentiation is applied since the measurements represent height differences with a 1-s time step. TLS measurements are modelled as white noise with a standard deviation of 0.5 mm/s.

For single differentiation, a one-sided expression is preferred over the central finite difference. While the central method is slightly more accurate, it results in signal loss over twice the differentiation interval. Therefore, the one-sided forward finite difference is used to approximate the first derivative:

$$\frac{df(t)}{dt} \approx \frac{x[n+k] - x[n]}{\tau_A}. \tag{6}$$

As a remark, the first-order finite difference approximation adopted in Eqs. (5) and (6) is sufficient for the standard deviations and frequency ranges involved. However, it is worth mentioning that if greater accuracy was required, a higher-order representation should be used to better describe the derivative operator (Bruton 2000).

For the inertial sensors – i.e. the quantum gravimeter and platform accelerometer – a low-pass filter is applied, with a 5-s moving average currently adopted. The optimal low-pass filter selection will be investigated during real-data operations, as it needs to be tuned based on the analysed data (Bidel et al. 2020, 2023).

A 0.1 μm/s^2 noise level for the quantum gravimeter is assumed. Regarding the platform tilt, an acceleration difference model is adopted (see Olesen 2002 for more details). The model depends on GNSS-based kinematics and accelerations measured by the inertial accelerometer on the platform along the East and North directions (perpendicular to gravity). It is assumed that the horizontal acceleration and position will not exceed 8 mm/s^2 and 10 mm, respectively.

All operations to compute the acceleration are assumed to work on a 5-s interval as the final objective is to achieve 1 μm/s^2 after 5 s of integration. The impact of the time interval is investigated in the following subsection.

4.2 Time Interval Impact: The GNSS Example

Before analysing all the sensor PSDs, it is worth understanding the impact of the time interval τ_A in computing the acceleration. The general behaviour can be deduced by looking at Fig. 3, which depicts the PSD of the error signal of the GNSS-based acceleration. Increasing τ_A from 5 to 50 s has two main effects.

On the one hand, it reduces the maximum PSD value. On the other hand, it narrows the frequency range, as the informative content for frequencies above $1/(2\tau_A)$ is lost according to the Nyquist-Shannon sampling theorem.

Frequencies above the Nyquist frequency cannot be correctly represented, and any components exceeding this threshold will undergo aliasing, which introduces erratic, misleading fluctuations in the PSD without any physical relevance to the signal.

In Fig. 3, the aliasing effect occurs for the orange line at frequencies above 0.01 Hz, which corresponds to the Nyquist frequency for the curve with τ_A= 50 s. The energy from the aliased frequencies is reflected back into the displayed frequency range on the plot.

Therefore, depending on the needs, a longer or shorter time interval may be preferred. Here, a 5-s τ_A is selected to cover a broader frequency range. Like in Fig. 2, the green-shaded area represents the acceptable range, while the light-blue-shaded area highlights the focus on regional geoid modelling. This layout is used for all subsequent PSD-related figures.

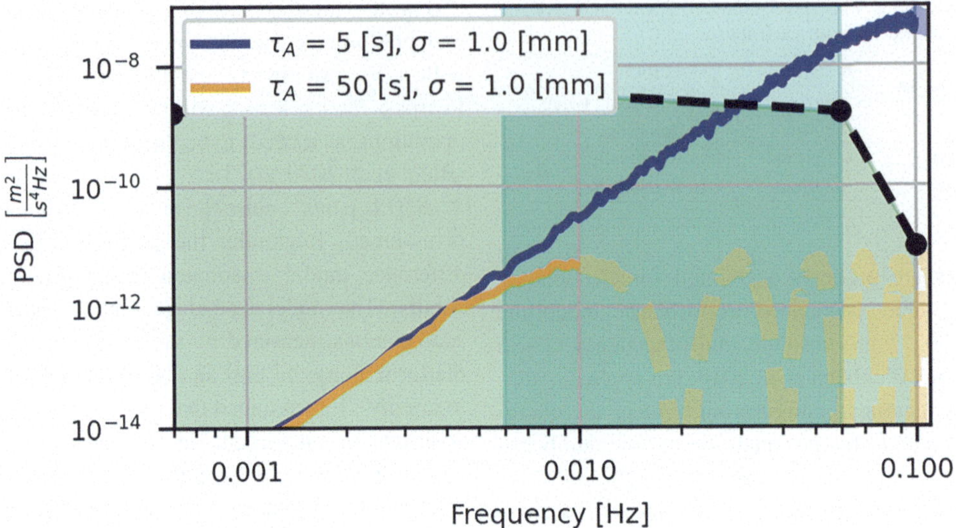

Fig. 3 GNSS error PSD varying τ_A

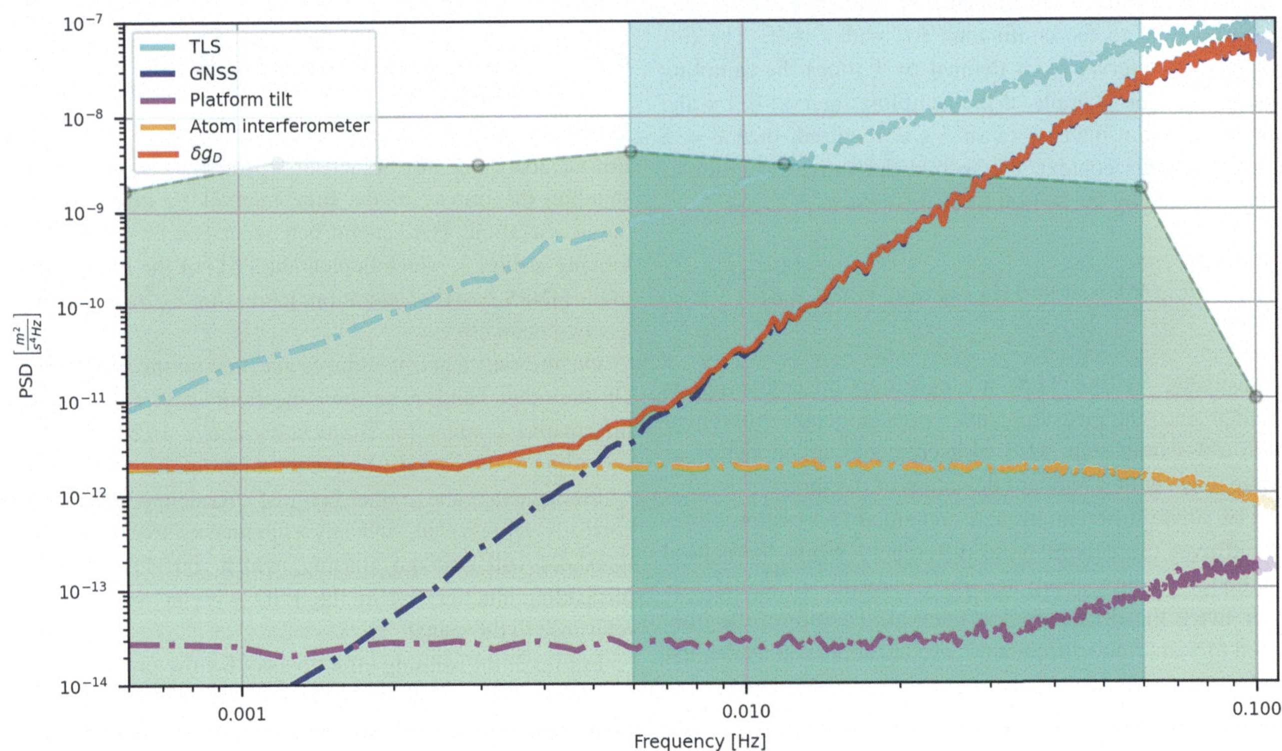

Fig. 4 PSD of individual sensor errors and of the reconstructed gravity error (continuous red curve)

4.3 PSD of~Sensor Errors

Figure 4 shows the PSD of the error signal of each sensor. The green-shaded area indicates the acceptance area; that is, the PSD of the errors should remain below the limits defined above. If the PSD of a sensor error exceeds the black-coloured dashed line, the related sensor does not meet the project goals.

The platform tilt's PSD (magenta line) is obtained by assuming a GNSS solution precision of 1 mm in the North and East directions. Additionally, the inertial acceleration in the North and East directions is assumed to be on the order of 8 mm/s^2. It is worth noting that this is a reasonable assumption, and higher values would interfere with achieving the objectives.

As it can be seen on the right side of Fig. 4, the kinematic sensors (i.e. the TLS – light-blue line – and GNSS – dark-blue line – sensors) limit the frequency range to below approximately 0.03 Hz.

It is worth noting that the kinematic sensors are responsible for limiting the successful achievement of the AeroQGrav goals, assuming that the atom interferometer performs as expected under nominal conditions (shown in Fig. 4 by the orange line). Validation with real data will be required to assess this theoretical analysis.

Considering Eq. (4) and summing up all the error components, the gravity signal error is reconstructed in Fig. 4 (selecting GNSS, and not TLS, for the kinematic error contribution). The red curve depicts the PSD of the gravity error δg_{D}. It can be observed that the PSD is too large for frequency values above 0.03 Hz. This limitation is primarily due to the GNSS solution, which is less restrictive than the TLS sensor.

However, having multiple sensors on board provides the opportunity to apply sensor fusion techniques that can overcome the limitations specific to each sensor.

5 Sensor Fusion

5.1 Approach

The sensor fusion approach adopted in AeroQGrav corresponds to what is called the direct method (Johann et al. 2019), which has also been called the accelerometer approach (Kwon and Jekeli 2001) since the inertial accelerations are compared directly to the kinematic accelerations. The direct method follows a cascaded approach (Becker et al. 2016) and involves multiple consecutive processing steps. An alternative is the so-called indirect method (Jekeli 2000), which uses a single-step extended Kalman filter (EKF) to fuse IMU and GNSS measurements and estimate position, velocity, and acceleration together.

Figure 5 shows the sensor fusion strategy employed in AeroQGrav. The GNSS measurements are processed by a positioning engine, and the resulting geometric position is fused together with the TLS and LDV measurements to obtain the kinematic trajectory in the NED frame. The trajectory and the velocity, obtained through numerical differentiation, are processed to compute the Eötvös effect. The latter corrects the kinematic acceleration obtained by double differentiation of the reconstructed position. The Eötvös $a_{\mathrm{Eöt}}$, kinematic $\ddot{x}_{\mathrm{D}}$, and sensed a_{D} accelerations are passed through a low-pass filter before being input to a Kalman filter. This process ultimately reconstructs the gravity signal in the Down direction of the NED frame, g_{D}.

A key factor for successful sensor fusion is ensuring proper time synchronisation.

5.2 Time Synchronisation

The importance of time synchronisation is well recognised in sensor fusion and has been addressed by several authors (e.g. Ding et al. 2008; Becker et al. 2016; Jellum et al. 2022). Previous studies have shown that synchronisation within 2 ms is necessary to align IMU and GNSS data accurately (Becker et al. 2016). In airborne gravimetry, the precise timing of measurements is critical for correcting the inertial acceleration, including both the kinematic component and the Eötvös effect (Bidel et al. 2020).

Synchronisation can be cumbersome when the computer clock controlling the gravimeter has an unknown delay compared to GNSS or when there are jitters between recorded time stamps and measurement times. Moreover, often, there are missing data points, and interpolation is needed. Therefore, a consistent and robust time synchronisation strategy is required.

In AeroQGrav, all sensors will be aligned to the GNSS time provided by a Septentrio Mosaic-T receiver fed by a 10 MHz oven-controlled crystal oscillator (OXCO) with a 5×10^{-12} stability at 1-s integration time. Figure 6 depicts the time synchronisation approach adopted in the project. Each measurement unit will receive a constant wave as a clock reference from the 10 MHz oscillator and a pulse-per-second (PPS) signal with an accuracy of 1 ns. Additionally, a position, referenced to a time stamp in the NMEA format, will be used to ensure consistent and accurate time tagging of each sensor measurement.

This setup allows each sensor to accurately time-tag its measurements without ambiguities by assigning the NMEA time stamp to the correct second, thanks to the PPS. Fine-grained timing is achieved by referencing the 10 MHz oscillator, ensuring stability. This approach, combined with the low-pass filtering procedure, ensures accurate handling of high-dynamic motion.

5.3 The GNSS and~TLS Fusion

The fusion of GNSS and TLS leverages the strengths of both sensors. GNSS provides an absolute reference at a lower update frequency (e.g. every 50 s), while TLS offers a relative reference at 1 Hz. This combination enables precise ellipsoidal height estimates with a standard deviation of 0.5 mm at 1 Hz. After processing through double differentiation, the final acceleration measurements are obtained. It is important to note that the approach is only valid within the update interval of the absolute reference (i.e. GNSS). Although a 50-s update interval is considered here, real-world data should be used to optimise the interval.

Figure 7 shows how sensor fusion extends the effective frequency range where measurement requirements are met,

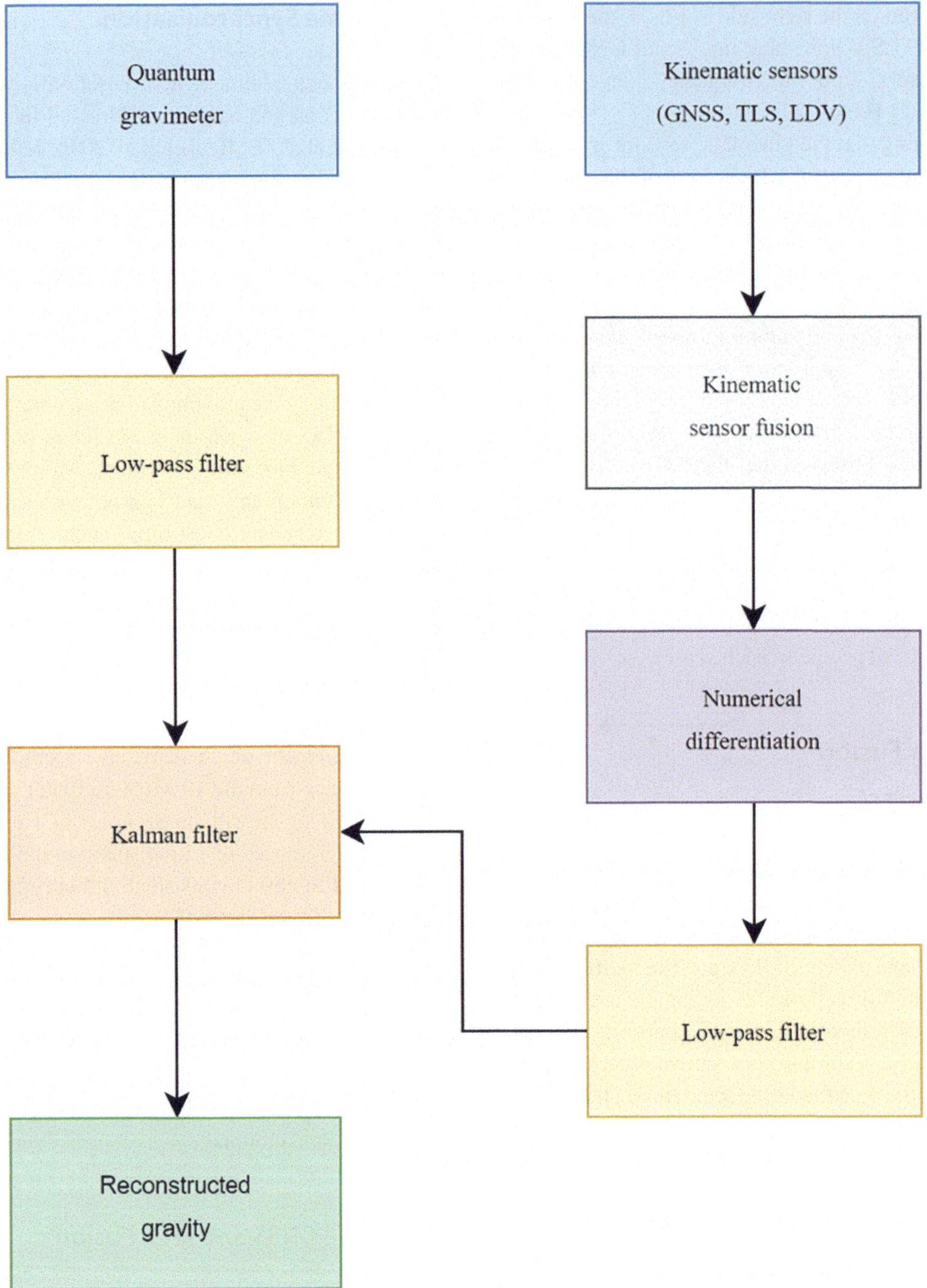

Fig. 5 Sensor fusion strategy

as represented by the green line. By combining GNSS and TLS, δg_{D} errors remain below 10 μm/s^2 for frequencies up to approximately 0.05 Hz. In this context, the TLS relative precision, while already high, effectively sets the limit on the overall performance of the fused solution when GNSS noise is sufficiently low. The additional fusion with the LDV sensor is expected to cover the remaining frequency range up to 0.1 Hz (yellow area), thus meeting all AeroQGrav objectives.

6 Conclusion and~Outlook

The AeroQGrav geodetic applications and requirements have been presented and described in both the spatial and frequency domains. Focusing on frequencies ranging from 6×10^{-4} to 0.1 Hz, a spectral analysis of sensor errors has been conducted to highlight the limiting factors based on the geodetic requirements.

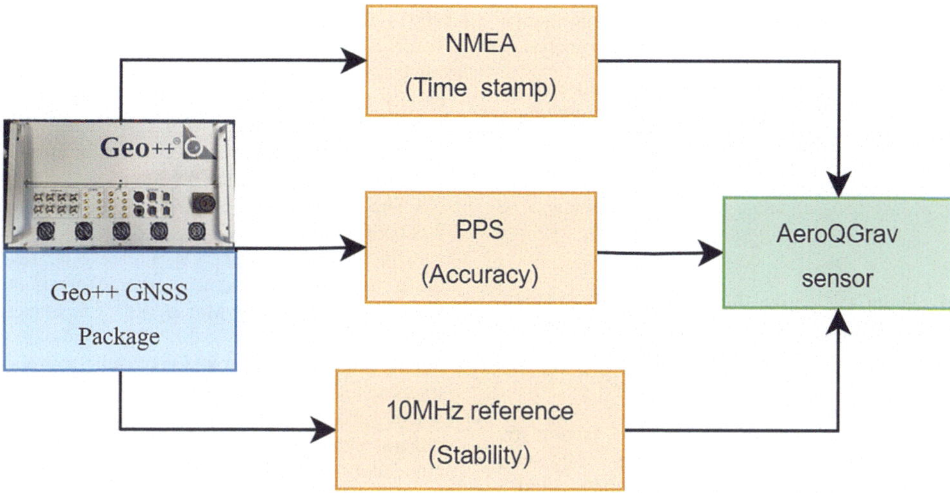

Fig. 6 Time synchronisation approach

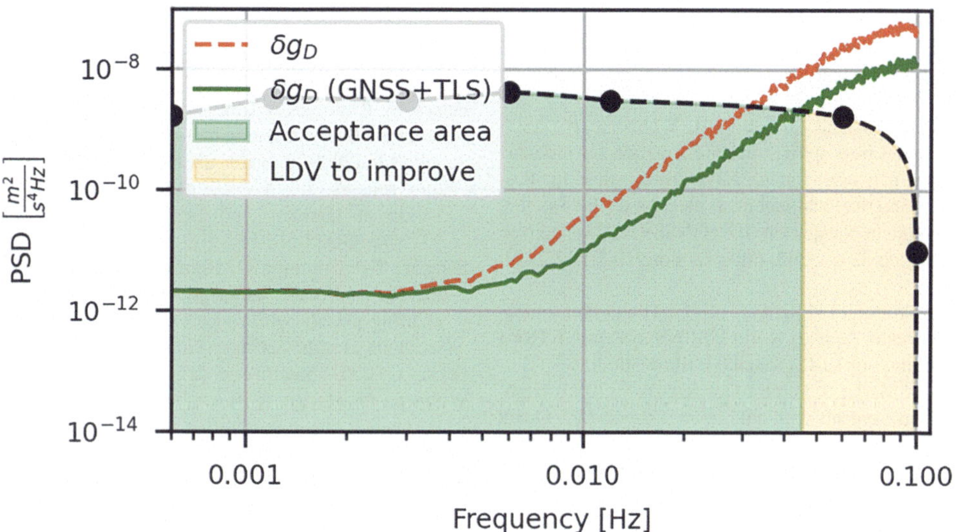

Fig. 7 PSD of the AeroQGrav error including the GNSS and TLS fusion (green line)

A general approach has been employed to model the sensors contributing to the error budget for reconstructing the gravity signal. The analysis focuses on the sensor noise, characterised as white noise with zero mean.

Particular attention has been given to the kinematic sensors GNSS, TLS, and LDV, which play a crucial role in separating gravitational from kinematic acceleration. It has been shown that, provided that the atom interferometer functions correctly, the kinematic sensors are the factor that limits the system's actual performance to meet the requirements only up to about 0.03 Hz.

A sensor fusion direct approach has been introduced to mitigate errors in the reconstructed gravity signal and broaden the frequency range in which the requirements are met. A GNSS-based robust time synchronisation has been presented to enable all sensors to time-tag their measurements consistently. It has been demonstrated that the fusion of GNSS and TLS ensures precision below 10 $\mu m/s^2$ for frequencies up to approximately 0.05 Hz.

The next step is to integrate the LDV, currently under testing, into the AeroQGrav sensor fusion. The adoption of the LDV is expected to cover the remaining frequencies up to 0.1 Hz, ultimately achieving 1 $\mu m/s^2$ with a 5-s time interval.

In the near future, initial flight campaigns will take place to evaluate the TLS and GNSS sensor fusion. This interaction will pave the way for achieving high-precision kinematic acceleration and integrating the atom interferometer for the final gravity signal reconstruction.

Acknowledgements The AeroQGrav partners, the German Federal Agency for Cartography and Geodesy (BKG) and the Institute of Flight Guidance (IFF) at the Technical University of Braunschweig (TUB), are acknowledged for providing valuable insights and information regarding geodetic applications and previous flight campaign analyses.

Funding The research work presented here has been funded by the German Federal Ministry of Research, Technology and Space (BMFTR), FKZ 13 N16519.

References

Becker D, Becker M, Olesen AV, Nielsen JE, Forsberg R (2016) Latest results in strapdown airborne gravimetry using an iMAR RQH unit. In: Proceedings of the 4th IAG symposium on terrestrial gravimetry: static and mobile measurements (TG-SMM 2016), pp 19–25

Berman PR (1997) Atom interferometry. Springer

Bidel Y, Zahzam N, Bresson A, Blanchard C, Cadoret M, Olesen AV, Forsberg R (2020) Absolute airborne gravimetry with a cold atom sensor. J Geod 94:1–9. https://doi.org/10.1007/s00190-020-01350-2

Bidel Y, Zahzam N, Bresson A, Blanchard C, Bonnin A, Bernard J et al (2023) Airborne absolute gravimetry with a quantum sensor, comparison with classical technologies. J Geophys Res Solid Earth 128(4). https://doi.org/10.1029/2022JB025921

Bruton AM (2000) Improving the accuracy and resolution of SINS/DGPS airborne gravimetry, vol 20145. University of Calgary, Calgary

Darugna F, Wübbena T, Wübbena G, Albers H, Wübbena JB (2023) Improving GNSS-based tropospheric delay estimation for airborne quantum gravimetry: first results using NWM forecasting. In: Proceedings of the 36th international technical meeting of the Satellite Division of the Institute of Navigation (ION GNSS+ 2023), Denver, Colorado, Sept 2023, pp 3233–3248. https://doi.org/10.33012/2023.19266

Ding W, Wang J, Li Y, Mumford P, Rizos C (2008) Time synchronization error and calibration in integrated GPS/INS systems. ETRI J 30(1):59–67. https://doi.org/10.4218/etrij.08.0106.0306

Geiger R, Landragin A, Merlet S, Pereira Dos Santos F (2020) High-accuracy inertial measurements with cold-atom sensors. AVS Quant Sci 2(2):024702

Guo D, Chen X, Xue Z, He H, Xing L, Ma X, Niu X (2023) High-accuracy quasi-geoid determination using Molodensky's series solutions and integrated gravity/GNSS/leveling data. Remote Sens 15(22):5414

Hirt C, Schmitz M, Feldmann-Westendorff U, Wübbena G, Jahn CH, Seeber G (2011) Mutual validation of GNSS height measurements and high-precision geometric-astronomical leveling. GPS Solut 15(2):149–159

Hofmann-Wellenhof B, Moritz H (2005) Physical geodesy. Springer, Wien, New York

Jekeli C (2000) Inertial navigation systems with geodetic applications. Walter de Gruyter GmbH & Co. KG

Jekeli C (2003) Statistical analysis of moving-base gravimetry and gravity gradiometry. The Ohio State University, Report, 446

Jellum ER, Bryne TH, Johansen TA, Orlandić M (2022) The syncline model-analyzing the impact of time synchronization in sensor fusion. In: 2022 IEEE conference on control technology and applications (CCTA). IEEE, pp 1446–1453. https://doi.org/10.1109/CCTA49430.2022.9966179

Johann F, Becker D, Becker M, Forsberg R, Kadir M (2019) The direct method in strapdown airborne gravimetry – a review. Z Geod Geoinform Land Manag (ZFV) 5. https://doi.org/10.12902/zfv-0263-2019

Kwon JH, Jekeli C (2001) A new approach for airborne vector gravimetry using GPS/INS. J Geod 74(10):690–700

Okada K (2021) A historical overview of the past three decades of mineral exploration technology. Nat Resour Res 30:2839–2860. https://doi.org/10.1007/s11053-020-09721-4

Olesen AV (2002) Improved airborne scalar gravimetry for regional gravity field mapping and geoid determination. University of Copenhagen, Copenhagen

Pavlis NK, Holmes SA, Kenyon SC, Factor JK (2012) The development and evaluation of the Earth Gravitational Model 2008 (EGM2008). J Geophys Res Solid Earth 117(B4). https://doi.org/10.1029/2011JB008916

Peters A, Chung KY, Chu S (1999) Measurement of gravitational acceleration by dropping atoms. Nature 400(6747):849–852

Schubert C (2023) Towards absolute airborne gravimetry using quantum sensors. In: XXVIII general assembly of the International Union of Geodesy and Geophysics (IUGG). https://doi.org/10.57757/IUGG23-5019

Schwarz KP, Colombo O, Hein G, Knickmeyer ET (1992) Requirements for airborne vector gravimetry. In: From Mars to Greenland: charting gravity with space and airborne instruments: fields, tides, methods, results. Springer, New York, pp 273–283

Seeber G (2003) Satellite geodesy. Walter de Gruyter, Berlin

Wübbena G, Bagge A, Boettcher G, Schmitz M, Andree P (2001) Permanent object monitoring with GPS with 1 millimeter accuracy. In: Proceedings of the 14th international technical meeting of the Satellite Division of the Institute of Navigation (ION GPS 2001), pp 1000–1008

Wübbena G, Schmitz M, Warneke A (2019) Geo++ absolute multi frequency GNSS antenna calibration. In: Presentation at the EUREF Analysis Center (AC) workshop, Oct 2019, Warsaw, pp 16–17

Evaluation of Novel Airborne Gravity Levelling Methods

Felix Johann, Hannes Eisermann, Antonia Ruppel, and Graeme Eagles

Abstract

In dynamic gravimetry, i.e. airborne and shipborne gravimetry, levelling methods are used to refine gravity disturbance results based on neighbouring trajectories. In the traditional crossover adjustment, line biases are estimated using gravity disturbance residuals at trajectory line crossings as input to a least-squares adjustment. In an alternative method, the results along the complete trajectory are used to estimate the gravity disturbance field in the survey area and line biases in a one-step least-squares adjustment applying spherical radial basis functions. This makes the bias estimation more robust since the observations are not restricted to a small number of residuals at crossings strongly affected by random errors. Adjustment becomes applicable to a wider range of campaigns including irregular trajectories without many crossings. Within the scope of this work, existing methods that estimate line biases are extended to bias estimation based on trajectory segments with inter-bias interpolation. The extended method can be particularly useful for irregular trajectories without a sufficient number of line crossings. The introduced levelling methods are evaluated at the example of three airborne campaigns: a fixed wing survey at Germany with a very dense grid, a fixed wing survey in East Antarctica with varying line separation, and a helicopter survey on Svalbard with highly irregular trajectories. It is shown that the levelling method based on spherical radial basis functions improves the precision in all evaluated campaigns, even when a traditional crossover levelling is not possible.

Keywords

Adjustment · Airborne gravimetry · Levelling · Radial basis functions · Strapdown

1 Introduction

In dynamic gravimetry, the gravity field is observed from a moving platform like an aircraft (airborne gravimetry) or a ship (shipborne gravimetry). In contrast to highly precise static terrestrial gravimetry, dynamic gravimetry facilitates the rapid coverage of large and remote regions. Satellite gravimetry enables the acquisition of gravity data on a global scale; however, due to the distance to the geoid, the spatial resolution is finite. Consequently, dynamic gravimetry has become a widely adopted approach in numerous applications, including regional geoid determination, geological research, exploration, and glaciology.

F. Johann (✉)
Physical and Satellite Geodesy, Technical University of Darmstadt, Darmstadt, Germany
e-mail: johann@psg.tu-darmstadt.de

H. Eisermann · G. Eagles
Alfred Wegener Institute for Polar and Marine Research, Bremerhaven, Germany
e-mail: hannes.eisermann@awi.de; graeme.eagles@awi.de

A. Ruppel
Federal Institute for Geosciences and Natural Resources (BGR), Hannover, Germany
e-mail: antonia.ruppel@bgr.de

J. T. Freymueller, L. Sànchez (eds.), *International Symposium on Gravity, Geoid and Height Systems 2024 (GGHS2024)*, International Association of Geodesy Symposia 158, https://doi.org/10.1007/1345_2025_291

Typically, a precision of gravity results between approximately 0.5 and 2 mGal (Studinger et al. 2008; Forsberg and Olesen 2010; Yuan et al. 2020; Johann 2023) (1 mGal = 10^{-5} m/s^2) is achievable after "end-matching" (Kwon and Jekeli 2001), i.e. by tying the results of relative gravity measurements to known terrestrial gravity values at the airfield/harbour between the flights/cruises. However, significant systematic errors can remain in the end-matched results due to sensor drifts of the gravimeters and possibly other unknown effects. These systematic deviations can be mitigated through levelling procedures, also referred to as "adjustment".

A straightforward and widely used levelling strategy is the crossover levelling. Assuming a static gravity field, the gravity residuals at obtained trajectory intersections, i.e. the "crossover points" (COs), are used to estimate drift parameters in a least-squares adjustment. A gravity bias can be estimated per flight/cruise or per approximately straight trajectory line (henceforth referred to as "line"). Thereby, the redundancy at the COs is used to enhance the precision of the gravity results. The seven campaigns described by Becker et al. (2016) and Johann et al. (2020) exhibit an average precision enhancement of 48%. When at least two COs are available per line, a linear drift can be estimated per line in addition to the line bias (Glennie and Schwarz 1999; Hwang et al. 2006; Zhang et al. 2017). In scenarios where the network contains a low number of COs, the risk of achieving overly optimistic precision values after adjustment can be reduced by incorporating correction factors into the crossover residuals at lines with few COs (Becker 2016). Nevertheless, in order to avoid a distortion of the resulting gravity network, a line-wise crossover levelling should only be applied if many COs per line are available. This outlines the primary limitation of crossover levelling: The adjusted results are dependent on a limited number of crossover residuals with the risk of result distortion. Hence, a reliable line-wise adjustment is only feasible for campaigns with a high number of COs per line; that is, many cross lines are required increasing the campaign duration and costs.

Instead of COs, the spatial neighbourhood of gravity observations can serve as input for gravity levelling, assuming similar low wavelength gravity at nearby points. For the levelling of magnetic field observations, various approaches based on directional filtering and (weighted) spatial median/average filters exist (Mauring and Kihle 2006; Ishihara 2015), as well as field modelling with subsequent line-wise drift estimation (White and Beamish 2015).

While the aforementioned approaches do not reflect the physical properties of the Earth's gravity field, Vyazmin (2020) uses strapdown gravimeter observations to estimate the sensor errors of the inertial measurement unit (IMU) simultaneously with the disturbance potential of the gravity field, modelled with spherical radial basis functions (SRBFs). The gravity disturbance is then computed as the gradient of the disturbance potential. A substantial limitation of this approach is that, since the resulting gravity disturbance is uniform at both adjacent lines at a CO, precision evaluation via crossover evaluation or repeated line analysis is not possible (Vyazmin et al. 2021). Li (2021) uses (unlevelled) airborne gravity disturbance data to estimate the gravity disturbance field (rather than the disturbance potential field). Within the same least-squares adjustment, he estimates gravity disturbance biases for each line. Like in the crossover levelling, the final gravity disturbance results are obtained by removing the estimated line biases from the unlevelled gravity disturbance. The computation of a precision indicator based on the adjusted crossover residuals is possible. The primary advantage of this method over CO levelling is that it uses the complete gravity disturbance trajectory for levelling. Hence, the method mitigates the risk of network distortion and avoids the necessity for a substantial number of crossing lines for levelling. Furthermore, a 2-D gravity disturbance field is estimated simultaneously within the least-squares adjustment.

This chapter presents the functional model of the CO and SRBF levelling approaches and their validation at the example of three airborne gravimetry campaigns with completely different trajectory designs. The existing levelling approaches are extended by estimating biases for equidistant trajectory segments instead of straight lines to allow for an application for campaigns with highly irregular trajectories. Section 2 presents the extended CO and SRBF levelling approaches. Section 3 introduces the three airborne gravimetry campaigns. The results of Sect. 3 are discussed in Sect. 4.

2 Methods

In this study, the levelling approaches are applied to results of gravity disturbance, i.e. the difference between gravity and normal gravity $\boldsymbol{\gamma}$, which are obtained with the direct method of strapdown gravimetry. This method is based on the subtraction of the 3-D accelerometer observations of an IMU, i.e. the so-called specific force $\boldsymbol{f}$, from the kinematic acceleration of the vehicle, yielding the gravity disturbance

$$\delta \boldsymbol{g}^i = \ddot{\boldsymbol{r}}^i - \boldsymbol{f}^i - \boldsymbol{\gamma}^i \tag{1}$$

in an inertial frame i. The kinematic acceleration is obtained as the second numerical derivative of the position, determined using Global Navigation Satellite Systems (GNSS). If the gravity disturbance is required to be expressed in a navigation frame, e.g. North, East, and Down, the specific force must be rotated from the body-fixed to the navigation frame. Additionally, the Eötvös correction considers (fictitious) accelerations due to the rotating sensor frames. Details on the applied algorithm of the direct method of strapdown gravimetry can be found in the works of Johann et al. (2019) and Johann (2023). The commercial software NovAtel Waypoint Inertial Explorer is used for the GNSS positioning, applying a precise point positioning (PPP) approach with satellite correction data from the Centre for Orbit Determination in Europe (CODE) and for the GNSS/IMU integration to compute the vehicle orientation. The remaining calculations are performed with MATLAB software developed at the Technical University of Darmstadt (TUDa).

The CO and SRBF levelling approaches, where a bias is estimated per flight or line, are presented in detail in Sects. 2.1 and 2.2, respectively. Section 2.2 also outlines the advantages of modelling gravity field quantities using SRBFs. Section 2.3 introduces modifications to both levelling approaches that allow for the levelling of irregular flight trajectories.

2.1 Line-Wise CO Levelling

In the crossover levelling method used in this chapter, one bias is estimated per line. The model is based on Becker (2016) and Hwang et al. (2006). The residual $\chi_{A,B}$ at a CO with the observed gravity disturbances δg_A, δg_B at the adjacent lines A, B is the difference

$$\chi_{A,B} = \delta g_B - \delta g_A = \kappa_B - \kappa_A, \tag{2}$$

assuming error-free gravity disturbance except for the line biases κ_A, κ_B. Based on the linear functional model in Eq. (2), the line biases are estimated within a least-squares adjustment (Gauss-Markov model)

$$\boldsymbol{L} = \boldsymbol{A}\ \boldsymbol{X} + \boldsymbol{\varepsilon}, \quad \boldsymbol{X} = \left(\boldsymbol{A}^{\mathrm{T}}\boldsymbol{P}\boldsymbol{A}\right)^{-1}\boldsymbol{A}^{\mathrm{T}}\boldsymbol{P}\boldsymbol{L}, \tag{3}$$

with $\boldsymbol{L}$ being the observations, $\boldsymbol{A}$ being the design matrix, $\boldsymbol{X}$ being the parameters to be estimated, and $\boldsymbol{\varepsilon}$ being normally distributed random errors in the observations. Assuming uncorrelated observations with equal accuracy, the weight matrix $\boldsymbol{P}$ becomes the unit matrix. The observation vector contains the crossover residuals $\boldsymbol{\chi}$; the parameter vector contains the line biases as

$$\boldsymbol{L} = \begin{pmatrix} \boldsymbol{\chi} \\ 0 \end{pmatrix}_{(C+1)\times 1}, \quad \boldsymbol{X} = \boldsymbol{\kappa}_{M\times 1}, \tag{4}$$

with C, M being the number of COs and lines, respectively. A row of the complete design matrix

$$\boldsymbol{A}_{(C+1)\times M} = \left(\boldsymbol{A}_1^{\mathrm{T}}\ \boldsymbol{A}_2^{\mathrm{T}}\ \cdots\ \boldsymbol{A}_C^{\mathrm{T}}\ \boldsymbol{A}_{C+1}^{\mathrm{T}}\right)^{\mathrm{T}}, \tag{5}$$

corresponding to one CO j and the last row is obtained as

$$\boldsymbol{A}_j = \left(\boldsymbol{0}^{\mathrm{T}}\ -1\ \boldsymbol{0}^{\mathrm{T}}\ 1\ \boldsymbol{0}^{\mathrm{T}}\right)_{1\times M}, \quad \boldsymbol{A}_{C+1} = \boldsymbol{1}^{\mathrm{T}}. \tag{6}$$

The columns of -1 and 1 correspond to the indices of the adjacent lines of the CO. The last row avoids a rank defect by introducing the pseudo-observation that the sum of all line biases is zero.

2.2 Line-Wise SRBF Levelling

The approach of the SRBF-based levelling is based on Li (2021) and is executed after the computation of the unlevelled gravity disturbance. The gravity disturbance results of a global gravity model evaluated up to degree and order $N_{\min}$ may be subtracted from the gravity disturbance before SRBF levelling. For an epoch i and a line l, the gravity disturbance portion in the bandwidth between the orders $N_{\min}$, $N_{\max}$ is assumed to be

$$\delta g_{i,l} = \kappa_l + \delta g_{\mathrm{mod},i} = \kappa_l + \frac{GM}{R^2}\sum_{k=1}^{K}\alpha_k B_{i,k}, \tag{7}$$

i.e. the sum of the line bias κ_l, and an SRBF-based regional gravity disturbance model $\delta g_{\mathrm{mod},i}$. GM is the geocentric gravitational constant, R is the Earth's mean radius under spherical approximation, $B_{i,k}$ is an SRBF with a scaling factor α_k, and K is the total number of SRBFs. An SRBF is obtained as

$$B_{i,k} = \sum_{n=N_{\min}}^{N_{\max}} b_n \left(\frac{R}{r_i}\right)^{n+2} (n+1)\sqrt{2n+1}\ \overline{P}_n\left(\cos\psi_{i,k}\right). \tag{8}$$

As kernel b_n, the Shannon kernel $b_n = 1$ is used in this study. For details on kernel types, see Bentel et al. (2013) and Lieb et al. (2016). r_i is approximated as $r_i = R + h_i$, with h_i being the ellipsoidal height. The weighting based on the spherical distance

$$\psi_{i,k} = \overline{\boldsymbol{r}}_i^{\mathrm{T}}\overline{\boldsymbol{r}}_k, \tag{9}$$

with $\overline{\boldsymbol{r}}_i, \overline{\boldsymbol{r}}_k$ being the normed Cartesian geocentric coordinates, i.e. unit vectors, of the δg observation point i and the SRBF origin k, respectively (Klees et al. 2008), is computed using normed Legendre polynomials $\overline{P}_n$ (see e.g. Torge et al. 2023) for a degree n. If the factors in Eq. (8) would be adapted, other functionals of the gravity field like the potential or gradient could be estimated instead of the gravity disturbance field (Liu et al. 2020).

In a least-squares adjustment according to Eq. (3), Eq. (7) serves as linear observation equation, i.e. the obtained gravity disturbances δg_i for N epochs i along the trajectory are the observations, and the scaling factors α_k of the SRBFs k are estimated along with the line biases κ_l,

$$L = \begin{pmatrix} \delta g \\ 0 \end{pmatrix}_{(N+1)\times 1}, \quad X = \begin{pmatrix} \alpha \\ \kappa \end{pmatrix}_{(K+M)\times 1}. \tag{10}$$

The design matrix is set up as

$$\begin{aligned} &A_{(N+1)\times(K+M)} = \left(A_1^{\mathrm{T}}\ A_2^{\mathrm{T}} \cdots A_N^{\mathrm{T}}\ A_{N+1}^{\mathrm{T}} \right)^{\mathrm{T}}, \quad \text{with} \\ &A_i = \left(\tfrac{GM}{R^2} B_{i,1} \cdots \tfrac{GM}{R^2} B_{i,k}\ A_{i,\mathrm{end}} \right)_{1\times(K+M)}, \\ &A_{i,\mathrm{end}} = \left(\mathbf{0}^{\mathrm{T}}\ 1\ \mathbf{0}^{\mathrm{T}} \right), \quad A_{N+1} = \left(\mathbf{0}_{1\times K}\ \mathbf{0}_{1\times M} \right), \end{aligned} \tag{11}$$

where the last row introduces the same pseudo-observation like in Eq. (6). The column of 1 in A_i corresponds to the index of the current line in the parameter vector in Eq. (10).

The quality of the gravity field parameters estimated in SRBF levelling strongly depends on the selection of some processing parameters, such as the kernel functions (Bentel et al. 2013; Lieb et al. 2016), the distribution and number of SRBFs (Klees et al. 2008; Bentel et al. 2013; Lieb et al. 2016), the subsampling and ratio of the gravity disturbance observations in comparison to the number of SRBFs (Klees et al. 2008; Li 2018), and the SRBF bandwidth (Klees et al. 2008; Lieb et al. 2016). For details on an appropriate parameter selection, the reader is referred to the mentioned references. However, due to the large number of possibilities and combinations, some trial and error is required to find optimal results for modelling the gravity (disturbance) field in a particular campaign or target region (Ma 2024). It has been found that the computation area, i.e. the area covered by SRBF origins, needs to be larger than the observation area, and the latter should be larger than the target area, because the result quality decreases near the outer observation boundary (Bentel et al. 2013; Lieb et al. 2016; Liu et al. 2024).

2.3 Modification: Segment-Wise Levelling

The dependence on flight lines in the levelling methods introduced in Sects. 2.1 and 2.2 has drawbacks. Typically, the duration of the lines varies, and they are not evenly distributed throughout a flight. When parts of a flight or even the whole flight do not have straight lines, they cannot be levelled with the traditional line-based methods. This chapter introduces a new type of levelling that divides the flights into equidistant segments that are independent of possibly irregular lines (Fig. 1). Instead of estimating a bias per line, biases are estimated in-between of the segments. The number of segments can be selected by the user based on the expected non-linear variation of the gravity disturbance error.

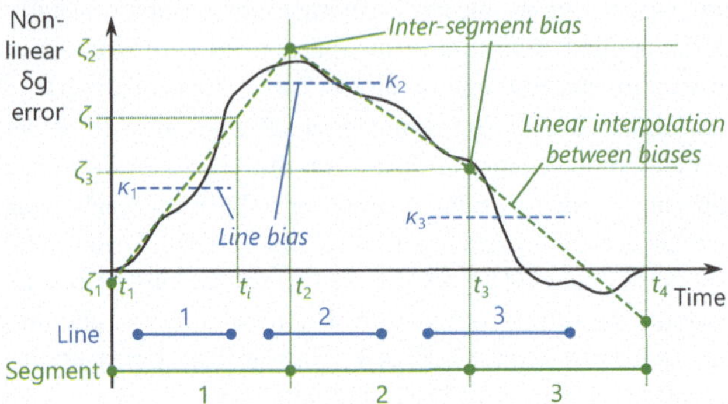

Fig. 1 Line-wise (blue) and segment-wise (green) levelling methods on the example of a non-linear gravity disturbance error during a fictitious flight with three lines, divided into three segments. In the line-wise levelling type, the error is approximated by a bias that is estimated for each approximately straight flight line, omitting trajectory parts between the lines. In the segment-wise levelling type, the trajectory is divided into equidistant segments, and the error is interpolated between biases before and after each segment

The gravity disturbance error

$$\begin{aligned} \zeta_i &= \zeta_1 + (\zeta_2 - \zeta_1)\frac{t_i - t_1}{T} = (1 - p_i)\ \zeta_1 + p_i \zeta_2, \\ &\text{with} \quad p_i := \frac{t_i - t_1}{T} = \frac{t_i - t_1}{t_2 - t_1}, \end{aligned} \tag{12}$$

at an epoch t_i of segment 1 is obtained by interpolation between the previous and subsequent biases ζ_1, ζ_2 at the epochs t_1, t_2 at the beginning and end of the line. The segment duration T is equal for all segments.

For the segment-wise type, the residual at a CO becomes

$$\begin{aligned} \chi_{A,B} &= \delta g_B - \delta g_A = \zeta_B - \zeta_A \\ &= (1 - p_B)\,\zeta_{B,1} + p_B \zeta_{B,2} + (p_A - 1)\,\zeta_{A,1} - p_A \zeta_{A,2} \end{aligned} \tag{13}$$

by replacing the line biases κ_A, κ_B in Eq. (2) by the gravity disturbance errors ζ_A, ζ_B of Eq. (12), which are interpolated between the neighbouring inter-segment biases $\zeta_{A,1}$, $\zeta_{A,2}$, $\zeta_{B,1}$, $\zeta_{B,2}$. The indices A, B represent both adjacent trajectory segments at the CO; the indices 1, 2 represent the beginning and end of the particular segment. While the observations L remain unchanged compared to Eq. (4), the parameter vector becomes $X = \boldsymbol{\zeta}_{(S+1)\times 1}$. Since one bias is estimated before and after each segment, the length of X is the number of segments S plus one. To enable a least-squares adjustment according to Eq. (3), the design matrix is set up like in Eq. (5), with $S + 1$ columns, but a row j of the design matrix becomes

$$A_j = \left(\mathbf{0}^{\mathrm{T}}\ p_A - 1\ -p_A\ \mathbf{0}^{\mathrm{T}}\ 1 - p_B\ p_B\ \mathbf{0}^{\mathrm{T}} \right)_{1\times M}, \tag{14}$$

translated from Eq. (13). The indices of the non-zero columns correspond to the indices of the parameters $\zeta_{A,1}$, $\zeta_{A,2}$, $\zeta_{B,1}$, $\zeta_{B,2}$. The last row consists of ones, like in Eq. (6).

The segment-wise type can also be applied to the SRBF levelling. The observation equation then becomes

$$\delta g_{i,s} = (1 - p_i)\,\zeta_{s,1} + p_i \zeta_{s,2} + \delta g_{\mathrm{mod},i}, \quad (15)$$

where, in comparison to Eq. (7) of the CO levelling, the line bias κ_l is replaced by the gravity disturbance error of Eq. (12), with $\zeta_{s,1}$, $\zeta_{s,2}$ being the neighbouring biases at the beginning and the end of the current segment s. The observation vector $\boldsymbol{L}$ remains unchanged compared to Eq. (10); the parameter vector becomes $\boldsymbol{X} = \left(\boldsymbol{\alpha}^{\mathrm{T}}\ \boldsymbol{\zeta}^{\mathrm{T}}\right)^{\mathrm{T}}_{(K+S+1)\times 1}$. The design matrix of Eq. (11) is modified according to Eq. (13) as

$$\begin{aligned} \boldsymbol{A}_{(N+1)\times(K+S+1)} &= \left(\boldsymbol{A}_1^{\mathrm{T}}\ \boldsymbol{A}_2^{\mathrm{T}} \cdots \boldsymbol{A}_N^{\mathrm{T}}\ \boldsymbol{A}_{N+1}^{\mathrm{T}}\right)^{\mathrm{T}}, \quad \text{with} \\ \boldsymbol{A}_i &= \left(\tfrac{GM}{R^2} B_{i,1} \cdots \tfrac{GM}{R^2} B_{i,k}\ \boldsymbol{A}_{i,\mathrm{end}}\right)_{1\times(K+S+1)}, \\ \boldsymbol{A}_{i,\mathrm{end}} &= \left(\boldsymbol{0}^{\mathrm{T}}\ 1 - p_i\ p_i\ \boldsymbol{0}^{\mathrm{T}}\right), \\ \boldsymbol{A}_{C+1} &= \left(\boldsymbol{0}_{1\times K}\ \boldsymbol{1}_{1\times(S+1)}\right). \end{aligned} \quad (16)$$

3 Validation in Dynamic Campaigns

The line- and segment-wise CO and SRBF levelling methods introduced in Sect. 2 were applied to the gravity disturbance results of three airborne campaigns. The Odenwald 2018 campaign ("ODW2018") and its results are presented in Sect. 3.1. With a dense line network and significant errors in the unlevelled gravity disturbance, ODW2018 is well suited for CO and SRBF levelling. The RIISERBATHY/GEA-VI campaign ("GEA-VI"), which is presented in Sect. 3.2, includes different areas with varying line network density. The trajectory of the CASE 23-Aerogeophysics ("CASE 23") campaign lacks straight lines and has a very low number of COs. It is therefore a good example for testing the segment-wise SRBF levelling method. The main characteristics and SRBF settings of all three campaigns are summarised in Table 1.

In all campaigns, the iMAR iNAV-RQH-1003 (iMAR Navigation 2012; Johann 2023) owned by TUDa was used as strapdown gravimeter. In the GEA-VI and CASE 23 campaigns, it was additionally equipped with a temperature-stabilising housing, the iTempStab-AddOn. In the SRBF levelling, the SRBF origins were homogeneously distributed based on Reuter grids (Eicker 2008). The other SRBF modelling settings, including the origin separation, the buffer radius defining the area covered by the origins, and the minimal and maximal model degrees in Eq. (8), were selected for the campaigns individually as stated in Table 1, depending on the covered area, the flight velocity and altitude, and the variability of the gravity disturbance field. If the lower wavelengths of a global gravitational model were subtracted from the unlevelled gravity disturbance, the EGM2008 (Pavlis et al. 2012) was used. While the choice of the above SRBF settings required careful tuning, the addition and choice of a global model were found to lead to only small result changes. Table 2 shows the crossover precision based on the root mean square error (RMSE), which is obtained by dividing the RMS of the residuals at the COs by $\sqrt{2}$. CO residuals with height differences of both adjacent trajectories of more than 150 m were removed from the computation to ensure comparability of the precision indicators between different campaigns. The number of segments was selected in a process optimising the CO RMSE. The optimal number of segments decreases with a decreasing non-linear gravity disturbance error variation.

Table 1 Campaign characteristics (upper part) and SRBF settings (lower part) (ODW2018 data based on Johann et al. (2020))

Properties/settings	ODW2018	GEA-VI	CASE 23
Mean ellipsoidal height [m]	947	1,005	486
Mean velocity [m/s]	54	76	47
Turbulence (RMS-g) [mm/s^2]	128	240	608
Filter length (−6 dB) [s]	120	130	160
Half-wave. resolution [km]	3.3	4.9	3.7
Variability $\lvert\delta g\rvert$ [mGal/km]	2.7	1.6	2.1
Number of COs	222	184	32
Segments per flight	55	2	1
SRBF separation [km]	7.5	50	20
SRBF buffer [km]	20	360	60
SRBF degree range	400–2,400	200–800	450–850
Global model subtracted	No	Yes	Yes

Table 2 Crossover point RMSE (all values in mGal)

Levelling	ODW2018		GEA-VI		CASE 23	
	CO	SRBF	CO	SRBF	CO	SRBF
None	4.05		1.71		2.16	
Line-wise	0.64	0.63	1.01	1.53	–	–
Segment-wise	0.53	0.83	1.02	1.24	–	2.08

3.1 Dense Regular Grid: Odenwald 2018

In March 2018, a test flight (ODW2018) was conducted with a light aircraft Cessna 206 "Stationair 6" at the low mountain range Odenwald near Frankfurt, Germany. Details on the campaign are presented by Johann et al. (2020) and Johann (2023). The presence of strong non-linear drifts in the unlevelled gravity disturbance results (Fig. 2) and the high CO RMSE of 4.05 mGal are attributed to environmental temperatures close to the freezing point in the morning and lack of temperature stabilisation. However, the large number of COs per line permits a robust line bias estimation in the

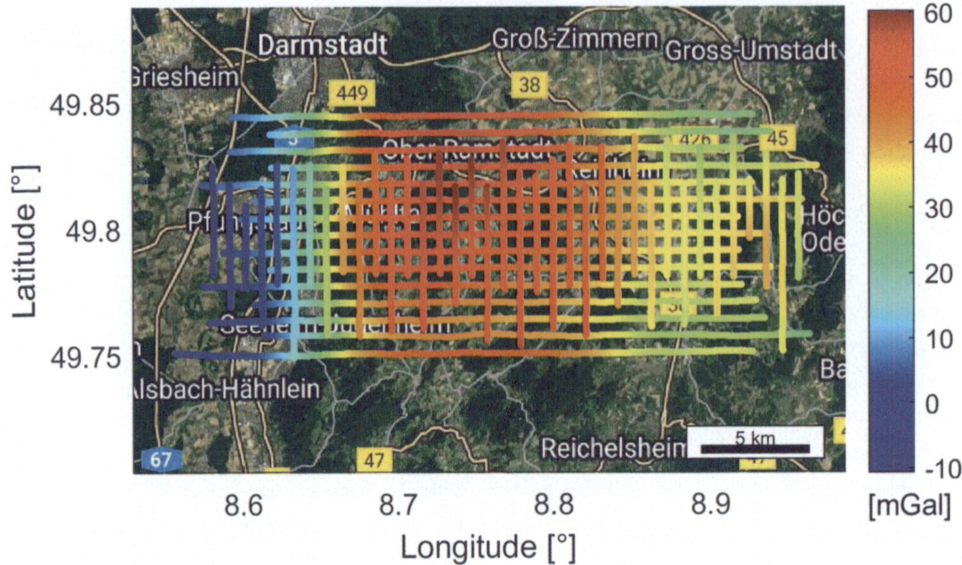

Fig. 2 Gravity disturbance [mGal] along the lines of ODW2018 before levelling (map data: Google, Landsat/Copernicus)

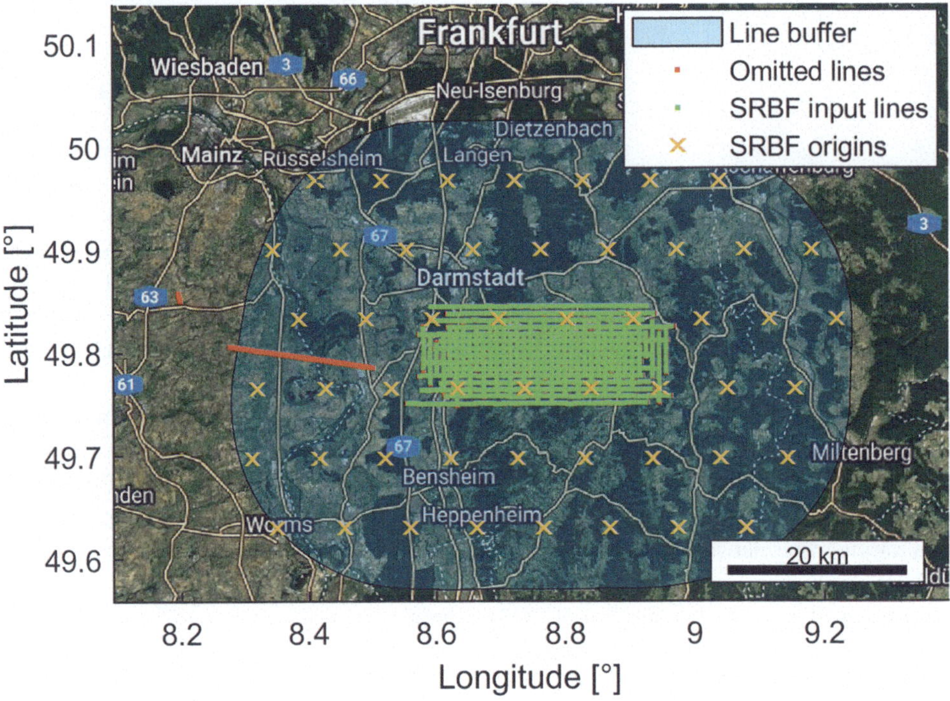

Fig. 3 ODW2018 lines and SRBF origins within a buffer area (map data: Google, Landsat/Copernicus)

CO levelling, resulting in a strongly improved RMSE of 0.64 mGal (see Table 2). When the novel segment-wise CO adjustment (see Sect. 2.3) is applied, dividing the flight into 55 segments, the RMSE further improves to 0.53 mGal.

Figure 3 depicts the distribution of the SRBF origins in and around the survey area. All Reuter grid positions located beyond a buffer area surrounding the flight lines were omitted. Isolated lines west of the grid were removed beforehand. After line-wise SRBF levelling, the precision is on par with CO levelling (see Table 2). The estimated line biases are illustrated in Fig. 4. While the obtained biases are very similar, the SRBF method allows for the estimation of additional outer lines with an insufficient number of CO for CO adjustment. For the segment-wise type, the CO levelling outperforms the SRBF levelling (see Table 2).

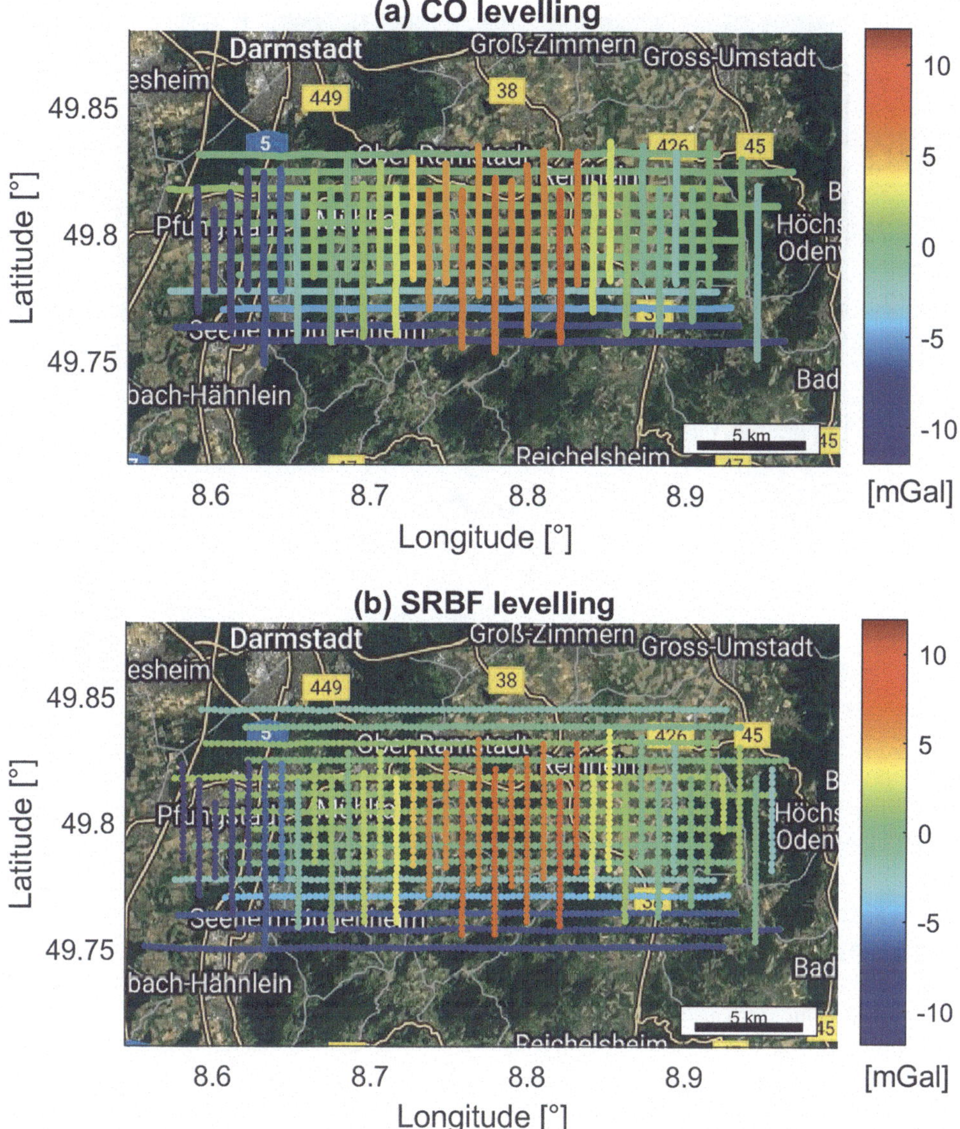

Fig. 4 Estimated line biases [mGal] of ODW2018 (map data: Google, Landsat/Copernicus): (**a**) CO levelling; (**b**) SRBF levelling

3.2 Large Polar Grid: RIISERBATHY/GEA-VI 2022/23

In the Antarctic summer of 2022/23, the Alfred Wegener Institute for Polar and Marine Research (AWI) and the Federal Institute for Geosciences and Natural Resources (BGR) conducted an airborne campaign, designated RIISERBATHY/GEA-VI (GEA-VI), at the Riiser-Larsen Ice Shelf, East Antarctica. Based at the Neumayer III station in the Northeast, ten survey flights were conducted with a Basler BT-67, resulting in a dense trajectory net in the Northeast and lines widening towards the Southeast (Fig. 5). Details on the campaign and a comparison of the strapdown gravimeter to an additionally installed stable-platform gravimeter can be found in the work of Johann et al. (2025a). The strapdown results are available in the study of Johann et al. (2025b). Without levelling, the CO RMSE is 1.71 mGal. After line-wise CO levelling, the RMSE becomes 1.01 mGal (see Table 2). Given the negligible line-to-line drift observed in this campaign (Johann et al. 2025a), the segment-wise CO levelling performs best (1.02 mGal) when only two segments per flight are used.

Applying line-wise SRBF levelling results in a reduction of formerly large CO residuals in the dense trajectory area in the Northeast, while CO residuals at the wider line spacing in the Southwest generally increase (Fig. 6a, b). The RMSE slightly improves to 1.53 mGal. The segment-wise SRBF levelling performs better over the complete survey area (Fig. 6c), resulting in an RMSE of 1.24 mGal.

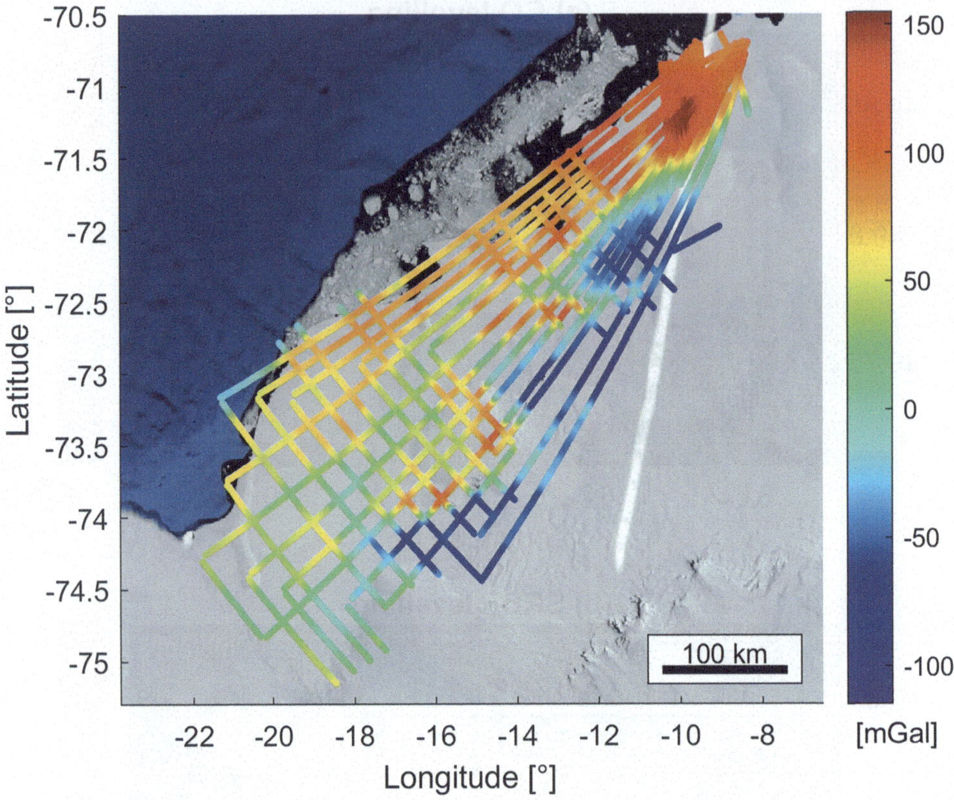

Fig. 5 Gravity disturbance [mGal] along the lines of GEA-VI before levelling (map data: Google, U.S. Geological Survey)

3.3 Irregular Trajectory: CASE 23-Aerogeophysics

While the aforementioned campaigns used fixed-wing aircraft, a helicopter (Airbus H125) was deployed during the expedition CASE 23-Aerogeophysics (CASE 23) in August 2023 across Nordenskiöld Land, Svalbard, led by the BGR. Like at GEA-VI, the gravimeter was equipped with temperature stabilisation and was turned on also between the flights to minimise the influence of thermally induced drifts. Due to unfavourable weather conditions, all four flights followed the valleys of Nordenskiöld Land instead of flying along straight flight lines (Fig. 7); the flight altitude followed the terrain height. After the removal of only the narrowest turns from the results, 32 COs remained, with an RMSE of 2.16 mGal. Due to the lack of straight lines and with the low number of COs, line-wise levelling and CO levelling were not possible (see Table 2). Consequently, the sole viable levelling method was the segment-wise SRBF levelling, which resulted in a slightly improved RMSE of 2.08 mGal when two biases were estimated per flight. The magnitude of the interpolated bias was low (Fig. 8), particularly when contrasted to the ODW2018 campaign that did not incorporate temperature stabilisation (see Fig. 4).

4 Discussion

The application of the levelling methods introduced in Sect. 2 to three airborne campaigns with very different trajectory layouts and local gravity fields unveils differing properties of the methods. The ODW2018 flight comprises almost ideal conditions for all levelling methods because it features a dense line network with many COs per line and levelling has a large improvement potential due to the large thermally induced gravity disturbance errors. All methods lead to strong improvements (see Table 2). The good performance of the segment-wise CO levelling suggests that the error in the un-levelled results is not line dependent.

The disparity in the enhancement of CO residuals across the Northeast and Southwest regions of GEA-VI indicates that the selection of SRBF settings like the SRBF origin distribution and spacing may not be equally well suited to trajectory parts with dense and wide line spacing. For the irregular trajectory of CASE 23 with rapid altitude changes, attributable to the terrain following flight mode, an RMSE for the unlevelled results of 2.16 mGal is deemed satisfactory. The SRBF levelling is the only method applicable to this challenging trajectory with a low number of COs.

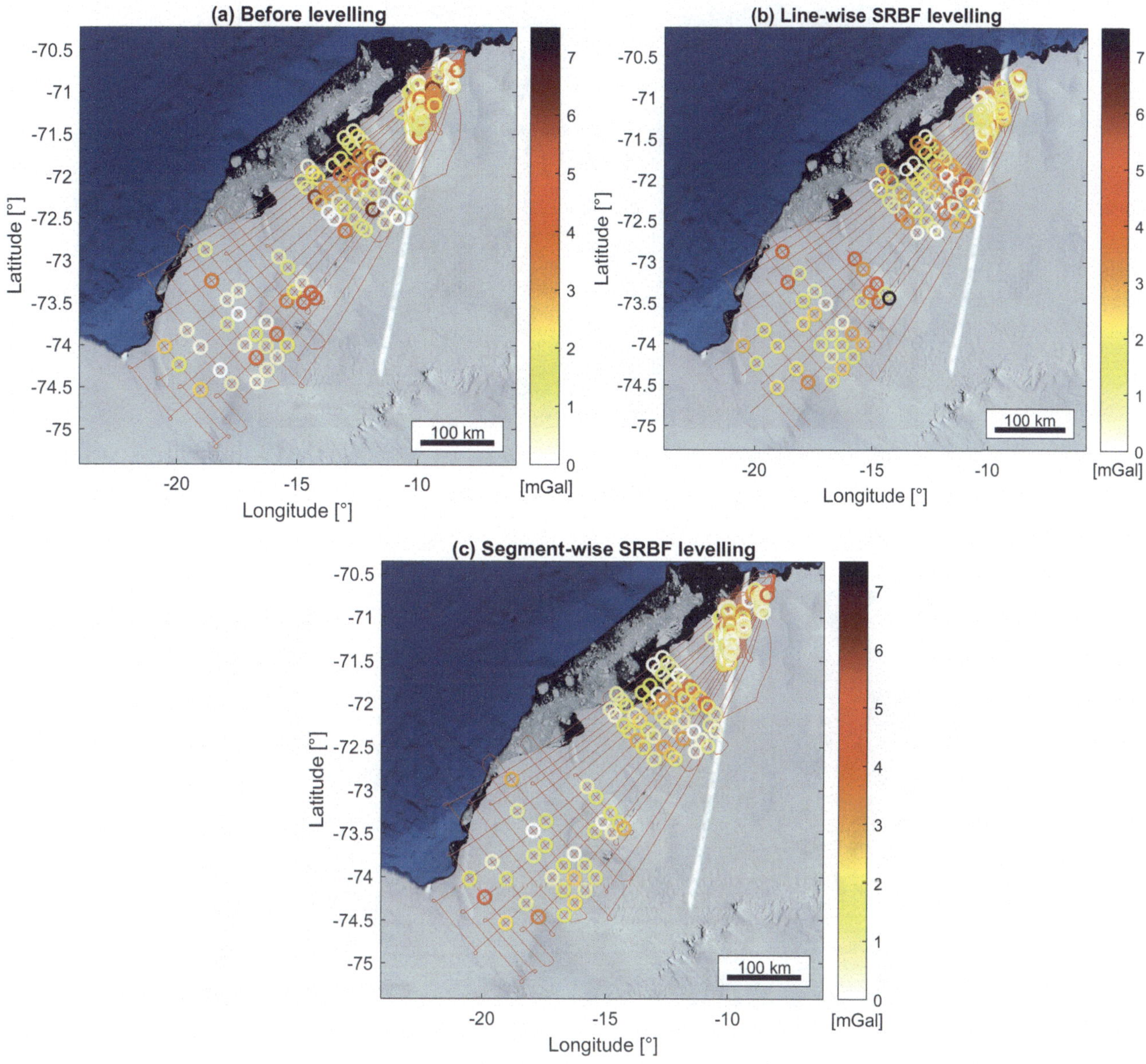

Fig. 6 Crossover point residuals [mGal] at GEA-VI (map data: Google, U.S. Geological Survey): (**a**) before levelling; (**b**) line-wise SRBF levelling; (**c**) segment-wise SRBF levelling

5 Conclusions and Outlook

CO and SRBF levelling methods were presented and evaluated. Biases were estimated for each flight line, which is the default approach in gravimetry, and, alternatively, between equidistant trajectory segments, which is a newly introduced approach. All levelling methods were shown to improve a CO RMSE from 4.05 mGal to values ranging between 0.53 and 0.83 mGal. In most cases, the CO levelling performed slightly better than the SRBF levelling. However, the SRBF method uses the gravity disturbance along the entire trajectory instead of being restricted to a few CO residuals. Hence, it was shown to enable bias estimation even for lines without COs. The new segment-wise SRBF method was shown to be capable of slight improvements in a campaign with a heterogeneous trajectory and a very low number of crossover points where a classical line-wise CO levelling could not have been applied.

In subsequent studies, the implementation of a topographic reduction prior to SRBF levelling may be considered to distinguish between sensor errors and gravity variations induced by the topography. Especially in campaigns with a high field variability and broad line spacing, enhancements in the SRBF levelling results are expected. In campaigns with a varying line spacing, a more sophisticated SRBF origin

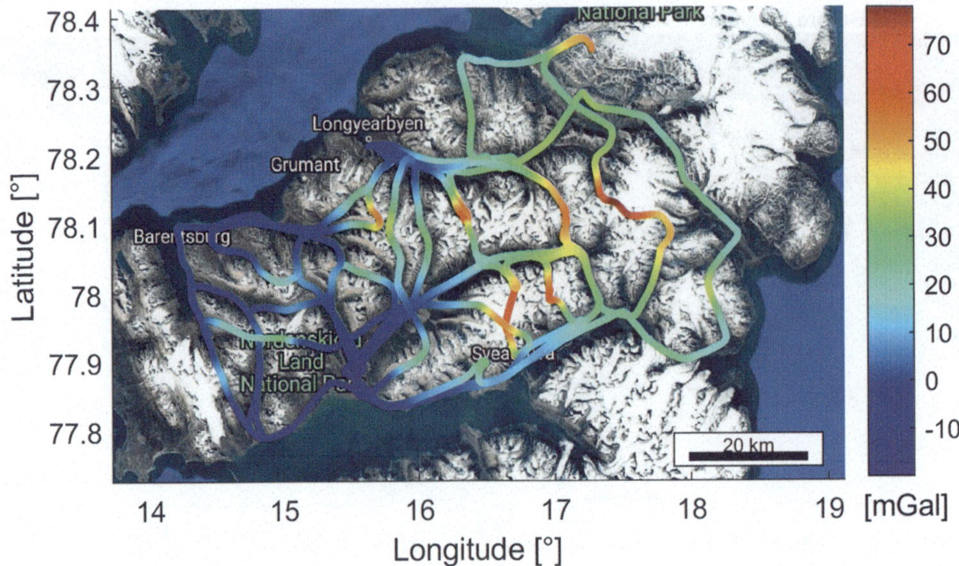

Fig. 7 Gravity disturbance [mGal] along the complete trajectory of CASE 23 before levelling (map data: Google, Landsat/Copernicus)

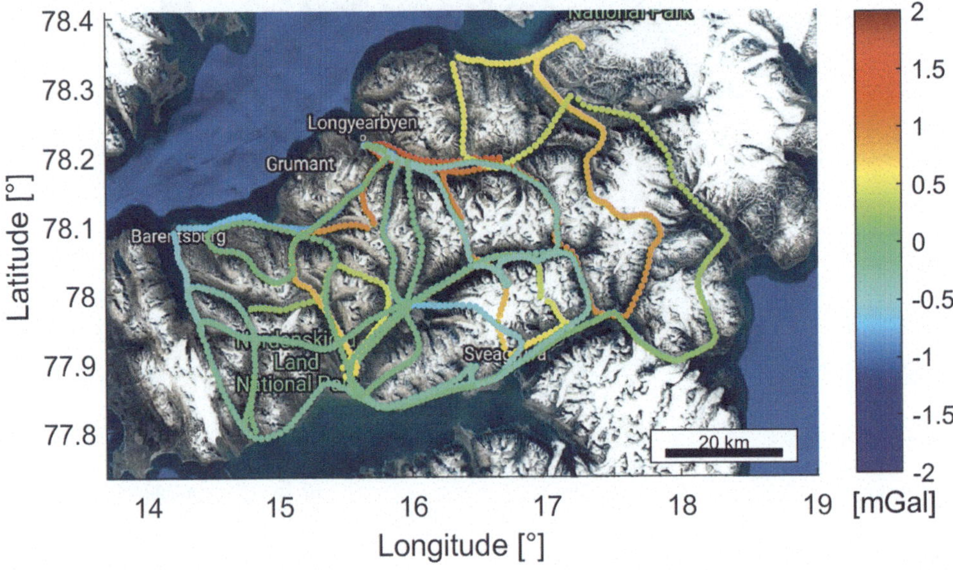

Fig. 8 Interpolated inter-segment biases [mGal] from SRBF levelling along the trajectory of CASE 23 (map data: Google, Landsat/Copernicus)

distribution, e.g. with more origins close to the trajectory (Klees et al. 2008), may prove advantageous. A comprehensive investigation into the impact of SRBF settings on the gravity disturbance results may also be recommended.

Acknowledgements The Odenwald 2017 campaign was conducted by TUDa in cooperation with DTU Space, the Technical University of Dresden, and the Technical University of Munich. The RIISERBATHY/GEA-VI campaign 2022/23 was conducted by the Alfred Wegener Institute for Polar and Marine Research (AWI) and the German Federal Institute for Geosciences and Natural Resources (BGR) in cooperation with TUDa. The CASE 23-Aerogeophysics campaign was organised by the BGR as part of CASE 23, with a cooperation with TUDa for the airborne project. H. Eisermann is funded through the INSPIRES program at the Alfred Wegener Institute.

References

Becker D (2016) Advanced calibration methods for strapdown airborne gravimetry. Technische Universität Darmstadt

Becker D, Becker M, Olesen AV, Nielsen JE, Forsberg R (2016) Latest results in strapdown airborne gravimetry using an iMAR RQH unit. In: 4th IAG symposium on terrestrial gravimetry. State Research Center of the Russian Federation, pp 19–25

Bentel K, Schmidt M, Rolstad Denby C (2013) Artifacts in regional gravity representations with spherical radial basis functions. J Geod Sci 3(3):173–187. https://doi.org/10.2478/jogs-2013-0029

Eicker A (2008) Gravity field refinement by radial basis functions from in-situ satellite data. Universität Bonn

Forsberg R, Olesen AV (2010) Airborne gravity field determination. In: Xu G (ed) Sciences of geodesy-I: advances and future directions. Springer, Berlin, pp 83–104

Glennie CL, Schwarz KP (1999) A comparison and analysis of airborne gravimetry results from two strapdown inertial/DGPS systems. J Geod 73:311–321. https://doi.org/10.1007/s001900050248

Hwang C, Hsiao YS, Shih HC (2006) Data reduction in scalar airborne gravimetry: theory, software and case study in Taiwan. Comput Geosci 32:1573–1584. https://doi.org/10.1016/j.cageo.2006.02.015

IMAR Navigation (2012) iNAV-RQH-1003: inertial measurement system for advanced applications. https://www.imar-navigation.de/downloads/NAV_RQH_1003_en.pdf. Accessed 17 May 2024

Ishihara T (2015) A new leveling method without the direct use of crossover data and its application in marine magnetic surveys: weighted spatial averaging and temporal filtering. Earth Planets Space 67(1):1–14. https://doi.org/10.1186/s40623-015-0181-7

Johann F (2023) Magnetic calibration and GNSS processing in strapdown dynamic gravimetry. Technical University of Darmstadt

Johann F, Becker D, Becker M, Forsberg R, Kadir M (2019) The direct method in strapdown airborne gravimetry – a review. Z Geod Geoinf Land Manag 144(5):323–333. https://doi.org/10.12902/zfv-0263-2019

Johann F, Becker D, Becker M, Ince ES (2020) Multi-scenario evaluation of the direct method in strapdown airborne and shipborne gravimetry. In: International Association of Geodesy symposia. Springer, Berlin, pp 1–8

Johann F, Eisermann H, Eagles G (2025a) A comparison and combination of stable platform and strapdown airborne gravimeters. J Appl Geophys 241:1–12. https://doi.org/10.1016/j.jappgeo.2025.105826

Johann F, Eisermann H, Eagles G (2025b) Airborne strapdown gravity disturbance data (iNAV) across Riiser-Larsen Ice Shelf, East Antarctica, from 2022/23 (RIISERBATHY Project) [dataset]. PANGAEA. https://doi.org/10.1594/PANGAEA.974371

Klees R, Tenzer R, Prutkin I, Wittwer T (2008) A data-driven approach to local gravity field modelling using spherical radial basis functions. J Geod 82(8):457–471. https://doi.org/10.1007/s00190-007-0196-3

Kwon JH, Jekeli C (2001) A new approach for airborne vector gravimetry using GPS/INS. J Geod 74:690–700. https://doi.org/10.1007/s001900000130

Li X (2018) Using radial basis functions in airborne gravimetry for local geoid improvement. J Geod 92(5):471–485. https://doi.org/10.1007/s00190-017-1074-2

Li X (2021) Leveling airborne and surface gravity surveys. Appl Geomat 13(4):945–951. https://doi.org/10.1007/s12518-021-00402-2

Lieb V, Schmidt M, Dettmering D, Börger K (2016) Combination of various observation techniques for regional modeling of the gravity field. J Geophys Res Solid Earth 121(5):3825–3845. https://doi.org/10.1002/2015JB012586

Liu Q, Schmidt M, Pail R, Willberg M (2020) Determination of the regularization parameter to combine heterogeneous observations in regional gravity field modeling. Remote Sens 12(10):1–25. https://doi.org/10.3390/rs12101617

Liu Q, Schmidt M, Sánchez L, Moisés L, Cortez D (2024) High-resolution regional gravity field modeling in data-challenging regions for the realization of geopotential-based height systems. Earth Planets Space 76(1):1–24. https://doi.org/10.1186/s40623-024-01981-1

Ma Z (2024) Gravity field modeling in mountainous areas based on band-limited SRBFs. J Geod 98(5). https://doi.org/10.1007/s00190-024-01852-3

Mauring E, Kihle O (2006) Leveling aerogeophysical data using a moving differential median filter. Geophysics 71(1):5–12. https://doi.org/10.1190/1.2163912

Pavlis NK, Holmes SA, Kenyon SC, Factor JK (2012) The development and evaluation of the Earth Gravitational Model 2008 (EGM2008). J Geophys Res Solid Earth 117(4):1–38. https://doi.org/10.1029/2011JB008916

Studinger M, Bell R, Frearson N (2008) Comparison of AIRGrav and GT-1A airborne gravimeters for research applications. Geophysics 73(6). https://doi.org/10.1190/1.2969664

Torge W, Müller J, Pail R (2023) Geodesy, 5th edn. De Gruyter Oldenbourg, Berlin, Boston

Vyazmin VS (2020) New algorithm for gravity vector estimation from airborne data using spherical scaling functions. https://doi.org/10.1007/1345_2020_113

Vyazmin V, Golovan A, Bolotin Y (2021) New strapdown airborne gravimetry algorithms: testing with real flight data. In: 28th Saint Petersbg international conference on integrated navigation systems (ICINS 2021). https://doi.org/10.23919/icins43216.2021.9470826

White JC, Beamish D (2015) Levelling aeromagnetic survey data without the need for tie-lines. Geophys Prospect 63(2):451–460. https://doi.org/10.1111/1365-2478.12198

Yuan Y, Gao J, Wu Z, Shen Z, Wu G (2020) Performance estimate of some prototypes of inertial platform and strapdown marine gravimeters. Earth Planets Space 72(1). https://doi.org/10.1186/s40623-020-01219-w

Zhang X, Zheng K, Lu C, Wan J, Liu Z, Ren X (2017) Acceleration estimation using a single GPS receiver for airborne scalar gravimetry. Adv Space Res 60:2277–2288. https://doi.org/10.1016/j.asr.2017.08.038

DroneSOM—Towards a Drone-Based Gravity Observation System

Bjørnar Dale, David Becker, Tim Enzlberger Jensen, and René Forsberg

Abstract

In this article we present the challenges, practical limitations together with results and current status of two drone-based gravity systems developed within an EU EIT Raw Materials funded project DroneSOM. Both systems have a maximum take-off weight below 25 kilograms simplifying the application procedure for obtaining beyond-visual-line-of-sight permits from relevant aviation authorities. The two systems apply the strapdown gravimetry technique using a navigation-grade inertial measurement unit and geodetic GNSS antenna as sensor payload. The fixed-wing system is designed towards mapping of larger areas, while the quadcopter system can obtain a higher spatial resolution through its lower flight speed. We also tested hovering-based observations for point-based observations of the Earth's gravity field. Results of two quadcopter campaigns obtain a root mean square of 2–3 mGal along repeated flight lines after accounting for IMU cross-coupling errors. Comparison of turbulence estimator with traditional aircraft campaigns, and the large reduction in repeat line root mean square when including cross-coupling errors show that IMU-errors are the dominant factor compared to GNSS for drone-based gravity measurements.

Keywords

Airborne gravimetry · Drones · IMU · Strapdown gravimetry

1 Introduction

The application of small Unmanned Aerial Vehicles (UAVs) or drones to geophysical mapping has recently received much attention due to efficient data collection and lower cost (Grover et al. 2021; Beirens et al. 2023; Vyazmin and Golovan 2023). Such a system infers requirements on the size and power consumption of sensors and the performance of the UAVs, pushing for innovative solutions. Mapping of the local gravity field is relevant to applications such as subsurface mapping, hydrology, volcanology and mineral surveying for defence technology and green transition (IEA 2021). A drone-based gravity system allows for higher spatial resolution gravity mapping than with conventional airborne carriers, while increasing efficiency compared to terrestrial survey techniques.

Gravity measurements from moving platforms have traditionally been carried out using platform-based instruments which are not suited for UAV applications. Recently, the use of Inertial Measurement Units (IMUs) in a strapdown configuration has seen its use in airborne gravimetry (Becker et al. 2015; Golovan and Vyazmin 2023). Such a strapdown system is much smaller, requires less power than conventional systems and has demonstrated superior resilience to dynamic flight conditions leading to increased data coverage (Jensen et al. 2019, 2025). DTU Space owns such an iNAT-RQH IMU from iMAR Navigation GmbH (Jensen

B. Dale (✉) · T. E. Jensen · R. Forsberg
Department of Space Research and Space Technology, Technical University of Denmark (DTU Space), Kongens Lyngby, Denmark
e-mail: baada@space.dtu.dk

D. Becker
AlgoNav UG, Darmstadt, Germany

J. T. Freymueller, L. Sànchez (eds.), *International Symposium on Gravity, Geoid and Height Systems 2024 (GGHS2024)*, International Association of Geodesy Symposia 158, https://doi.org/10.1007/1345_2025_303

2024), which was recently employed on drone-based experiments within the DroneSOM (Drone Geophysics and Self-Organizing Maps) project funded by EU EIT Raw Materials.

The DroneSOM project goal was to develop both a fixed-wing and a quadcopter drone-based gravity observation platform with corresponding processing software. Existing drone-based gravity experience at DTU Space was based on the first ever beyond-visual-line-of-sight (BVLOS) drone gravity test campaign offshore Wales in 2019, using a fixed-wing UAV Prion-Mk3 platform with a wingspan of 3.8 m and a 43.3 kg take-off weight (TOW), in cooperation with UAVE LTD and iMAR Navigation GmbH (Grover et al. 2021). A single 70 km flight line was flown twice forward and backwards over a sedimentary basin in the Irish Sea, initially showing an accuracy of 2.5 mGal root-mean-square (rms), and after estimating an improved lever arm component from the survey data, a rms accuracy of 1.1 mGal was obtained (D. Becker, AlgoNav, Personal Communication). Although the test was successful, lessons learned were to look for new and smaller sensors, and develop a system with maximum take-off weight (MTOW) of below 25 kg to ease the regulatory BVLOS application procedure.

This paper reports on the practical challenges and developments of the two drone-based gravity systems within DroneSOM. Section 2 describes the drone design and sensor integration, followed by Sect. 3 which outlines the data processing, before Sect. 4 describes the various field activities and show result from two drone campaigns. At last, Sect. 5 summarises the current status of both systems and planned future drone-based gravity activities beyond DroneSOM at DTU Space.

2 Drone Design and Sensor Integration

The two developed drone systems, shown in Fig. 1, serve different application purposes due to their respective flight characteristics as described in Table 1. A fixed-wing drone can cover larger areas due to its higher flight speed and operation time, while the quadcopter system has a lower flight time, but can on the other hand have a very low speed, or even hover. Hovering-based observations or constant flight speed should give equivalent result, and successful retrieval of in-flight hovering-based observations can open new applications for airborne gravimetry by point-wise observations of the Earth's gravity field. Table 1 shows the different flight characteristics of the two systems. In moving-base gravimetry, the platform movement controls the spatial resolution of the derived gravity estimates, determined by the flight speed and filter length (Jensen 2022). To suppress high-frequency noise, both in the GNSS and IMU measurements, a low-pass filter is applied to the raw gravity disturbance measurements. With a flight speed of 25–35 m/s for a fixed-wing system and typical filters with full-width-half maximum (FWHM) of 75–90 seconds, this implies an absolute minimum survey line length of 2.5–3.0 km, corresponding to the spatial resolution of the derived result, is necessary to obtain any gravity estimates without degraded quality caused by influence of flight turns. Fixed-wing operations must hence be conducted as BVLOS flights to achieve sufficiently long flight lines. The heavy IMU payload results in a limited flight time and operational range of the quadcopter system, and it is hence not relevant for BVLOS applications. The MTOW requirement still applies in order to conduct a VLOS campaign within the European Union Aviation Safety (EASA) Open Class category.

Weight and size were the main limitations when developing a drone-based system compared to traditional airborne systems. For drones with MTOW below 25 kg, application for operational authorisation from national aviation authorities can use Predefined Risk Assessment (PDRA) developed by EASA. This simplifies the application process for obtaining VLOS and BVLOS permits. The initial phase of DroneSOM involved searching for available commercial quadcopter and fixed-wing drones capable of carrying a total sensor payload of 8 kg and with a MTOW below 25 kg. A commercial X55 quadcopter from ArcSky was acquired, while no publically available fixed-wing drone within the MTOW limitation was found, so a custom-made fixed-wing drone was developed by Dines Avitech, Denmark. For custom-built drones an additional EASA regulation of 3 meter maximum wing span applies. The two developed systems are shown in Fig. 1. Beirens et al. (2023) gives an overview of existing drone-based gravity campaigns, but apart from their study using a MEMS-based IMU, all drone systems were above 25 kg using not public (military) drones.

On the used ArcSky X55 quadcopter, the IMU was installed upside down using custom-made mountings. It is noted that orientation of installation is stated by the manufacturer to have no influence on the performance. In December 2023, the internal IMU GNSS receiver was upgraded to a multi-constellation Septentrio Mosaic-X5, making an external GNSS receiver redundant for the two quadcopter campaigns. During operation, all sensors are powered by small 3200 mAh LiPo batteries, which can be replaced in a few seconds on ground. Lab tests documented capacity of a single LiPo battery to be 1.5 hours. To ensure continuous power during operations, we installed two ideal diodes with an ORing controller allowing for battery change without power interruption. For all field campaigns, the IMU was continuously powered throughout the whole campaign.

3 Data Processing

In the processing of airborne gravimetry, the commercial Waypoint post-processing software from Hexagon/NovAtel

Fig. 1 The two drones used in DroneSOM; ArcSky X55 quadcopter system on the left, and the custom-built fixed-wing DroneSOM-01 drone on the right (a second fixed-wing DroneSOM-02 drone acting as backup)

Table 1 Technical specification and target parameters for the two drone-based gravity systems. Both systems used the same sensor payloads

	quadcopter	Fixed-wing
Manufacture/Type	ArcSky X55	Dines Avitech
TOW [kg]	24.8	24.9
Flight speed [m/s]	0–8	25–30
Flight time [min]	16–18	90
Range limitation [km]	2–4	40
GNSS rate [Hz]	5	5
IMU rate [Hz]	300	300
Permission requirement	Open class / VLOS	BVLOS
Survey area [km x km]	2 x 2	20 x 20
Survey design	Grid/point-based	Grid

was employed. The IMU observations were combined with GNSS observations obtaining an integrated IMU/GNSS navigation solution. The navigation solution is combined with the raw IMU observations to derive gravity estimates, using a MATLAB-based software applying the direct method as described in Johann et al. (2019).

The derived gravity estimates contain both longer wavelength gravity signal and high frequency noise from the IMU and GNSS observations. To suppress high-frequency noise, a 3-pass forward-backward Butterworth filter with typical FWHM of 75–90 seconds is applied.

As described in Becker et al. (2015), the main error source in traditional airborne strapdown gravimetry is temperature induced sensor drift. Compared to aircraft, the drone-environment will be prone to larger temperature variations due to lack of shielding. Hence, we expect significant temperature variation, requiring proper temperature handling in the data processing. Performing alignment and removal of linear drift through tie-measurements during battery changes minimizes the temperature induced effects. Applying the estimated temperature calibration curve, derived following the same approach as presented in Becker et al. (2015) and Johann et al. (2019), to static lab recordings with an internal sensor temperature ranging between 22–44°C results in a rms of 0.2–0.3 mGal for the corrected static measurements. During field operations the IMU sensor temperature was between 30–44°C.

4 Field Operations

Three field campaigns were carried out within the DroneSOM project. The first field campaign in Finland was carried out using the Dines DroneSOM-01 fixed-wing platform, while the second and third campaigns in Germany and Sweden were carried out using the ArcSky X55 quadcopter platform. An overview of the field campaigns is given in Table 2.

All three field campaigns and the Værløse demonstration test were successfully conducted in challenging weather conditions. The quadcopter system flew in total 21 hours, where the Swedish survey amounted to 8 hours covering 55 kilometer with wind gusts reaching 11 m/s demonstrating its robustness. The Erzgebirge survey, left part of Fig. 2, consisted of three repeat flights at 2 m/s along line 1 together with static hovering points observed for 120 seconds separated by 100 meters coinciding with the targeted spatial resolution. A gravity validation profile was measured along

Table 2 Overview of field campaigns carried out within the DroneSOM project

Campaign	Sokli	Sweden	Erzgebirge
Location	Finland	Sweden	Germany
Time period	Sep, 2023	Aug, 2024	Oct, 2024
Platform	Dines DroneSOM-1	ArcSky X55	ArcSky X55
Instrumentation	iNAT-RQH + Mosaic-X5	iNAT-RQH	iNAT-RQH
Area size	20 x 20 km	2 x 2 km	2 km lines
Flight speed	30 m/s	2 m/s	2 m/s
Target resolution	2–3 km	0.1 km	0.1 km
Repeat line segments			Yes

parts of line 1 using a Scintrex CG-6 gravimeter. Lines 2 and 3 were surveyed once at constant flight speed ensuring full coverage of the target area used for geophysical inversion within the DroneSOM project.

Apart from the short survey flights in Finland, the fixed-wing drone conducted extensive flight tests in Denmark documenting a flight endurance above the 1.5-hour project target and receiving the required BVLOS permit in Finland. The radio link and autopilot system are designed for 40 km communication range enabling survey lines with a length of up to 80 km. Due to geopolitical reasons, the originally mining-relevant survey area at Sokli, Finland, had to be moved further away from the Finish-Russian border. The fixed-wing drone was constructed using GNSS-transparent materials (wood and fibreglass), with an internally installed GNSS antenna. Post-processing of the GNSS data showed degraded results, caused by GNSS interference from either the drone material or onboard instruments. In future campaigns, the GNSS antenna will be installed externally on the drone body. Bad weather conditions and restrictions on finding a suitable runway limited the survey in Sokli to only a few lines of the planned pattern.

Figure 3 and Table 3 shows the rms results from the first gravity demonstration flight along three parallel lines in Værløse and the repeat flight line in Erzgebirge. We used four different GNSS solutions. In Table 3, (1) and (2) denotes the DGNSS and PPP solution processed with AlgoNav software, while (3) and (4) are two different PPP solutions processed with Waypoint post-processing software from Hexagon/NovAtel.

To maximise survey length under VLOS limitations and minimizing operational risk, hovering points were defined at both the start and end of each line. Assuming similar noise characteristics during hovering and constant speed flight, a definite line length can then be defined by selecting hovering time corresponding to the applied filter length. As seen in Table 3, the rms values are reduced with increased line cropping for all survey lines. This reduction in rms reflects different noise characteristics between surveying in hovering mode or at a constant flight speed, similar to flight turn effects experienced in conventional airborne gravimetry. For the drone campaigns this effect was unexpected as the hovering points were introduced to prevent degraded accuracy at line boundaries. Understanding the underlying different noise characteristics is necessary to utilize gravity retrieval in future drone-based gravity applications. This is a prerequisite in order to perform in-flight hovering-based observations.

The rms results are based on Kalman filter processing using one common set of IMU cross-coupling terms, determined as the median over all drone gravity campaigns with the terms modelled as additional constant states in the Kalman filter system. Estimating one common set of cross-coupling terms reduces the risk of over-fitting, even though estimates from the individual campaigns showed consistent estimates. The cross-coupling effect accounts for possible sensor misalignment or insufficient IMU sensor calibration,

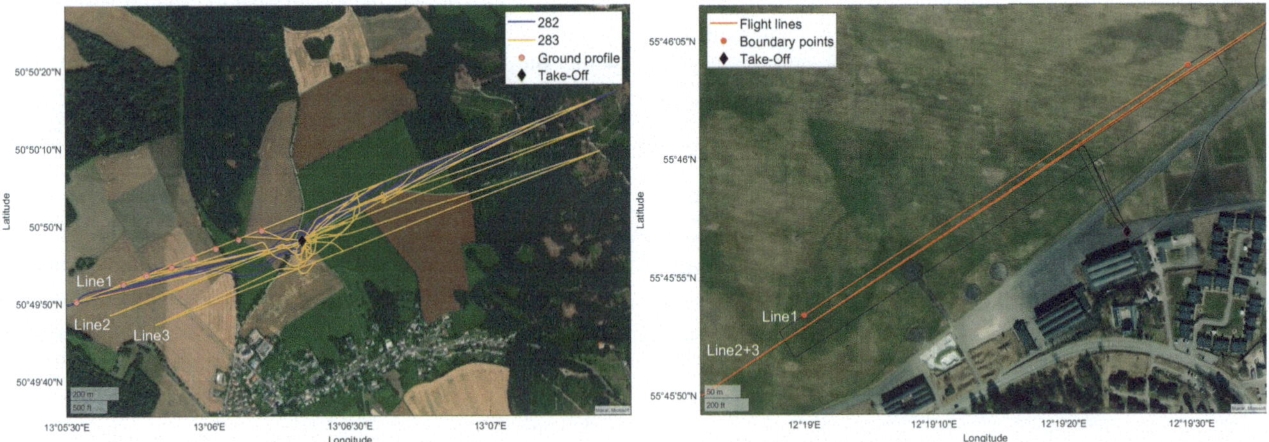

Fig. 2 The two quadcopter drone campaigns in Erzgebirge (left) and Værløse (right). For Værløse, the rms values are computed along the common distance of the three parallel lines determined by the boundary points of line 1

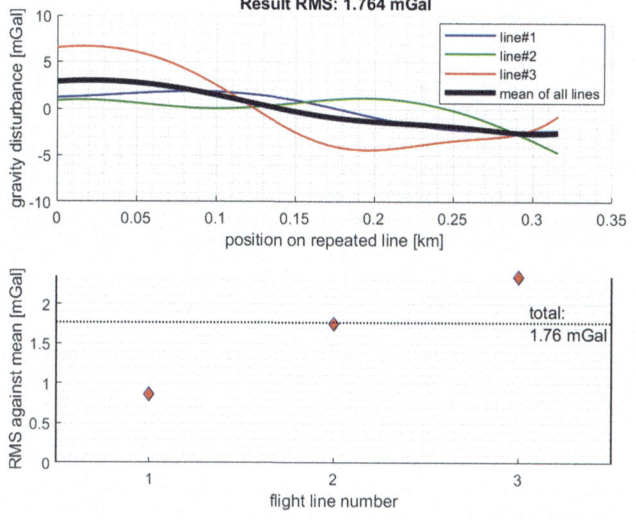

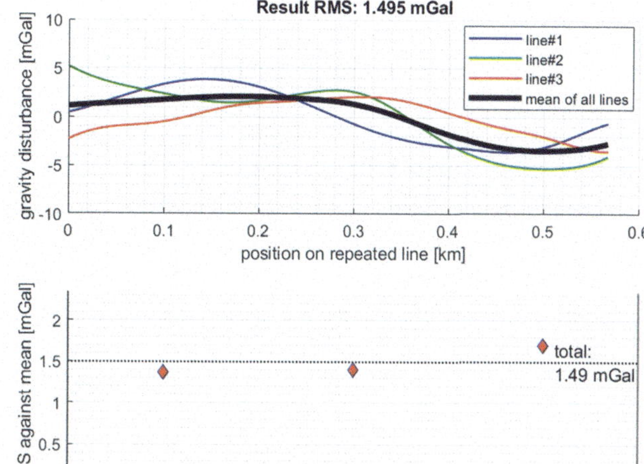

Fig. 3 Timeline plot of repeat line rms for Værløse (left) and Erzgebirge (right) using solution (4) from Table 3. For Værløse, the rms values are computed along the longest common distance of the three parallel lines determined by the boundary points of line 1. Lower part of plot shows deviation of the respective line rms compared to the overall rms of all lines

Table 3 rms values of gravity disturbance estimates along repeated survey lines both with (w) and without (w/o) applying IMU cross-coupling errors. 0, 25, 50 and 100 s denotes cropping time of start and end of flight line. (1), (2), (3) and (4) represent different DGNSS and PPP GNSS solutions discussed in the text

	Værlose						Erzgebirge					
	0s		25s		50s		0s		50s		100s	
	w/o	w	w/o	w	w/o	w	w/o	w	w/o	w	w/o	w
(1)	11.92	3.17	7.59	2.41	2.43	1.85	3.59	2.48	3.17	1.31	2.42	1.06
(2)	12.01	3.05	9.16	2.54	3.58	1.89	3.68	3.83	3.31	1.74	2.45	1.26
(3)	7.00	2.54	3.07	1.95	3.02	2.03	3.89	2.67	3.31	1.67	2.69	1.41
(4)	11.86	3.19	7.00	2.27	2.08	1.76	3.51	2.74	3.28	1.50	2.53	1.30

which normally is not accounted for in processing of traditional airborne gravimetry.

In general it can be seen that the rms values are lower for Erzgebirge than in Værløse. Værløse was the first demonstration flight with the original versions of mounting and sensor integration. Further developments, with tighter fittings and installation of rubber pads between the IMU and drone platform to mitigate high-frequency vibrations, were implemented before the system reached the final configuration used in both Sweden and Erzgebirge campaigns. In addition, a tuning of flight control parameters tailored to the heavy weight payload was implemented throughout the project phase. Further investigation is needed to quantify these effects.

The average rms-g is computed as a 50 seconds sliding moving standard deviation of the vertical GNSS acceleration and is used as a measure of flight turbulence (Becker 2016). For Værløse and Erzgebirge, the rms-values using GNSS solution (4) results in 202 and 280 mm/s^2, respectively. This is on-par and below the observed rms-g value in traditional airborne gravimetry campaign, for comparison see chapter 8.2 of Johann (2023). Both campaigns show large reductions in rms value when accounting for the estimated cross-coupling effects. By including cross-coupling errors, the rms reduces to a level of 2–3 mGal, and even further when increasing cropping at the start and end of the line. There are notable differences in the repeat line rms values between the four different GNSS solutions. However, in comparison with the computed rms-g and large reduction in rms value when including IMU cross-coupling errors, we state that IMU-errors are the dominant error source for drone-based gravity observation.

5 Conclusion and Outlook

We presented the development and current status of two drone-based gravity observation systems developed within DroneSOM. Through three drone campaigns, we tested practical limitations and capabilities for each system. The fixed-wing system suffered from GNSS signal interference caused by the drone material, expected to be resolved by external installation of the geodetic GNSS antenna, while the quadcopter system has reached a close to optimum practical payload integration. Considering practical limitations under VLOS operations, a survey area of 2 x 2 km with line spacing of 100 m can be surveyed within two days using the quadcopter system. Based on field operations from the three campaigns and a substantial number of test flights, we consider both systems to be robust and have significantly lowered the operational risk of installing a high-value payload as the iMAR-RQH IMU onboard.

In the reported surveys, we estimated repeat line rms both with and without IMU cross-coupling errors. Together with the computed rms-g values being on-par with or below traditional aircraft campaign values, the large reduction in repeat line rms when applying the estimated IMU cross-coupling parameters shows that IMU-errors are the dominant error sources in drone-based gravity observations. The rms results reduced further with additional cropping at the start and end of each survey line. From the survey design, this shows different noise characteristics between surveying in hovering mode and with constant flight speed. Further investigations are necessary to understand the reason for this difference.

The effort to make gravity measurements from lightweight fixed-wing and quadcopter platforms as an operational surveying technique is ongoing. Investigations to better understand inherent IMU errors in high dynamic environments are crucial to obtain this. Future lab calibrations are planned for determining these IMU sensor-specific errors in a controlled environment. Application of the derived parameters to traditional airborne campaigns is an area of natural interest. With better IMU sensor understanding, DTU plans to move forward with a future drone campaign targeting observation of climate change effects (glacier mapping) with validation to independent ground data. Present results with an rms of 2–3 mGal are unsatisfactory, and better calibration of the IMU sensors and internal offsets/misalignments is necessary to fully explore the potential accuracy of drone-based gravity measurements.

Acknowledgements DTU Space acknowledges the collaboration with DroneSOM partners GTK, Radai Oy, Finland, and Beak Consultants GmbH, Germany, for the help with field campaigns and project management. DTU Space acknowledge Lantmäteriet and BKG for access to their absolute gravity network during the field campaigns in Sweden and Germany.

References

Becker D (2016) Advanced calibration methods for strapdown airborne gravimetry. Ph.D. thesis. Technische Universität Darmstadt, Darmstadt. Available from: http://tuprints.ulb.tu-darmstadt.de/5691

Becker D, Nielsen JE, Ayres-Sampaio D, Forsberg R, Becker M, Bastos L (2015) Drift reduction in strapdown airborne gravimetry using a simple thermal correction. J Geodesy 89(11):1133–1144

Beirens B, Darrozes J, Ramillien G, Seoane L, Médina P, Durand P (2023) Using a SPATIAL INS/GNSS MEMS unit to detect local gravity variations in static and mobile experiments: First results. Sensors 23(16):7060

Golovan AA, Vyazmin VS (2023) Methodology of airborne gravimetry surveying and strapdown gravimeter data processing. Gyroscopy Navigat 14(1). https://doi.org/10.1134/S2075108723010029

Grover G, Forsberg R, Jensen T, Müller F, Hoss M, Bourne T, Wright O (2021) World's first fixed wing uav gravity data collection flight. less cost & less carbon. In: Second EAGE Workshop on Unmanned Aerial Vehicles, vol 2021, pp 1–3. European Association of Geoscientists & Engineers

IEA (2021) The role of critical minerals in clean energy transitions. Technical report. OECD Publishing, Paris. https://doi.org/10.1787/f262b91c-en

Jensen TE (2022) Spatial resolution of airborne gravity estimates in Kalman filtering. J Geodetic Sci 12(1):185–194

Jensen TE (2024) 4. iMAR iNAT-RQH-4001. https://doi.org/10.11583/DTU.25673604.v2

Jensen TE, Dale B, Stokholm A, Forsberg R, Bresson A, Zahzam N, Bonnin A, Bidel Y (2025) Airborne gravimetry with quantum technology: observations from iceland and greenland. Earth Syst Sci Data 17(4):1667–1684. https://doi.org/10.5194/essd-17-1667-2025

Jensen TE, Olesen AV, Forsberg R, Olsson PA, Josefsson Ö (2019) New results from strapdown airborne gravimetry using temperature stabilisation. Remote Sens 11(22). https://doi.org/10.3390/rs11222682

Johann F (2023) Magnetic Calibration and GNSS Processing in Strapdown Dynamic Gravimetry. Ph. D. thesis, Technische Universität Darmstadt, Darmstadt. Available from: https://tuprints.ulb.tu-darmstadt.de/handle/tuda/10828

Johann F, Becker D, Forsberg R, Kadir M (2019) geodaesie.info - The Direct Method in Strapdown Airborne Gravimetry – a Review. ZfV - Zeitschrift für Geodäsie, Geoinformation und Landmanagement (zfv 5/2019)

Vyazmin VS, Golovan AA (2023) Strapdown airborne gravimetry based on aircrafts and uavs: Postprocessing algorithms and new results. In: JT Freymueller, L Sánchez (eds), Gravity, Positioning and Reference Frames. REFAG2023, vol 156 of International Association of Geodesy Symposia, pp. 45–51. Springer, Cham

Scale Factor (In)stability of ZLS-B78: A Worst Case Scenario?

Christian Gerlach

Abstract

Among other applications, terrestrial gravimeters are used for geoscientific studies of local or regional mass transport. Thereby, requirements on observational accuracy and stability of the observational system are high, given that time variable signals have amplitudes of only some few tens of microGal ($1\,\mu\text{Gal} = 10\,\text{nm/s}^2$) or less and repeated surveys often span time frames of several years.

In 2013, the Bavarian Academy of Sciences and Humanities acquired the spring-type relative gravimeter ZLS-B78 and uses it since then for long term monitoring of gravity variations related to local mass transport in hydrological and glaciological studies. Measurements are mostly based on the electrostatic feedback system of the instrument. The corresponding linear scale factor was determined in various calibration experiments over a period of about 10 years. From these measurements we find a strong rate of change of the feedback scale of 0.2 % per year and a seasonal component with amplitudes of up to 0.1 %. Ignoring the linear rate would distort gravity differences measured over the full 50 mGal feedback range by about 0.1 mGal/year, which is larger than most of the time variable signals we want to observe. In general, instabilities of scale factors have been reported in the geodetic literature for different types of gravimeters, but most are at least one order of magnitude smaller and, to my knowledge, annual variations are not documented for any of these feedback scale functions. Therefore, ZLS-B78 may be considered a worst-case scenario. The results verify that the scale factor stability of each and every instrument must carefully be evaluated and that regular instrumental checks are essential for obtaining reliable results especially in long-term monitoring projects.

Keywords

Calibration · Feedback · Gravimetry · Scale factor · ZLS-Burris

1 Introduction

Gravimetry is used to observe static and time variable structures of the Earth gravity field by means of satellite and terrestrial methods. One field of application is monitoring mass transport in the hydrosphere, cryosphere or solid earth. In 2013, the geodesy group of the Bavarian Academy of Sciences and Humanities (BAdW) in Munich acquired a new ZLS-Burris terrestrial relative gravimeter (ZLS-B78) for studying gravity variations on alpine glaciers (Gerlach et al. 2017). With local gravity networks typically covering a range of up to 100 mGal and with an expected measurement accuracy of better than $10\,\mu\text{Gal}$, the scale factor of the instrument should be known to at least $1 \cdot 10^{-4}$. In the best case, the value is stable over time spans of several years. In order to verify these requirements, various calibration

C. Gerlach (✉)
Bavarian Academy of Sciences and Humanities, Munich, Germany
e-mail: gerlach@badw.de

J. T. Freymueller, L. Sànchez (eds.), *International Symposium on Gravity, Geoid and Height Systems 2024 (GGHS2024)*,
International Association of Geodesy Symposia 158, https://doi.org/10.1007/1345_2025_295

experiments were conducted, the results of which are the topic of the present paper.

Comparable studies have been published mostly for instruments of type Scintrex CG-3 or CG-5, but to my knowledge not much is reported about ZLS Burris instruments. Earlier studies using LaCoste and Romberg instruments often concentrate on scale factor variations (non-linearities) of the mechanical spring system (see, e.g., Atzbacher and Gerstenecker 1993), which is not relevant for the present study which deals with the scale factor of the electrostatic feedback system.

Studies on the scale of Scintrex instruments are presented by, among others, Timmen and Gitlein (2009), Ukawa et al. (2010), Oja (2016), Cheraghi et al. (2020) or Timmen et al. (2020). While Timmen and Gitlein (2009) find the scale factor of their CG-3M instrument to be stable to $1 \cdot 10^{-4}$ over almost three years of observations and without any indication of secular trends or seasonal components, the other authors find significant trends of the scale factor. Ukawa et al. (2010) find from three repeated calibrations performed over 8 years with three CG-3M instruments trends between epochs of up to $5 \cdot 10^{-5}$/year. Oja (2016) shows results from three CG5 instruments tested on different calibration lines in Estonia. For two of the instruments the calibration experiments cover a period of more than 10 years. The rate of change of the scale factor of these instruments is in the order of $4 \cdot 10^{-5}$/year with deviations from the linear trend of up to $2 \cdot 10^{-4}$. The third instrument confirms a scale rate of change in the same order of magnitude, however over a short period of two years only and with an opposite sign. Cheraghi et al. (2020) find from 12 years of observations with three CG-3M and three CG-5 instruments on a calibration line in Iran a linear trend of the scale factor of almost $1 \cdot 10^{-4}$/year for two of the CG-5 instruments, while the other instruments show smaller trends of less than $2 \cdot 10^{-5}$/year. Deviations with respect to the linear trend are mostly below $1 \cdot 10^{-4}$ but reach $5 \cdot 10^{-4}$ for one of the instruments. Extending the three years CG-3M calibration time series reported in Timmen and Gitlein (2009) to 14 years, Timmen et al. (2020) show a linear trend of the scale factor of $-4 \cdot 10^{-5}$/year and maximum deviations from linearity of around $5 \cdot 10^{-4}$. In the same study, four different CG-6 instruments are tested on the Vertical Gravimeter Calibration Line Hannover (VGCH) over variable time spans of 1 to 6 months, each instrument conducting between 2 and 4 measurement epochs. Scale factors of the same instrument differed typically by $2 \cdot 10^{-4}$ to $3 \cdot 10^{-4}$ between epochs; one instrument showed maximum differences of $7 \cdot 10^{-4}$. However, due to the small number of epochs and the short study period, reliable linear trends are not derived.

The feedback scale factor of ZLS-B25 was investigated by Jentzsch (2008). From three different sessions conducted on a 16 mGal calibration line he finds scale factors that agree to $4 \cdot 10^{-4}$ (which roughly corresponds to the expected measurement accuracy) if observed in similar dial settings but differ by $18 \cdot 10^{-4}$ if observed in significantly different dial settings, which indicates non-linearities in the feedback system. However, the study does not comprise systematic monitoring of the scale factor over longer time. Schilling and Gitlein (2015) report on repeated calibrations of ZLS-B64 over about 1.5 years. They conclude that the linear scale factor stays stable to within $3 \cdot 10^{-4}$. From graphical inspection of their Figure 4, one finds deviations from linearity of less than around $6 \cdot 10^{-4}$, showing signatures of an annual component of varying amplitude, which, however, is not further discussed by the authors. Timmen et al. (2020) extend the time series of scale factors of the same instrument to three years and find a small linear trend of $-1.5 \cdot 10^{-4}$/year. A much longer time series of eight years is available for ZLS-B20 which is operated at the Geodetic Observatory Pecny, Czech Republic (V. Pálinkáš, private communication). The feedback scale of ZLS-B20 is regularly calibrated in comparison to the superconducting gravimeter operated in Pecny. Each of the individual calibration experiments has an estimated precision of around $5 \cdot 10^{-4}$, the linear trend of the feedback scale is found to be at the level of $8 \cdot 10^{-4}$/year without any significant deviations from linearity.

These investigations confirm that feedback scale factors may vary in time and that the behaviour can differ from instrument to instrument. In most cases, the calibration is determined with an uncertainty of better than $5 \cdot 10^{-4}$, just above the requirement of $1 \cdot 10^{-4}$ mentioned further above. For some but not for all instruments there is an obvious drift of the scale factor. While for Scintrex instruments reported trends do not surpass a value of $1 \cdot 10^{-4}$/year, trends of the feedback scale factor may be larger at least for some ZLS-Burris instruments. In any case, the numbers suggest that the stability requirement of $1 \cdot 10^{-4}$ is not always met. To which degree this is also true for ZLS-B78 is the topic of the present paper.

In Sect. 2, the instrumentation and the different calibration scenarios are presented for the case of ZLS-B78. In Sect. 3, the time variations of the scale factor of its electrostatic feedback system are discussed.

2 Instrumentation and Calibration Experiments

The ZLS Burris gravity meters are relative instruments based on the LaCoste and Romberg gravity meter design (Jentzsch 2008). They employ a metal spring measuring system consisting of a horizontal lever arm carrying on one end the sensitive test mass. The Burris instruments are equipped with an electrostatic feedback system with a typical working range of 50 mGal. The calibration experiments discussed in the present paper only relate to the linear scale factor of this feedback system. Should a calibration line exceed the 50 mGal range, the line is separated into mutually overlapping segments, each of which are measured independently in different dial settings. In consequence, even though ZLS-B78 has a calibrated screw, the calibration function of the mechanical system is not relevant for our experiments and is thus not further discussed here. It is also worth mentioning that we assume linearity of the feedback system and do not further investigate scale factor variability as function of the feedback reading. This may be subject to a future study.

The absolute calibration factor of the feedback system is determined by comparison to absolute gravity values either on dedicated calibration lines or in combined gravity networks. As an alternative, the feedback system allows to observe apparent gravity differences by turning the dial of the mechanical system while the instrument remains at the same station. This provides a relative scaling between the mechanical and the electrostatic readings (Torge 1989, sec. 6.4). Absolute and relative scale factors are not directly comparable but time variations of both should be.

In all experiments reported here, the scale factor was derived as one of the unknowns in a least-squares gravity network adjustment. This also allows to estimate the formal error of the scale factor which mainly depends on the covered measurement range, the network geometry and the uncertainty of the datum points.

In its original version, ZLS-B78 was equipped with the UltraGrav software installed on an external unit (personal digital assistant, PDA) for controlling the feedback loop and registering the relative gravity readings (Jentzsch 2008). Thereby, gravity effects of solid Earth tides are computed using Longman's model (Longman 1959) and reduced from the observations automatically. Reductions for instrument height and atmospheric pressure variations are to be applied explicitly by the user during data preprocessing. In all experiments presented here, atmospheric pressure effects are reduced with a factor of $0.3\,\mu\text{Gal/hPa}$ (Torge 1989, sec. 10.2). Instrumental height is not of relevance for the relative calibration experiments (feedback versus dial) but for the absolute experiments in the field. In the latter case, measurements of the local vertical gravity gradient are available. Where local tidal models are available (Wank and Wettzell experiments, Sect. 2.1), these are used to replace the Longman model.

Calibration experiments performed during the first two years of operation (2013–2015) revealed that the linear scale factor of ZLS-B78 was highly unstable (Gerlach et al. 2017), so the instrument was returned to the manufacturer for repair. Afterwards, the instrument was equipped with the AGES software package (Schulz 2018) replacing the original UltraGrav for feedback control and data registration. AGES also provides a standardized maintenance procedure for checking instrumental parameters and levels. Such maintenance was conducted in irregular intervals but at least once a year, typically before fieldwork season, i.e., in spring or early summer. A detailed analysis of the time series of maintenance parameters was not conducted, but graphical inspection gives the impression that values either fluctuate around a time stable mean, or are drifting mostly linearly on a small rate.

2.1 Absolute Calibration

The following scenarios were used to determine the absolute feedback scale factor of ZLS-B78. In most cases, measurement campaigns with the relative gravimeter were performed close in time to the absolute campaigns thus minimizing degradation of the scale factor caused by time variations of the absolute reference values:

- **Vernagt:** In 2014 and 2022, A10 absolute gravity values were determined in the upper Ötz valley, Austria, by the German Federal Agency for Cartography and Geodesy (BKG) to complement BAdW's glacier monitoring program (Gerlach et al. 2017). Three stations are located in the area of the Vernagt glacier and cover a gravity range of 30 mGal. Two additional stations are located in the valley and cover a gravity range of 32 mGal. Since there was no relative link between both point groups, two independent estimates of the feedback scale were determined for the 2014 campaign, one from observations near the glacier (elevations between 2860 m and 3020 m) and one from relative connections between the two valley stations (elevations around 1935 m and 2015 m). In 2022, only the valley stations were observed with ZLS-B78, thus only one scale factor is estimated.
- **Wettzell:** Since about 10 years, BKG operates a gravity network of 11 field stations on and around the Geodetic Observatory in Wettzell, Germany, for monitoring hydrological gravity variations. Observations are taken twice a year (in spring and fall) using an A10 absolute gravimeter. For some of the spring-epochs (2015, 2017, 2019, 2020, 2021, 2024[1]), the network was also observed with ZLS-B78 allowing to monitor its scale factor in relation to A10 absolute values. Point distances in the network are up to 20 km. The gravity range is around 60 mGal. More details can be found in Gerlach et al. (2018).
- **Wank:** In 2004/05, a gravimeter calibration system was established on and in the vicinity of Germany's highest peak, Mt. Zugspitze (2962 m) in cooperation of TU Munich, Leibniz University Hannover (LUH) and BKG. One of the lines of this system connects the valley and

[1]The 2024 epoch is not yet processed, thus not included in the present paper.

summit stations of the cable car to Mt. Wank, a nearby peak about 10 km north-east of Mt. Zugspitze on the opposite side of the city of Garmisch-Partenkirchen. This calibration line covers an elevation difference of about 1000 m, and a gravity range of about 210 mGal. The line was remeasured with LUH's FG5 instrument in 2019 (Timmen et al. 2021) and found to represent a time stable calibration basis. Since then it is used to calibrate the feedback of ZLS-B78 in comparison to the FG5 absolute gravity values. For this scenario, the FG5 values are assumed to be stable, i.e., possible temporal variations are neglected. The comparison of the 2005 and 2019 FG5 observations agree to better than 1 μGal on the summit station, while 6 μGal differences are found at the valley station. However, considering the gravity range of 210 mGal, neglected temporal variations in the order of 10 μGal affect the calibration by only $5 \cdot 10^{-5}$, well below the accuracy requirement for the scale factor. For logistical reasons, an additional gravity value is established in every epoch at the intermediate station of the cable car using a Scintrex CG-5 instrument. Subsequently, the lower part of the line between valley and intermediate cable car station, which spans a range of around 85 mGal, is used for calibrating the ZLS instrument (again splitting the line in mutually overlapping segments; transport by foot).

2.2 Relative Calibration

In addition to absolute calibration experiments, relative scale factors were determined on a stable point in the BAdW office building. Thereby, the dial is set to an intermediate position such that the feedback reading is close to zero. Then feedback readings are taken at dial settings of 20 counter units left and right of the intermediate position. This results in an apparent gravity difference between the settings of around 40 mGal. Typically, each setting is repeated 4 to 6 times and the whole experiment takes between 1.5 and 2.5 hours. Residual tidal and atmospheric effects during such short time intervals are assumed to be negligible or to map to the largest extend into the daily instrumental drift estimate.

Repeating such an experiment over time (using identical dial settings), allows to track changes of the apparent gravity difference. Relating results of all epochs to the initial one allows to compute a relative scale factor that can be monitored over time. By definition, the scale factor of the initial epoch has a value of 1.0. After a certain period, say 1–2 years or so, the dial settings need to be changed due to the long term drift of the spring system and a new time series of relative scale factors needs to be started. Two such independent time series can be connected by determining at one epoch the relative scale factors in both of the independent dial settings. In doing so, a long time series of relative feedback scale factors can be determined.

3 Results

Figure 1 shows the feedback scale factor of ZLS-B78 determined between summer 2014 and fall 2024. The color codes of the dots refer to the various calibration experiments at Wettzell (blue), Vernagt glacier (green), Wank (orange) and the relative ones at BAdW (gray). Light gray dots indicate the relative scale factors referring to the initial epoch in late 2019 (initial value is 1.0 by definition). The offset of the relative scale factors with respect to the absolute ones is determined in the frame of a least-squares linear fit of the data, resulting in one single calibration function for the feedback system of ZLS-B78. Dark gray dots indicate the offset-corrected relative scaling factors.

During the first two years of operating ZLS-B78, it became quite clear that scale factors derived from different calibration experiments differed considerably without any functional dependency on time or feedback range. Values differed by almost 1 % and even calibration experiments carried out in different sessions on the same day were not consistent (cf. the two 2014 calibration experiments at Vernagt glacier in Fig. 1). After discussing the issue with the manufacturer, ZLS-B78 was returned for repair/maintenance in late 2015, (indicated by the left of the two vertical red lines in Fig. 1). Another maintenance became necessary in 2024 (indicated by the right vertical red line) as seals apparently were no longer tight.

During both of the maintenance operations the manufacturer made modifications to the electronics with possible effects on the feedback scale factor. This is quite evident comparing scale factors before and after 2016. It is less obvious for the second maintenance in 2024. Here, further interpretation requires several more calibration experiments in the coming months and years.

The following discussion will concentrate on the period 2017–2024 between the two maintenance operations. In this period, calibration factors show a significant linear drift. The blue line in Fig. 1 represents the linear regression curve. According to this, the scale factor rate of change amounts to 0.00248 ± 0.00007 per year. Applied to a gravity difference covering the full feedback range of 50 mGal, this rate amounts to 0.124 ± 0.003 mGal/yr. Such a large rate requires regular monitoring but is in principle of no major concern as long as it is constant. Deviations from linear regression, however, can seriously degrade the reliability of data collected with ZLS-B78.

Residuals with respect to linear regression are shown in Fig. 2. We find an obvious periodic behaviour with amplitudes of up to 0.001 and a period of almost exactly one year.

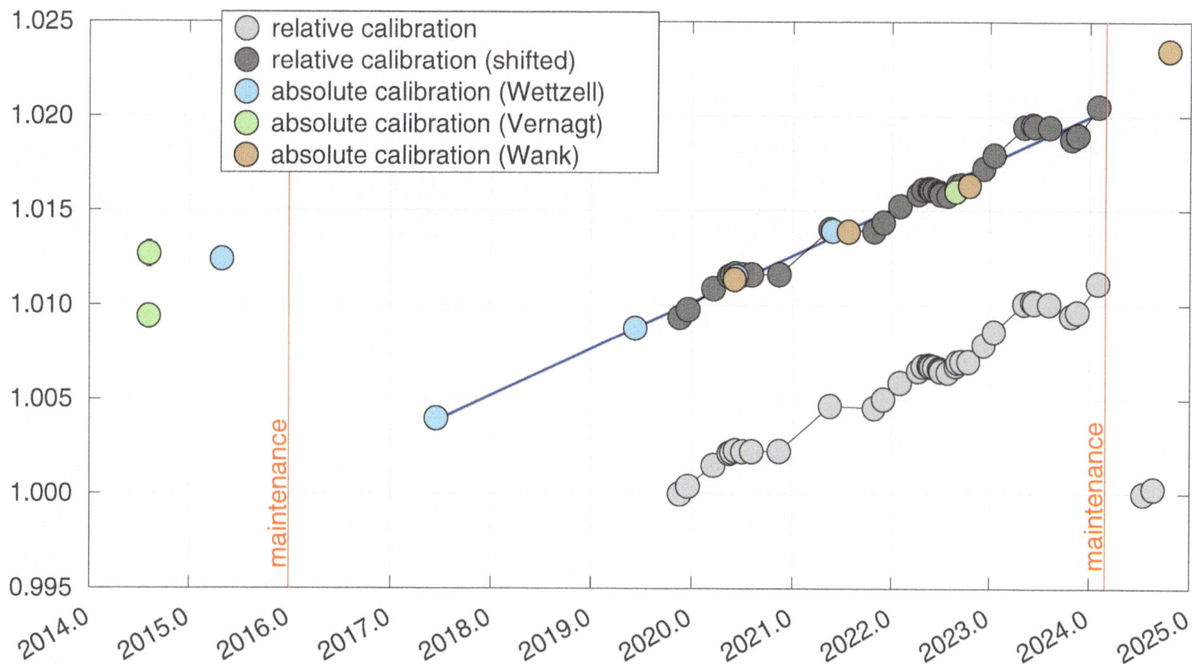

Fig. 1 Time series of feedback scale factors of ZLS-B78 from various calibration experiments along with a linear regression model (blue line). The red vertical lines indicate epochs of maintenance at the manufacturer

For illustrating this periodicity, the vertical dashed lines in Fig. 2 indicate the epochs of 1st of May (dark gray) and 1st of November (light gray) each year (between 2020 and 2024).

Due to the many relative calibration experiments (dark gray dots) the yearly signal is represented well. Also the absolute calibrations nicely fit into the yearly pattern of the relative calibrations. For all experiments, 1-σ formal error bars are provided. On the scale of Fig. 1 and also in case of the relative experiments (gray dots) in Fig. 2, error bars are mostly hidden by the dots, while they are clearly visible in Fig. 2 for the absolute experiments. They are largest for the Vernagt and Wettzell cases where A10 absolute gravimeters (10–12 μGal uncertainty) serve as calibration standards; adjusting a network of 12 stations helps to reduce the error bars for Wettzell as compared to the Vernagt experiment which is based on one single gravity difference only. Error bars are even smaller for the Wank experiment where an FG5 (1–2 μGal uncertainty) serves as calibration standard.

In many cases, absolute calibration experiments in the field are carried out in the same season. In principle, this is a good idea as it helps to minimize temporal variations of the reference values. Also in our case, in the period 2017 to 2021, the experiments in Wettzell (blue dots) and at Mt. Wank (orange dots) were carried out in almost the same season, late spring/early summer. For those experiments, the scale factors do not vary a lot around the linear trend (which is reduced in Fig. 2) and one could assume that variations are related to the accuracy of the estimated scale factors. Only when combining these absolute scale factors with the relative ones (gray dots) it gets clear that the variability reflects the annual component of the scale factor variability. This is nicely confirmed by the two absolute calibration experiments at Wank and Vernagt glacier, both carried out in late summer 2022, where the annual component reaches its maximum.

Fitting a periodic component to the time series gives an analytical approximation to the periodic scale variations and should help to reduce the scale factor variability. However, in our case it reduces the residuals by a factor of 2 only – not good enough for applications with requirements of $1 \cdot 10^{-4}$ or better. Therefore, further investigations are necessary to better describe the variability of the scale factor and to allow for a reliable reconstruction of scale factors between calibration epochs. As an initial speculation, annual components of ambient temperature or humidity could be possible candidates that affect the measurement system and cause variations of the scale factor. Effects of ambient humidity on spring-type gravimeters have been discussed before but to my knowledge only relating to the long-term instrumental drift (see, e.g., El Wahabi et al. 2001), not to the feedback scale factor.

Even though the annual pattern gives way to speculate about external drivers like meteorological conditions, I do not rule out the possibility of undetected instrumental effects. Time variability of the level system may lead to variations of the instrument's sensitivity with immediate implications for the scale factor. However, the sparse, but regular checks carried out with AGES's standard maintenance program, hopefully allow to keep related time variations under control.

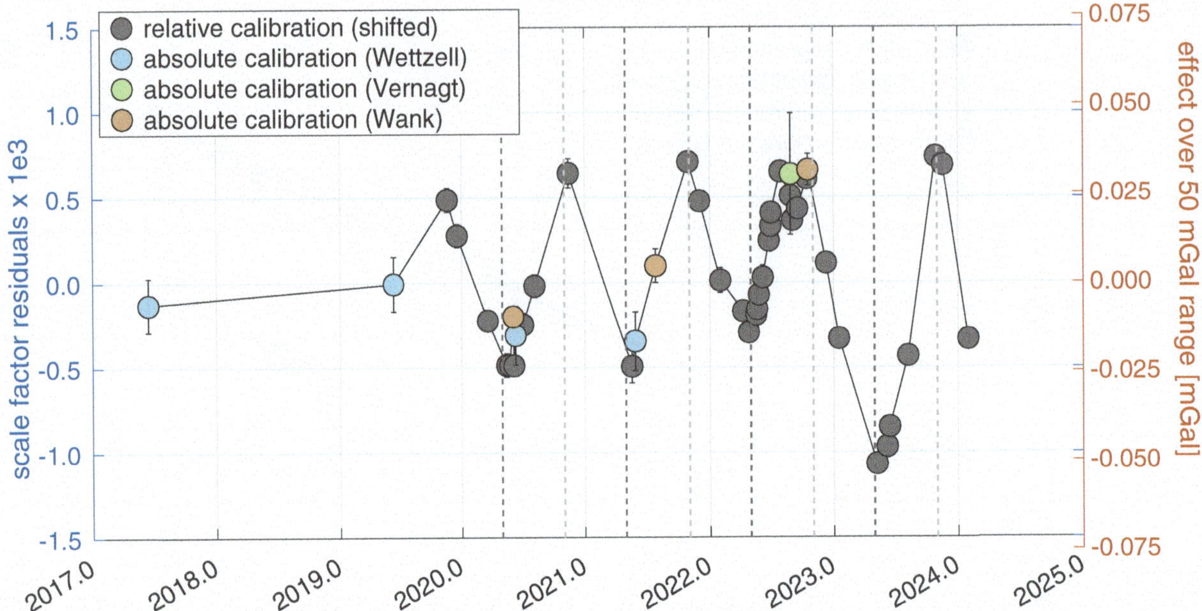

Fig. 2 Residuals of the scale factors shown in Fig. 1 with respect to the linear regression model. Residuals are evaluated only for the time span between the two maintenance epochs. The left axis (blue) indicates the magnitude of the residuals, the right axis (red) the corresponding gravity effect when applying the residuals over the 50 mGal full feedback range of ZLS-B78. Dashed vertical lines indicate beginning of May (dark gray) and November (light gray) every year in the period 2020–2024

I further assume, instrumental air pressure effects to be of negligible contribution because experiments conducted in quite different elevation ranges confirm the annual scale variability as provided by the relative scaling experiments and a dependence on elevation is not obvious.

4 Summary and Conclusions

Repeated calibration experiments of the LaCoste and Romberg-type spring gravimeter ZLS-B78 reveal a strong linear drift of the scale factor of the electrostatic feedback system. The drift amounts to 0.2 % per year and is modulated by an annual signal with amplitudes of up to 0.1 %. This is significantly larger than the value of $1 \cdot 10^{-4}$ or better, typically required in long-term monitoring projects. As is typical for spring-type gravimeters, each and every instrument can be expected to show a different behaviour, but in general it seems clear that one single calibration cannot be taken for granted over longer time frames, maybe not even on a monthly scale. Scale factor instabilities in the order of 0.01 % are reported for various spring-type gravimeters of type LaCoste and Romberg (Schilling and Gitlein 2015; Timmen et al. 2020, or V. Pálinkáš, personal communication) or Scintrex (e.g., Timmen and Gitlein 2009; Ukawa et al. 2010; Oja 2016; Cheraghi et al. 2020; Timmen et al. 2020). Some of these instruments show trends of the feedback scale factor in the order of 0.01–0.08 % per year. This is at least a factor of two below the trend observed for ZLS-B78. To my knowledge, no larger trends have been reported in the geodetic literature, so we may consider ZLS-B78 a worst-case scenario. However, of more concern than the linear trend (which we can easily deal with as long as it is constant) is the annual component of the scale factor variability. Further investigations on the possible cause of the annual component are necessary to allow for a reliable reconstruction of the scale factor for arbitrary epochs. In any case, it is obvious that regular calibrations are necessary to monitor scale factor variations. However, restricting repeated calibrations to the same season, with the idea of minimizing perturbations from seasonal gravity variations of the calibration base, is not necessarily sufficient to ensure complete control over the scaling factor. The temporal behaviour of the scale factor seems to be a specific characteristic of each and every instrument, so it is hard to give general advice. In case of significant variability of the feedback scale factor, it is at least advisable to do calibrations before and after every field campaign.

Acknowledgements Vojtech Pálinkáš and Przemyslaw Dykowski are gratefully acknowledged for discussion and valuable input during and after the GGHS2024 meeting. Special thanks to Vojtech for providing insight into the behaviour of the feedback scale of his ZLS-B20. Walter Zürn is acknowledged for pointing me to the work of El Wahabi et al. on environmental effects in long-term observations with relative spring gravimeters. I am also grateful to my former masterstudent Umang Dotel for conducting several of the relative experiments and to Ludger Timmen and another reviewer for their valuable comments.

References

Atzbacher K, Gerstenecker C (1993) Secular gravity variations: recent crustal movements or scale factor changes? J Geodynam 18(1):107–121. https://doi.org/10.1016/0264-3707(93)90033-3

Cheraghi H, Hinderer J, Saadat SA, Bernard J-D, Djamour Y, Tavakoli F, Arabi S, Azizian Kohan N (2020) Stability of the calibration of Scintrex relative gravimeters as inferred from 12 years of measurements on a large amplitude calibration line in Iran. Pure Appl Geophys 177(2):991–1004. https://doi.org/10.1007/s00024-019-02300-6

El Wahabi A, Ducarme B, Van Ruymbeke M (2001) Humidity and temperature effects on LaCoste & Romberg gravimeters. J Geodetic Soc Jpn 47(1):10–15. https://doi.org/10.11366/sokuchi1954.47.10

Gerlach C, Ackermann C, Falk R, Lothhammer A, Reinhold A (2017) Gravimetric investigations at Vernagtferner. In: Vergos G, Pail R, Barzaghi R (eds) International Symposium on Gravity, Geoid and Height Systems 2016, pp 53–60. International Association of Geodesy Symposia, Springer International Publishing. https://doi.org/10.1007/1345_2017_2

Gerlach C, Falk R, Reinhold A (2018) Feedback-Kalibrierung eines ZLS-Gravimeters in einem A10-Absolutschwerenetz mittlerer Ausdehnung. Allgemeine Vermessungsnachrichten (avn) 125(7):220–229.

Jentzsch G (2008) The automated Burris gravity meter – a new instrument using an old principle. In: Proceedings of Terrestrial Gravimetry: Static and Mobile Measurements, pp 21–28. St Petersberg, Russia

Longman IM (1959) Formulas for computing the tidal accelerations due to the Moon and the Sun. J Geophys Res 64(12):2351–2355. https://doi.org/10.1029/JZ064i012p02351

Oja T (2016) Unstable calibration factor of CG-5 relative gravimeter. In: Poster presented at the International Symposium on Gravity, Geoid and Height Systems 2016. Thessaloniki, Greece

Schilling M, Gitlein O (2015) Accuracy estimation of the IfE gravimeters Micro-g LaCoste gPhone-98 and ZLS Burris gravity meter B-64. In: Rizos C, Willis P (eds) IAG 150 years. International Association of Geodesy Symposia, vol 143, pp 249–256. Springer International Publishing. https://doi.org/10.1007/1345_2015_29

Schulz HR (2018) A new PC control software for ZLS-Burris gravity meters. Geodesy Geodynam 9(3):210–219. https://doi.org/10.1016/j.geog.2017.09.002

Timmen L, Rothleitner C, Reich M, Schröder S, Cieslack M (2020) Investigation of Scintrex CG-6 gravimeters in the gravity meter calibration system Hannover. Allgemeine Vermessungsnachrichten (avn) 127(4):155–162

Timmen L, Gerlach C, Rehm T, Völksen C, Voigt C (2021) Geodetic gravimetric monitoring of mountain uplift and hydrological variations at Zugspitze and Wank Mountains (Bavarian Alps, Germany). Remote Sens 13:918. https://doi.org/10.3390/rs13050918

Timmen L, Gitlein O (2009) The capacity of the Scintrex Autograv CG-3M no. 4492 gravimeter for absolute-scale surveys. Revista Brasileira de Cartografia 56(2). https://doi.org/10.14393/rbcv56n2-43505

Torge W (1989) Gravimetry. de Gruyter, Berlin/New York

Ukawa M, Nozaki K, Ueda H, Fujita E (2010) Calibration shifts in Scintrex CG-3M gravimeters with an application to detection of microgravity changes at Iwo-Tou Caldera, Japan. Geophys Prospect 58(6):1123–1132. https://doi.org/10.1111/j.1365-2478.2010.00869.x

On the Application of the Atmospheric Attraction Computation Service (Atmacs) in Absolute Gravimetry

Ezequiel D. Antokoletz, Hartmut Wziontek, Thomas Klügel, André Gebauer, Kyriakos Balidakis, and Henryk Dobslaw

Abstract

Classically, atmospheric corrections for terrestrial gravimetry are computed from the local air pressure record and a conventional admittance factor of $-3.0\,\mathrm{nm/s^2/hPa}$ is usually adopted to derive gravity effects. The reference level is based on a standard atmosphere in agreement with the Resolution N°1 of the International Association of Geodesy (IAG) of 2023, which defines the International Terrestrial Gravity Reference System (ITGRS). Roughly 90% of the atmospheric contributions to gravity variations are covered by this approach. To account also for the spatial distribution of air masses around a given station, the Atmospheric attraction computation service (Atmacs[1]) was established that relies on numerical weather models of the German Weather Service (Deutscher Wetterdienst – DWD). However, while Atmacs provides accurate time-variations of atmospheric corrections, the long-term stability is currently limited by occasional model improvements at DWD. Such model updates do not only include improvements in the spatial and/or vertical resolution but also modifications of the associated model orography, and may thus cause discontinuities in the time series of atmospheric corrections which may rise the tens of $\mathrm{nm/s^2}$ level and are typically altitude-dependent. In the present study, we aim to solve this problem by referring the time series to a common reference ensuring the compatibility with the conventions adopted for the ITGRS, and we show the advantage of using Atmacs for correcting absolute gravity observations.

Keywords

Absolute gravimetry · Atmacs · Atmospheric corrections · Terrestrial gravity time series

1 Introduction

In addition to Earth tides, high-precision terrestrial gravity measurements are mainly affected by mass-variations in the atmosphere. Classically, atmospheric corrections are based on a constant admittance factor for the surface atmospheric

[1]https://atmacs.bkg.bund.de/, last access: November 20, 2025.

E. D. Antokoletz (✉) · H. Wziontek · A. Gebauer
Federal Agency for Cartography and Geodesy (BKG), Leipzig, Germany
e-mail: Ezequiel.Antokoletz@bkg.bund.de; Hartmut.Wziontek@bkg.bund.de; Andre.Gebauer@bkg.bund.de

T. Klügel
Federal Agency for Cartography and Geodesy (BKG), Geodetic Observatory Wettzell, Bad Kötzting, Germany
e-mail: Thomas.Kluegel@bkg.bund.de

K. Balidakis
Federal Agency for Cartography and Geodesy (BKG), Frankfurt am Main, Germany

Helmholtz Centre for Geosciences (GFZ), Potsdam, Germany
e-mail: Kyriakos.Balidakis@bkg.bund.de

H. Dobslaw
Helmholtz Centre for Geosciences (GFZ), Potsdam, Germany
e-mail: dobslaw@gfz.de

J. T. Freymueller, L. Sànchez (eds.), *International Symposium on Gravity, Geoid and Height Systems 2024 (GGHS2024)*,
International Association of Geodesy Symposia 158, https://doi.org/10.1007/1345_2025_300

pressure measured at the station, i.e. (Torge 1989),

$$g_{atm} = a(P - P_0), \tag{1}$$

where a is an admittance factor, P is the observed atmospheric pressure and P_0 is a reference pressure related to a standard atmosphere. Although the admittance factor may vary for different locations, for absolute gravimetry, it is usually fixed to $a = -3.0\,\mathrm{nm/s^2/hPa}$, following the International Absolute Gravity Basestation Network (IAGBN): Absolute Gravity Observations Data Processing Standards by Boedecker (1988), recently confirmed through the International Terrestrial Gravity Reference System (ITGRS) Conventions 2020 (Wziontek et al. 2021). This fixed value accounts for both the vertical attraction of the air column above the gravimeter and the loading caused by it (Torge 1989). Within the ITGRS Conventions, the reference pressure for the standard atmosphere ISO 2533:1975 (DIN 5450) is adopted and computed by

$$P_0(H) = 1013.25(1 - 0.0065H/288.15)^{5.2559} \quad \mathrm{hPa} \tag{2}$$

where H is the physical height of the station in m. The standard atmosphere plays an essential role for the definition of gravity acceleration and the ITGRS, as stated through the Resolution N° 1 of the International Association of Geodesy (IAG) of 2023 (IAG Office 2025).

This simple and straightforward approach explains already about 90% of the total atmospheric contributions to gravity variations (Merriam 1992). More precise atmospheric corrections for terrestrial gravimetry can be achieved through the application of numerical weather models (e.g. Boy et al. 2002; Neumeyer et al. 2004; Klügel and Wziontek 2009). These allow not only to account for the global atmospheric masses variability, but also to consider the 3D distribution of air masses around the station. In this case, it has been demonstrated that these corrections are more efficient than the standard approach (e.g. Antokoletz et al. 2025) in particular cases where small signals e.g., from hydrology, need to be isolated.

The Atmospheric attraction computation service (Atmacs) of BKG provides on an operational basis, atmospheric corrections for high-precision gravity time series (Klügel and Wziontek 2009; Antokoletz et al. 2024). Atmacs benefits from global numerical weather simulations of the Icosahedral Nonhydrostatic (ICON) model of the German Weather Service (Deutscher Wetterdienst – DWD; Zängl et al. 2015) to compute atmospheric corrections at selected stations. Atmospheric models from DWD have a triangular grid and its current version (ICON 384) has an average spatial resolution of 13 km and a temporal resolution of 3 hours. The computation scheme in Atmacs is summarized in Fig. 1. Atmospheric attraction and deformation effects

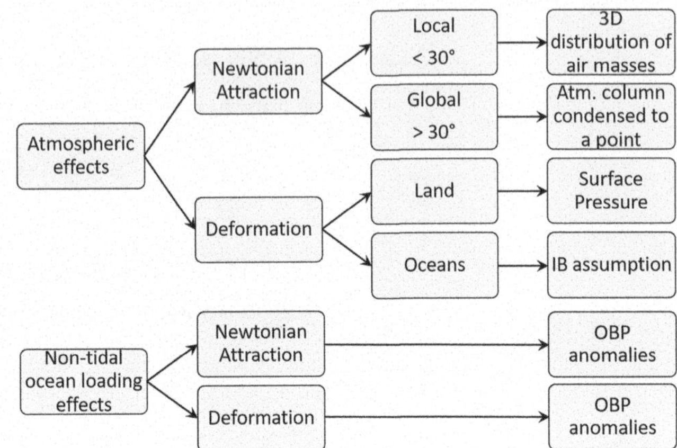

Fig. 1 Computation scheme of atmospheric and non-tidal ocean loading effects in Atmacs (from Antokoletz et al. (2024), licensed under CC-BY 4.0)

are computed separately. The computation of attraction effects is divided into a local and global regions, where within the local area (<30°), the 3D distribution of air masses is accounted for. Outside this area (global), the mass of each atmospheric column is condensed to a point. Further details on the atmospheric modelling can be found in Klügel and Wziontek (2009) and Antokoletz et al. (2024). Deformation contributions to loading are computed using the surface atmospheric pressure in a Green's functions approach following Farrell (1972) and Love load numbers for PREM (Jentzsch 1997). Over oceans, an Inverse Barometer response is assumed (Antokoletz et al. 2023). Moreover, Antokoletz et al. (2024) have recently updated the service by including non-tidal ocean loading effects based on ocean-bottom pressure (OBP) anomalies of the Max-Planck-Institute for Meteorology Ocean Model (MPIOM; Jungclaus et al. 2013) forced by atmospheric fields from the Operational European Centre for Medium-Range Weather Forecasts (ECMWF) model, as processed by the Helmholtz Centre for Geosciences (GFZ; Shihora et al. 2023).

Currently, atmospheric corrections from Atmacs cannot be applied for absolute gravity observations as the compatibility with equation (1) is not ensured. In this study, we review the reference level of atmospheric corrections in Atmacs and its capability to also provide atmospheric corrections for absolute gravity measurements.

2 Reference Level of Atmospheric Corrections in Atmacs

In Atmacs, mass anomalies at each model cell are computed with respect to the standard atmosphere ISO 2533:1975 (DIN 5450). For the computation of local attraction effects, mass anomalies with respect to vertical density profiles of the

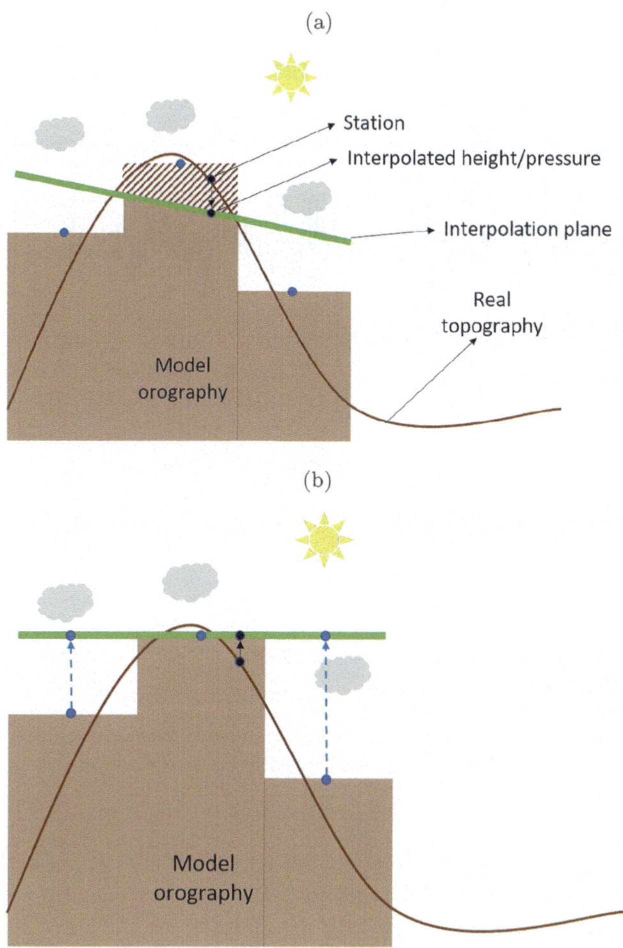

Fig. 2 Scheme of the actual (**a**) and new (**b**) interpolation approach of height and surface atmospheric pressure in Atmacs

standard atmosphere are derived. Outside this area, as well as for deformation effects, mass anomalies are derived from the surface atmospheric pressure and the reference pressure given by equation (2) assuming hydrostatic equilibrium (Antokoletz et al. 2024). Here, the position of each mass anomaly and its height is based on the model orography. As the model orography does not coincide exactly with the real topography, the height of the computation point is not given by the actual height of a station but interpolated from the model orography through a plane interpolation defined by the cells around the computation point. Moreover, besides attraction and deformation effects, surface atmospheric pressure at the computation point is routinely computed as it can be used to further improve the high-frequency variability of the corrections following a remove-restore approach. Surface atmospheric pressure at the computation point is interpolated in the same way as it is for height. The interpolation scheme is depicted in Fig. 2a.

For a given station within a cell, all cells around it are selected to define the interpolation plane, taking into account the height of each cell and the surface atmospheric pressure. The interpolation plane is defined through the three vertices of the cell containing the station and transformed into a local coordinate system (x_i, y_i, z_i), where i is each of the three vertices. The z component for each vertex is either height or surface pressure and it is obtained through a mean value of all cells sharing that vertex. Once the plane is defined, it is evaluated at the position of the computation point to obtain the required quantity.

Although this approach is commonly used to interpolate several atmospheric variables in atmospheric modelling (e.g. Reinert et al. 2020), it presents several disadvantages for computing gravity effects. First of all, if model changes occur due to e.g., improvements of the horizontal/vertical resolution, there will be discontinuities in the interpolated time series, either because of a change of the height of the cell containing the station or because of changes in the orography in the surroundings of the station which highly affects the inclination of the interpolation plane. Such step-like features are noticeable in height and, consequently, on the average surface pressure. To illustrate this, Fig. 3 shows the interpolated surface atmospheric pressure at the station Medicina, Italy compared to the local air pressure recordings (Wziontek et al. 2017). A step of about 3 hPa can be noted, which is related to a model change and enhancement of the spatial resolution from the Global Model (GME) 256 to GME 384 (Majewski et al. 2002). Medicina is a station located in a rather flat area but as the interpolation plane relies on the cells in the surroundings and its spatial resolution, the presence of the Apennine mountains have an impact on the tilting of the interpolation plane and, therefore, on the interpolated pressure. However, when looking at the total gravity effect from Atmacs, no step-like feature can be noticed, proving the internal consistency of the Atmacs computation.

On the other hand, as schematized in Fig. 2a, the interpolated height for a given station may be either above or below the model orography. In the near-field, atmospheric masses may be far away from the computation point or even below it, causing a miss- or overestimation (depending on the position of the mass with respect to the station) of the attraction effects in the local area. Finally, as the model orography represents a smooth description of the real topography, the interpolated height is not compatible with the real height of the station, which turns to be important for the definition of the reference level for absolute gravimetry when computing the reference pressure with equation (2). Since pressure is highly correlated with height, this is also reflected by the mean value (bias) of the pressure differences in the top panel of Fig. 3.

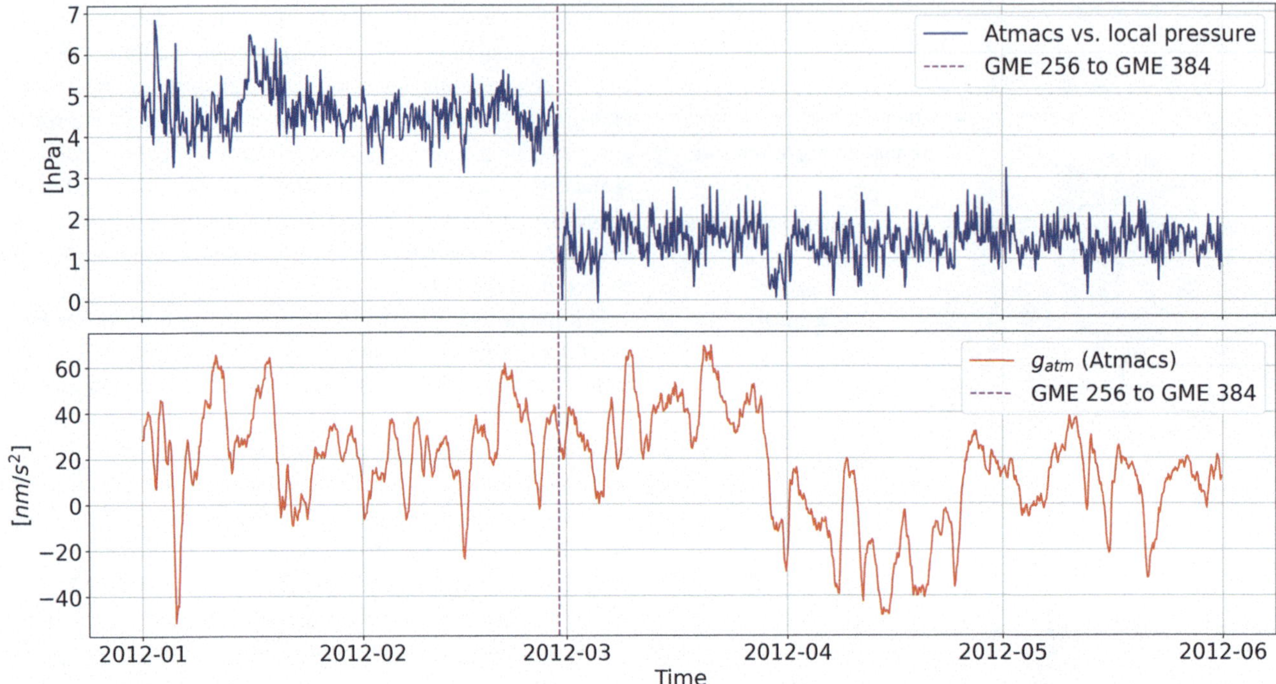

Fig. 3 Top: Difference between the local pressure recordings and the interpolated surface atmospheric pressure at the station Medicina. Bottom: Total atmospheric correction as obtained from Atmacs. The vertical dashed line corresponds to the epoch where the model change from GME 256 to GME 384 occurred

3 New Interpolation Scheme

To overcome the drawbacks described above, two modifications to the interpolation approach are proposed and schematized in Fig. 2b. First of all, in the new scheme, no interpolation of the height is performed and the height of the computation point is simply the height of the cell where the point lies on, based on the model orography. This simplification allows to consistently account for the distance between the air masses and the computation point, avoiding any miss- or overestimation of the attraction effects in the local area. In case a station is below the surface (e.g., in a tunnel), a certain depth can also be considered.

As for the interpolation of the surface atmospheric pressure, prior defining the interpolation plane, in the new approach the pressure of the neighbouring cells is shifted to the same height of the station (H_T) by means of the standard atmosphere (Equation (2)) as

$$\Delta P = P_0(H) - P_0(H_T), \quad (3)$$

where H is the height of the respective cell to be shifted. By this, the interpolation plane is horizontal and refers to a constant height, i.e., the height of the computation point given by the model orography. However, this height may change due to model changes as already described. To avoid this, at a final step, the interpolated pressure is shifted again to the real height of the station using equation (3). This not only helps to overcome step-like features but also to ensure its compatibility with the conventions for absolute gravimetry. Future model changes by DWD can also be handled in this same way, providing the long-term stability of the entire time series.

4 Results and Discussion

4.1 Surface Pressure Interpolation

In order to analyse whether the new interpolation approach allows to overcome the problems previously described, atmospheric surface pressure derived from Atmacs was compared with local pressure recordings at different stations. As illustration, Fig. 4 depicts this comparison for three stations contributing to the International Geodynamics and Earth Tide Service (IGETS) with different heights: Medicina, Italy (MC), Wettzell, Germany (WE) and Apache Point, USA (AP) with heights $H = 28.0$ m, $H = 606.6$ m and $H = 2788.0$ m, respectively. As already demonstrated, the interpolated pressure from Atmacs as described in Section 2 (Atmacs 1.0) present step-like features related to the model change from GME 256 to GME 384, which may vary up to tens of hPa for stations located in mountain areas. In contrast, with the new interpolation approach (Atmacs 2.0), no step-like features can be identified.

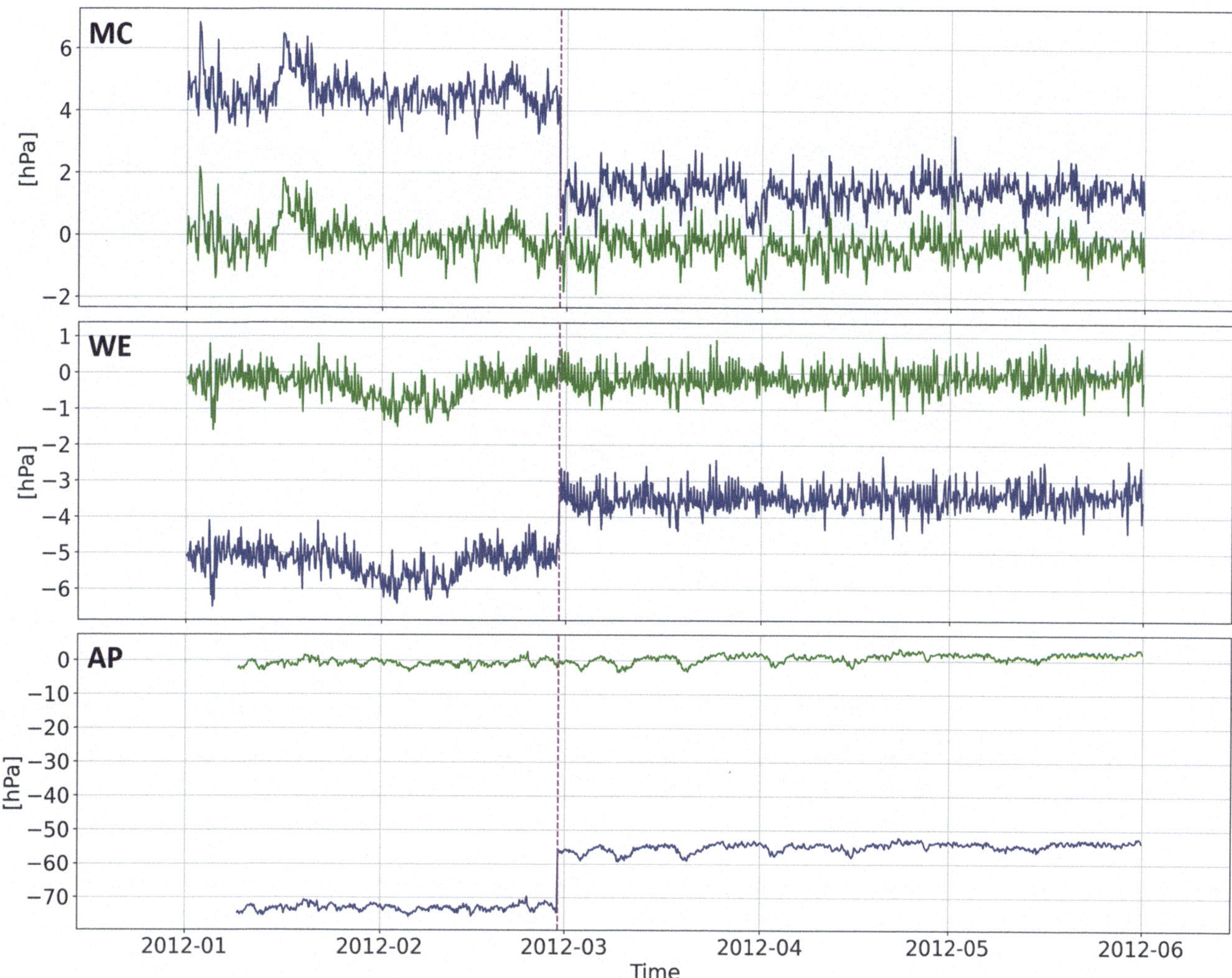

Fig. 4 Difference between the local atmospheric pressure recording and the interpolated pressure from Atmacs at stations MC, WE and AP. Blue lines correspond to the current interpolation approach and green lines to new one. The vertical dashed line corresponds to the model change from GME 256 to GME 384

Table 1 Bias introduced by the interpolated pressure from Atmacs. For Atmacs 1.0, biases before and after the model change from GME 256 to GME 384 are presented

	Bias [hPa]	
Station	Atmacs 1.0	Atmacs 2.0
MC	4.6 / 1.4	−0.3
WE	−5.3 / −3.5	−0.2
AP	−73.1 / −54.8	0.3

Table 1 shows the biases introduced by both interpolation approaches. For Atmacs 1.0, large biases of up to 70 hPa ($\approx$ 170 nm/s^2) can be noticed as the interpolated pressure refers to the model orography and not to the real height of the station. Such biases would cause a consequent bias in the absolute level if applied to absolute gravimetry which is well above the typical accuracies of absolute gravimeters ($<10^{-8}$). With the new interpolation approach (Atmacs 2.0), only small biases (<1 nm/s^2) are found as the interpolated pressure refers to the real height of the station. These biases can be considered negligible compared to the accuracy mentioned above and, by this, the compatibility of Atmacs 2.0 with the classical correction based on the admittance factor is ensured as the reference level remains the same.

4.2 Impact on Superconducting Gravity Time Series

To see the impact of the new interpolation approach on gravity observations, a superconducting gravity (SG) time series recorded at the Conrad Observatory (CO), Austria (Meurers et al. 2021) was selected. As this station is located in the Eastern Alps, the complexity of the topography in the surroundings makes it an ideal study case to evaluate the new interpolation approach. Moreover, Meurers (2024) points out the limitations of the atmospheric corrections

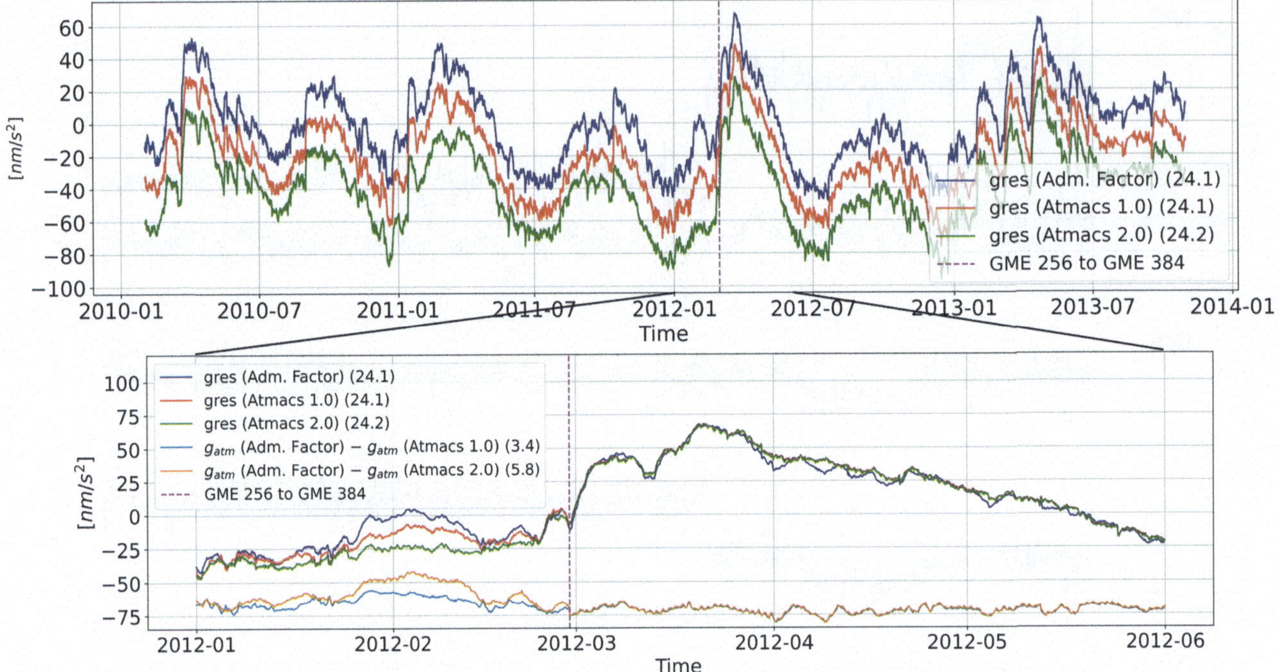

Fig. 5 Gravity residuals at CO after tidal analysis of superconducting gravity time series using different atmospheric corrections: classic admittance factor (blue), Atmacs 1.0 (red) and Atmacs 2.0 (green). In the first panel, an arbitrary offset has been applied. The bottom panel shows a zoom for a few months in 2012 where the model change occurred (vertical line). The differences between the classic admitance factor and Atmacs 1.0 (light blue) and Atmacs 2.0 (orange) are also shown. Legends also show the standard deviation of the residuals in nm/s^2

based on numerical weather models. For this study, SG time series from CO comprehend the Level 1 data sets from the IGETS database (Boy et al. 2020) with a length of four years (2010–2014). Harmonic contributions due to Earth, ocean and atmospheric tides where removed through a tidal analysis with the ETERNA-x package, version ETA34-X-V81 (Schüller 2019a,b). Polar motion effects were computed based on the EOP C04 20 pole coordinates series of the International Earth Rotation and Reference Systems Service (IERS) using an amplitude factor of 1.16 (Wahr 1985). Atmospheric corrections were then applied based on three approaches:

- the conventional admittance factor $-3.0\,nm/s^2/hPa$, as given by equation (1);
- Atmacs with the interpolation approach as described in Section 2 (Atmacs 1.0); and
- Atmacs with the new interpolation approach described in the previous section (Atmacs 2.0).

Atmospheric corrections were applied prior to tidal analysis to avoid the presence of tidal residuals in the remaining gravity signal. Therefore, gravity residuals from three tidal analyses are evaluated. Atmospheric tides and its coupling with Earth's tides are not discussed here as its modelling should be tackled separately and it is subject of future analysis.

Figure 5 depicts the gravity residuals obtained using the three different atmospheric corrections. Although they show no difference in the overall standard deviation, gravity residuals reduced with Atmacs show smoother variability. This becomes more visible in a zoomed view of the first six months of 2012 (lower panel of Fig. 5). Here it is worth noticing the period prior the model change from GME 256 to GME 384, where the gravity residuals obtained using Atmacs 2.0 shows the lowest variability. For this period, the atmospheric corrections differ up to almost $30\,nm/s^2$. Such differences turn to be important as incomplete atmospheric corrections may significantly distort the interpretation of other signals, like the gravity change in response of precipitation events and the subsequent vertical and lateral transport of the rain water through the soil, aquifers, and surface water channels.

Additionally, to analyse the variability of the gravity residuals at different periods, a variance analysis was performed. To do so, a set of subsequent Butterworth band-pass filters were applied to the gravity residuals to analyse the standard deviation in 26 narrow spectral ranges. Figure 6 shows the results from the variance analysis. As expected, the use of Atmacs helps to further reduce the variability of the gravity residuals, especially for periods between 3 to 60 days. It is also noticeable that, although small, the new

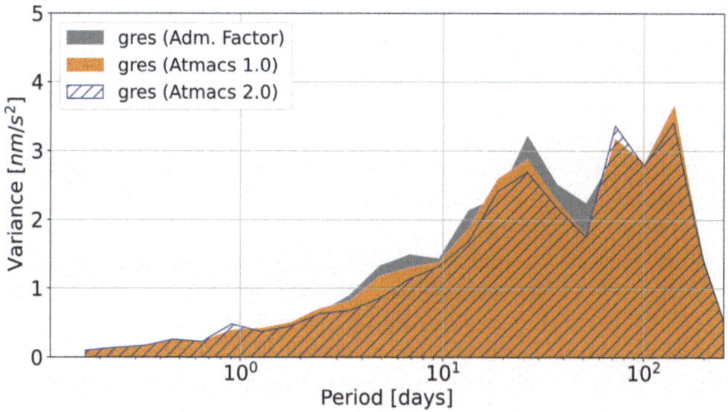

Fig. 6 Variance analysis of the gravity residuals at CO using the different atmospheric corrections

interpolation approach in Atmacs shows a further reduction of the variability for almost all periods.

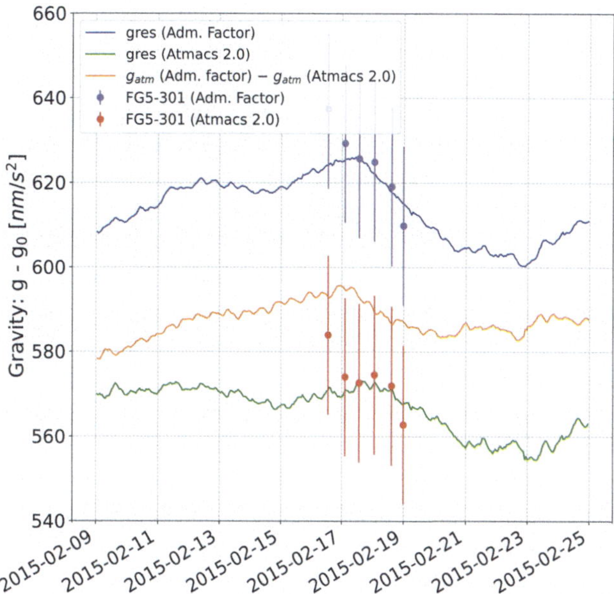

Fig. 7 Gravity reference function at station Wettzell, Germany with different atmospheric corrections: classic admittance factor (blue) and Atmacs 2.0 (green). Absolute gravity measurements performed with the FG5-301 with different atmospheric corrections: classic admittance factor (lilac) and Atmacs 2.0 (dark red). The difference between both atmospheric corrections is also shown (orange). An arbitrary offset of $50\,\mathrm{nm/s^2}$ is applied to the time series. $g_0 = 9808369000.0\,\mathrm{nm/s^2}$

4.3 Application to Absolute Gravity (AG) Measurements

Although with continuous SG time series the efficiency can be well illustrated, the impact of Atmacs 2.0 on periodical absolute gravity measurements performed with the FG5-301 (BKG 2025) at the station Wettzell during February, 2015 is demonstrated. They are also compared to the gravity reference function derived from the combination of the SG030 time series and FG5 measurements during the period November 2010 to October 2023. Both time series were processed with two different atmospheric corrections: the classic admittance factor and Atmacs 2.0. Atmospheric corrections from Atmacs were applied at the drop level through an interpolation of the time series and combined with the surface atmospheric pressure as recorded by the FG5.

Figure 7 depicts the gravity reference function and the set of AG measurements as processed with the different atmospheric corrections. Earth tides and ocean tide loading were removed through a local tidal model and polar motion effects in the same way as in the previous section. Although small, the AG measurements corrected with the classic admittance factor show a variability of about $15\,\mathrm{nm/s^2}$ over a few days, also present in the gravity reference function. Such variability is related to an insufficient atmospheric correction as reflected in the difference between the atmospheric corrections. These variations can be well corrected with Atmacs and, by this, most of the AG measurements are on the same level, presenting a better agreement with each other. These results show not only that absolute gravity measurements can benefit from better atmospheric corrections as provided by Atmacs but also enable a more realistic estimate of the long-term reproducibility of the instrument (Pálinkáš et al. 2013).

5 Conclusions

The long-term stability of atmospheric corrections from Atmacs is currently affected by changes of DWD's numerical weather models. Although these changes are occasional, they may cause discontinuities in the interpolated time series of surface atmospheric pressure provided in order to improve the high-frequency variability of the atmospheric corrections. To overcome this, a new interpolation approach was proposed and evaluated for several stations. Results not only show that discontinuities due to model changes are solved but also help to consistently account for attraction effects in the surroundings of the computation point. To illustrate this, gravity recordings at the Conrad Observatory, Austria were reduced with different atmospheric corrections and compared. The new interpolation approach helps to further decrease the variability of gravity residuals compared not only to the classic admittance factor correction but also to the current version of Atmacs. Atmacs 2.0 was also applied to a set of absolute gravity measurements performed at the station Wettzell, Germany. Although small, Atmacs not only provides better agreement between the measurements along several days but could also provide a better assessment of the long-term reproducibility of absolute meters.

Atmospheric corrections derived with the new interpolation approach show now compatibility with the ITGRS Conventions, recently adopted for the definition of International Terrestrial Gravity Reference System (ITGRS) by the Resolution N°1 of IAG of 2023. By this, once this new approach is implemented, Atmacs could also be used to correct absolute gravity observations without affecting the absolute level defined through the standard atmosphere. Precise atmospheric corrections are essential for further interpretation of the gravity signal in terms of e.g., hydrology (Güntner et al. 2017) or glacial isostatic adjustment (GIA; Olsson et al. 2019).

Finally, these results motivate a reanalysis of Atmacs to provide enhanced atmospheric corrections with the new interpolation approach and to obtain consistent time series over decades. Such reanalysis will be made available the Atmacs website and its operational implementation will also allow to overcome future model changes in DWD.

Acknowledgements The authors appreciate the effort of the operators of the SGs and FG5-301 to maintain and operate the instruments involved in this study. We also thank the Editor, Dr. George Vergos, and the two anonymous reviewers for their valuable comments and suggestions. EA thanks Tobias Bauer (BKG, Germany) and Claudia Tocho (UNLP, Argentina) for the helpful discussion regarding the atmospheric models of DWD and the preparation of the manuscript. KB was funded by the DFG via the Collaborative Research Cluster TerraQ (SFB 1464, project-ID 434617780).

Author Contribution TK, HW and EA worked in the development of the atmospheric modelling. EA and HW worked on the new interpolation approach and all authors participated in the discussion. EA drafted and coordinated the work on the manuscript and all authors contributed to the discussion and final version.

Data Availability Atmospheric corrections are available through the Atmospheric attraction computation service (Atmacs): http://atmacs.bkg.bund.de/. Superconducting gravity time series are available through the IGETS Database: https://isdc.gfz.de/igets-data-base/. Absolute gravity measurements are available through the International Absolute Gravity Database (AGrav): https://agrav.bkg.bund.de/.

Disclosure of Interests The authors declare that they have no conflict of interest.

References

Antokoletz ED, Wziontek H, Dobslaw H, Balidakis K, Klügel T, Oreiro FA, Tocho CN (2023) Combining atmospheric and non-tidal ocean loading effects to correct high precision gravity time-series. Geophys J Int 236(1):88–98. https://doi.org/10.1093/gji/ggad371

Antokoletz ED, Wziontek H, Klügel T, Balidakis K, Dobslaw H (2024) Update of the atmospheric attraction computation service (Atmacs) for high-precision terrestrial gravity observations. Springer, Berlin, Heidelberg. https://doi.org/10.1007/1345_2024_239

Antokoletz ED, Boy JP, Wziontek H, Balidakis K, Klügel T, Dobslaw H, Tocho CN (2025) Comparison of atmospheric and non-tidal ocean loading corrections for high-precision terrestrial gravity time series. Pure Appl Geophys. https://doi.org/10.1007/s00024-025-03804-0

Boedecker G (1988) International Absolute Gravity Basestation Network (IAGBN). Absolute gravity observations data processing standards station documentation. Bull d'information-Bureau Gravimétrique Int (63):51–56

Boy JP, Gegout P, Hinderer J (2002) Reduction of surface gravity data from global atmospheric pressure loading. Geophys J Int 149(2):534–545. https://doi.org/10.1046/j.1365-246X.2002.01667.x

Boy JP, Barriot JP, Förste C, Voigt C, Wziontek H (2020) Achievements of the first 4 years of the international geodynamics and earth tide service (IGETS) 2015–2019. Springer, Berlin, Heidelberg. https://doi.org/10.1007/1345_2020_94

Bundesamt für Kartographie und Geodäsie (2025) FG5-301 absolute gravity measurements. Absolute Gravity Database AGrav. https://doi.org/10.18168/agrav.m.FG5.301

Farrell W (1972) Deformation of the Earth by surface loads. Rev Geophys 10(3):761–797. https://doi.org/10.1029/RG010i003p00761

Güntner A, Reich M, Mikolaj M, Creutzfeldt B, Schroeder S, Wziontek H (2017) Landscape-scale water balance monitoring with an iGrav superconducting gravimeter in a field enclosure. Hydrol Earth Syst Sci 21(6):3167–3182. https://doi.org/10.5194/hess-21-3167-2017

IAG Office (2025). IAG Geodesist's Handbook 2024. International Association of Geodesy (IAG). https://doi.org/10.82507/IAG-GH2024

Jentzsch G (1997) Earth tides and ocean tidal loading. Tidal Phenomena, 145–171. https://doi.org/10.1007/BFb0011461

Jungclaus J, Fischer N, Haak H, Lohmann K, Marotzke J, Matei D, Mikolajewicz U, Notz D, Von Storch J (2013) Characteristics of the ocean simulations in the Max Planck Institute Ocean Model (MPIOM) the ocean component of the MPI-Earth system model. J Adv Model Earth Syst 5(2):422–446. https://doi.org/10.1002/jame.20023

Klügel T, Wziontek H (2009) Correcting gravimeters and tiltmeters for atmospheric mass attraction using operational weather models. J Geodynam 48(3):204–210. https://doi.org/10.1016/j.jog.2009.09.010

Majewski D, Liermann D, Prohl P, Ritter B, Buchhold M, Hanisch T, Paul G, Wergen W, Baumgardner J (2002) The operational global icosahedral–hexagonal gridpoint model GME: Description and high-resolution tests. Month Weather Rev 130(2):319–338. https://doi.org/10.1175/1520-0493(2002)130<0319:TOGIHG>2.0.CO;2

Merriam JB (1992) Atmospheric pressure and gravity. Geophys J Int 109(3):488–500. https://doi.org/10.1111/j.1365-246X.1992.tb00112.x

Meurers B (2024) Correction of high-frequency (> 0.3 mhz) air pressure effects in gravity time-series. Geophys J Int 237(1):14–30. https://doi.org/10.1093/gji/ggae030

Meurers B, Papp G, Ruotsalainen H, Benedek J, Leonhardt R (2021) Hydrological signals in tilt and gravity residuals at conrad observatory (Austria). Hydrol Earth Syst Sci 25(1):217–236. https://doi.org/10.5194/hess-25-217-2021

Neumeyer J, Hagedoorn J, Leitloff J, Schmidt T (2004) Gravity reduction with three-dimensional atmospheric pressure data for precise ground gravity measurements. J Geodynam 38(3):437–450. https://doi.org/10.1016/j.jog.2004.07.006

Olsson PA, Breili K, Ophaug V, Steffen H, Bilker-Koivula M, Nielsen E, Oja T, Timmen L (2019) Postglacial gravity change in Fennoscandia–three decades of repeated absolute gravity observations. Geophys J Int 217(2):1141–1156. https://doi.org/10.1093/gji/ggz054

Pálinkáš V, Lederer M, Kostelecký J, Šimek J, Mojzeš M, Ferianc D, Csapó G (2013) Analysis of the repeated absolute gravity measurements in the Czech Republic, Slovakia and Hungary from the period 1991–2010 considering instrumental and hydrological effects. J Geodesy 87(1):29–42. https://doi.org/10.1007/s00190-012-0576-1

Reinert D, Prill F, Frank H, Denhard M, Baldauf M, Schraff C, Gebhardt C, Marsigli C, Zängl G (2020) DWD database reference for the global and regional icon and icon-eps forecasting system. DWD 2023. https://www.dwd.de/DWD/forschung/nwv/fepub/icon_database_main.pdf

Schüller K (2019a) Installation Guide. Manual-03-ETA34-X-V80-Installation-Guide. https://eterna.bkg.bund.de/et34-x-docu-v80.7z

Schüller K (2019b) User's Guide. Manual-02-ET34-ANA-V80. https://eterna.bkg.bund.de/et34-x-docu-v80.7z

Shihora L, Balidakis K, Dill R, Dobslaw H (2023) Assessing the stability of AOD1B atmosphere–ocean non-tidal background modelling for climate applications of satellite gravity data: long-term trends and 3-hourly tendencies. Geophys J Int 234(2):1063–1072. https://doi.org/10.1093/gji/ggad119

Torge W (1989) Gravimetry. De Gruyter, Berlin, Boston. https://www.degruyter.com/view/title/10966

Wahr JM (1985) Deformation induced by polar motion. J Geophys Res Solid Earth 90(B11):9363–9368. https://doi.org/10.1029/JB090iB11p09363

Wziontek H, Wolf P, Nowak I, Richter B, Rülke A, Schwahn W, Wilmes H, Zerbini S (2017) Superconducting gravimeter data from Medicina - Level 1. GFZ Data Services. https://doi.org/10.5880/igets.mc.l1.001

Wziontek H, Bonvalot S, Falk R, Gabalda G, Mäkinen J, Pálinkás V, Rülke A, Vitushkin L (2021) Status of the international gravity reference system and frame. J Geodesy 95(1):1–9. https://doi.org/10.1007/s00190-020-01438-9

Zängl G, Reinert D, Rípodas P, Baldauf M (2015) The ICON (ICOsahedral Non-hydrostatic) modelling framework of DWD and MPI-M: Description of the non-hydrostatic dynamical core. Quart J Roy Meteorol Soc 141(687):563–579. https://doi.org/10.1002/qj.2378

Part II

Geoid Modelling and Height Systems

Contribution of the NGGM and MAGIC Constellation to Geodesy with an Emphasis on the IHRF

E. M. G. Mamagiannou and G. S. Vergos

Abstract

The Mass-Change and Geosciences International Constellation (MAGIC) mission is a collaborative effort between the National Aeronautics and Space Administration (NASA) and the European Space Agency (ESA), which will consist of four Earth observation satellites arranged into two pairs, with each agency contributing two satellites. Under NASA's supervision, the GRACE-C (P1) pairs will be developed, while ESA will oversee the development of the Next Generation Gravity Mission (NGGM) (P2). MAGIC aims to deliver mass-change products with superior attributes compared to its predecessors, the Gravity Recovery and Climate Experiment (GRACE) and the Gravity Recovery and Climate Experiment Follow-On (GRACE-FO), including higher spatial resolution, sub-weekly temporal resolution, shorter latency (a few days products), and heightened accuracy. For instance, MAGIC is designed to achieve an exceptional 1 mm accuracy in geoid measurements, along with a spatial resolution of 500 km within a 3-day timeframe or, alternatively, a spatial resolution of 150 km within a 10-day timeframe. In this study, we focus on the foreseen 30-day Level2a products of the NGGM and MAGIC constellation as Spherical Harmonical Coefficients (SHCs) of the disturbing potential. To simulate the NGGM/MAGIC time series, unfiltered GRACE and GRACE-FO release 06 Global Geopotential Models (GGMSs) from 2002 to 2023 are used. These are appropriately filtered, translated into temporal gravity field data products, and used as simulated MAGIC/NGGM observations, to allow us to explore hypothetical scenarios related to future forecasted gravity field and mass-change products. The focus is placed on estimating how the performance, and more specifically, the accuracy of the simulated MAGIC/NGGM data contributes to the estimation of the gravitational potential and geoid heights at a specific International Height Reference Frame (IHRF) core site. Singular Spectrum Analysis (SSA) is employed to complete the time-series in case voids are present. We explore various scenarios for both the tandem mission duration and observational accuracy. Each scenario offers unique insights into potential outcomes, offering an understanding of the anticipated performance of the MAGIC/NGGM missions. Conclusions on the achievable accuracy and its contribution to the realization of the IHRF are drawn along with the impact on the constellation and of NGGM itself as an operational service with emphasis on understanding Earth's dynamic processes and delivering high-precision gravity field data.

E. M. G. Mamagiannou (✉) · G. S. Vergos
Laboratory of Gravity Field Research and Applications - GravLab, Department of Geodesy and Surveying, Aristotle University of Thessaloniki, Thessaloniki, Greece
e-mail: elismama@topo.auth.gr

J. T. Freymueller, L. Sànchez (eds.), *International Symposium on Gravity, Geoid and Height Systems 2024 (GGHS2024)*, International Association of Geodesy Symposia 158, https://doi.org/10.1007/1345_2025_298

Keywords

Geodesy · GRACE/GRACE-FO · IHRF · NGGM/MAGIC · Spherical Harmonical Coefficients

1 Introduction

To contribution of satellite gravimetry missions, such as Global Ocean Circulation Explorer (GOCE) (Drinkwater et al. 2003), Gravity Recovery and Climate Experiment (GRACE) (Tapley et al. 2004), Gravity Recovery and Climate Experiment Follow-On (GRACE-FO) (Landerer et al. 2020), and the future GRACE continuity (GRACE-C), Next Generation Gravity Mission (NGGM), and Mass-Change and Geosciences International Constellation (MAGIC) (Daras et al. 2024) is pivotal. They contribute to the determination of the static and most importantly of the time-variable part of gravity-field related quantities, such as geoid heights, gravitational and gravity potential, physical heights, etc., which are of paramount importance for the realization of the International Height Reference System (IHRS) into the International Height Reference Frame (IHRF). The IHRS was established in 2015 by the International Association of Geodesy (IAG) as the official global standard for the precise and consistent determination of physical (orthometric) heights (Drewes et al. 2016) The IHRS is a geopotential-based reference system that co-rotates with the Earth and uses geopotential numbers (C_P) as its defining coordinate component. These geopotential numbers represent the difference between the gravitational potential at a point (P) on the Earth's surface (W_P) and a globally adopted constant reference potential (W_0), defined on the geoid (Ihde et al. 2017; Sánchez et al. 2021).

$$C_P = W_0 - W_P. \tag{1}$$

The IHRF is the physical realization of the IHRS, consisting of a globally distributed network of ~167 sites which implement collocated monitoring methods, i.e., permanent GNSS, gravity, levelling connection to nearby tide gauge stations and/or levelling connection to each national vertical datum. Each site is defined by two components, i.e., its geometric coordinates (X_P, Y_P, Z_P), referenced to the International Terrestrial Reference Frame (ITRF) (Altamimi et al. 2023), and its corresponding physical coordinate in the form of its gravity potential value (W_P), determined at that position. The combination of these geometric and physical components enables the consistent computation of the geopotential numbers (C_P) (Sánchez et al. 2024a).

During its initial realization (Ihde et al. 2017) the IHRF accounts only for the static gravity field, estimating (W_P) at specific epochs. However, the Earth's gravity field is inherently time-variable due to dynamic mass redistribution processes from hydrological, cryospheric, oceanic and solid earth forcing (Pail et al. 2015). Ignoring these variations introduces inconsistencies, especially when compared with the ITRF, which explicitly incorporates time-dependent changes in station positions through velocity components $(\dot{X}_P, \dot{Y}_P, \dot{Z}_P)$.

To ensure full consistency between the IHRF and ITRF, current efforts are focused, apart from the precise definition of potential values at the IHRF sites, on investigating the temporal evolution of the gravitational potential $(\dot{W}_P)$ across all IHRF reference stations. This inclusion marks a shift towards a dynamic realization of the IHRF, combining both static and time-varying elements of the gravity field. Such an approach would allow for the precise and time-resolved estimation of gravity potential values and associated heights, including orthometric heights (H_P), geoid heights (N_P), and ellipsoidal heights (h_P).

This close relationship is illustrated in Fig. 1, where geometric and physical quantities at point P and their temporal changes at a later epoch P' are shown. Temporal variations in gravitational potential (W_P), geopotential number (C_P), and surface deformation affect both the observed height components and the consistency of the reference system over time (Bíró 1983). Until today, the only available data source to infer variations with time of the potential and of physical heights, at a global scale, is offered by the time-series of the satellite missions of GRACE and GRACE-FO, only. Nevertheless, the spatial resolution of their monthly fields of ~200–500 km and accuracy of 2.5 cm at 450 km spatial scales (monthly) and 1 cm/year at 350 km (long-term trend), is a limiting factor (Daras et al. 2023). This is due to the fact that their current constellation and measurement principle does not manage to resolve higher frequencies of the gravity field spectrum, resulting in a large omission error. This is expected to improve with the future mission of NGGM and in combination with GRACE-C with the MAGIC constellation, as improved accuracies to higher spatial scales are expected, targeting 0.1 cm at 400 km and 1 cm at 200 km (monthly) and 0.01 cm/year and 0.1 cm/year for the long-term trend, respectively (Daras et al. 2024).

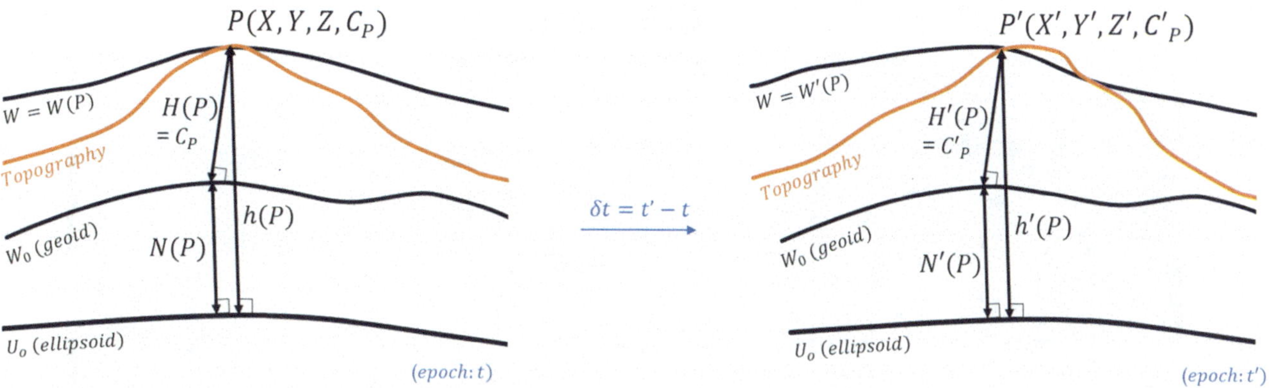

Fig. 1 Geometric variation (due to positional displacement) and dynamic variation (due to mass redistribution) of height components (N, H, h) over time. (Adaptation after Bíró (1983))

Note that in Fig. 1, the W_o corresponds to the conventional constant value adopted for the IHRF. Although W_o remains fixed, the geometry of the equipotential surface associated with that value (geoid) can vary over time due to temporal mass redistribution within the earth system. Thus, the geoid height (N) may change over time (N′) even though W_o itself does not (see Fig. 1).

2 Data and Study Area

Given the aforementioned prospect for improved, both in spatial resolution and accuracy, monthly gravity fields, the aim of this work is to examine how the forthcoming MAGIC constellation and NGGM satellite mission can contribute to the dynamic realization of the IHRF. As a representative example, we focus on the AUT1 reference station located in Thessaloniki, Greece, which is an IHRF core site and a EUREF station. The global distribution of the IHRF reference sites, including AUT1, is shown in Fig. 2.

To simulate future satellite observations, we used monthly Level-2 GRACE and GRACE-FO data from the CSR Release 06 dataset, covering the period from 2002 to 2023. These datasets include Spherical Harmonic Coefficients (SHCs) that represent the Earth's gravitational potential. Corrections were applied to the low-degree coefficients C_{20} and C_{30} following the recommendations of (Loomis et al. 2020) To reduce correlated noise, the SHCs were processed using a 350 km Gaussian spatial filter. Based on the filtered coefficients, geodetic products, such as gravitational potential (W) and geoid height (N) have been derived.

To fill gaps in the GRACE/GRACE-FO time series and ensure temporal continuity, Singular Spectrum Analysis (SSA) (Yi and Sneeuw 2021) was applied. Various methodologies have been proposed for reconstructing or extending GRACE-derived mass-change signals, which can generally be classified into three main groups. The first group includes data-substitution techniques, where complementary gravity observations, such as Swarm, are employed to fill temporal gaps (Richter et al. 2021; Forootan et al. 2020). The second group comprises data-adaptive approaches, such as empirical orthogonal function (EOF) analysis and singular spectrum analysis (SSA), which extract dominant temporal modes directly from the GRACE observations without relying on auxiliary datasets (Yi and Sneeuw 2021). The third group encompasses data-driven strategies, including machine-learning and deep-learning frameworks that use hydrometeorological variables (e.g., precipitation, soil moisture, or temperature) to model mass variations (Forootan et al. 2014; Gyawali et al. 2022; Mandal et al. 2025). Recent developments have shown that deep-learning methods can effectively capture nonlinear patterns but require extensive training datasets and parameter optimization. In contrast, SSA was selected in this study as a non-parametric, self-adaptive technique that reconstructs the signal solely from the available observations, ensuring methodological transparency and stability. Its straightforward implementation and ability to maintain the temporal behavior of the geopotential field make it particularly well suited for non-stationary time series. SSA was used to both address the period of missing observations between GRACE and GRACE-FO (mid 2017–mid 2018) as well as for monthly solutions that have not been delivered, resulting in a total of 35 months of the time-series. The method decomposes the available time series into distinct components representing the long-term trend, seasonal and sub-seasonal oscillations, and residual noise. The dominant components were then reconstructed and extrapolated beyond the observed period by continuing their characteristic temporal patterns, rather than by a simple linear or harmonic fit. This process allows the forecast to preserve the amplitude, phase, and variability of the original GRACE/GRACE-FO signal. The resulting reconstructed and gap-free time series of 254 months served as the basis for simulating the NGGM and MAGIC mission scenarios.

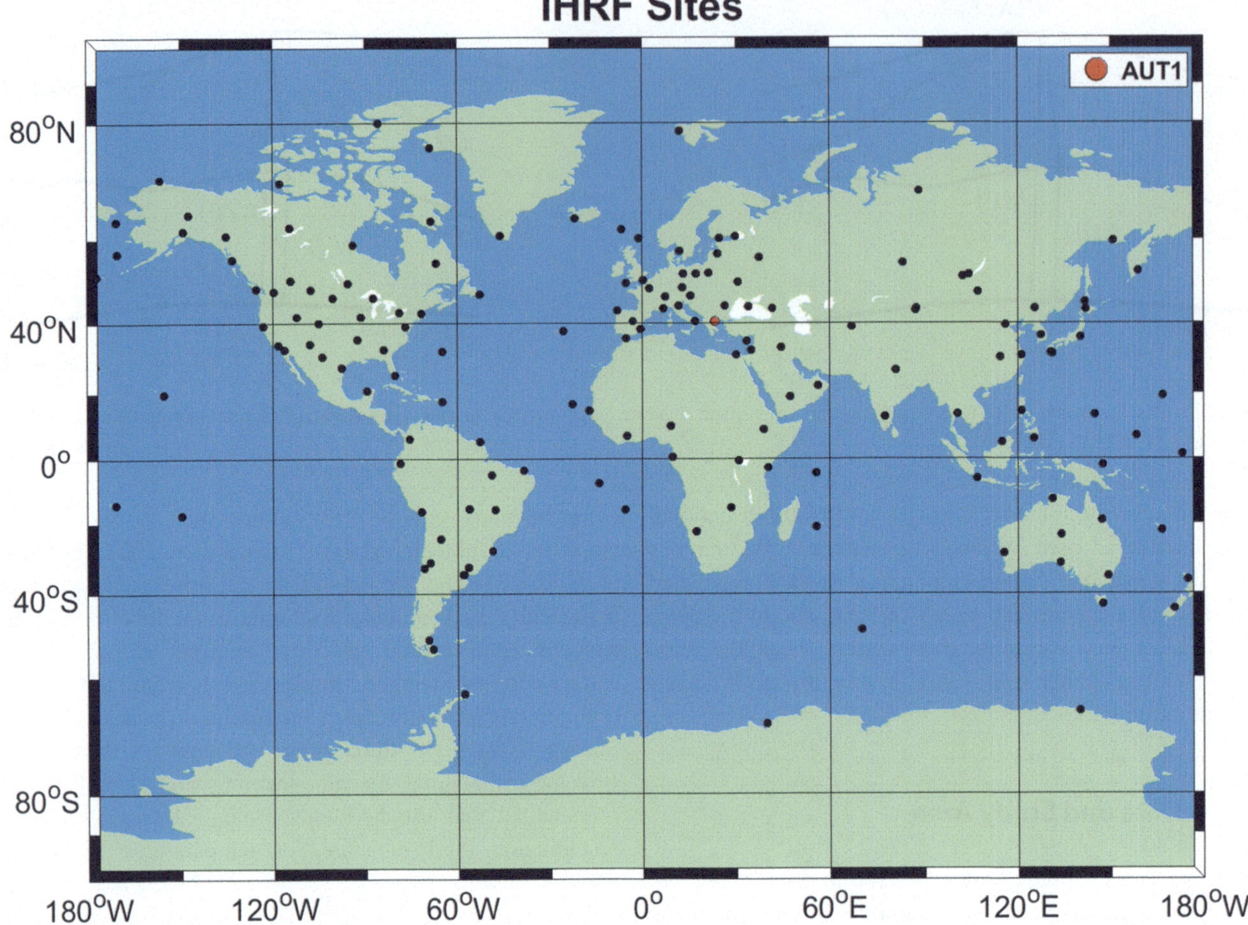

Fig. 2 The current (April 2025) list of IHRF sites

In the various simulations examined herein, it is assumed that the start of the existing GRACE/GRACE-FO time-series coincides with the planned launch of NGGM in 2032 (see Fig. 3), following various scenarios for NGGM life duration, GRACE-C and NGGM tandem, i.e., MAGIC constellation realization and forecasting in the future. Therefore in essence, forecasting up to a period of 13 years is examined, after NGGM nominal lifetime of 7 years.

3 Tested Scenarios in Simulations

To assess the potential contribution of the NGGM mission and MAGIC constellation to the dynamic realization of the IHRF, a series of simulation scenarios was developed using the reconstructed, gap-free GRACE/GRACE-FO time series. Three missions' duration scenarios were considered as: (1) 3 years of overlapping NGGM and MAGIC observations, (2) 5 years of NGGM and 4 years of MAGIC, with a 4-year overlap, and (3) 7 years of NGGM with 4 years of MAGIC and the same overlap period as scenario (2). It should be noted that the aforementioned scenarios aim to investigate the influence to geodesy and the IHRF of the nominal mission planning (scenario (3)), of a shorter NGGM life span but with the same as the nominal overlap with GRAE-C (scenario (2)) and a short NGGM mission with 3 years overlap with GRACE-C (scenario (1)). In all scenarios, we aim to investigate the extend in time, i.e., for how many years after the missions end, reliable potential and geoid heights estimates can be derived, simulating the case that no new similar missions are launched.

Furthermore, each scenario was evaluated under three accuracy (@) configurations, following (Daras et al. 2023) and the provided target requirements (Tables 11 and 12, *ibid.*) and science and applications traceability matrix for NGGM as well as those for the MAGIC constellation (Haagmans and Tsaoussi 2020): NGGM at @1.8 cm and MAGIC at @1.0 cm, which form the nominal scenario; NGGM at @2.2 cm and MAGIC at @1.2 cm, i.e. increased uncertainties by 20%; and NGGM at @2.6 cm and MAGIC at @1.45 cm, i.e. increased uncertainties by 45%. It should be noted that the aforementioned uncertainties refer to the

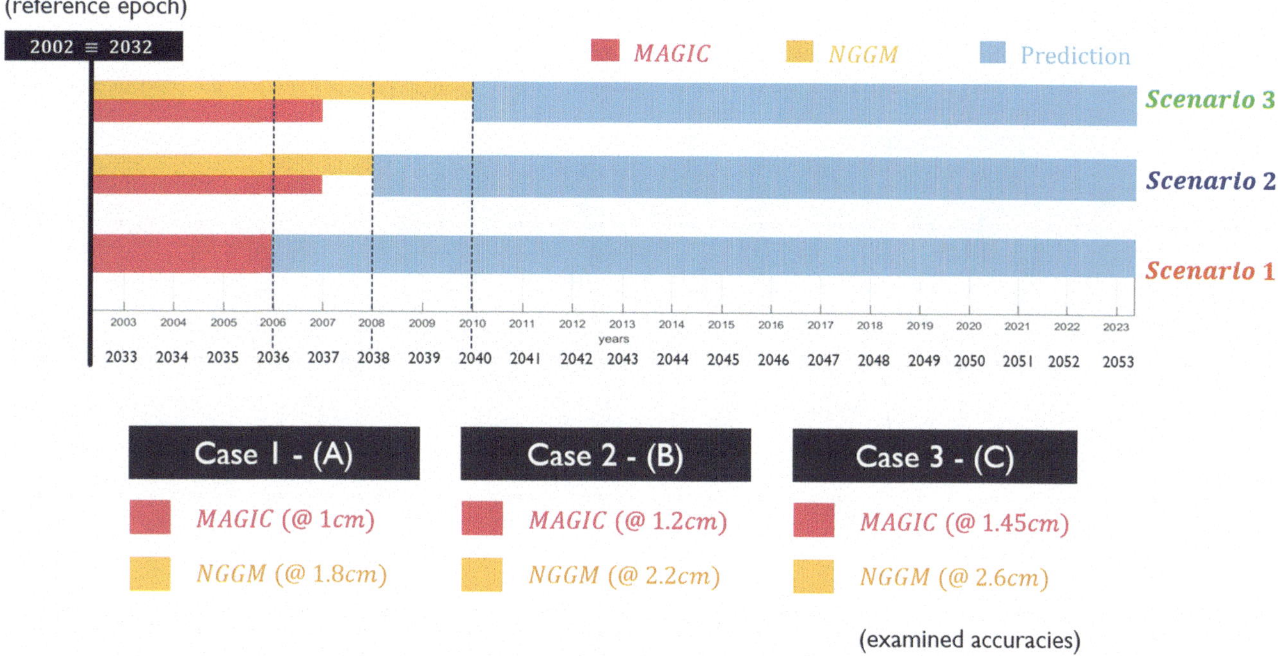

Fig. 3 Timeline of GRACE/GRACE-FO data usage for simulating NGGM/MAGIC observations

foreseen monthly NGGM and MAGIC products at a spatial resolution of 200 km corresponding to SHCs complete to degree and order (d/o) 100, which will be the ones of main use to geodetic applications. For each scenario/case combination, simulated values of gravitational potential and geoid height anomalies were computed at the AUT1 IHRF station, with the full set of tested scenarios and their corresponding simulation timelines being shown in Fig. 3. In all cases, the nominal GRACE/GRACE-FO timeseries referred to in the prequel are considered as the ground truth.

4 Results

4.1 Gravitational Potential Anomalies (ΔW)

Gravitational potential anomalies (ΔW), i.e., monthly estimates relative to a long-term mean (2004.0–2009.9), at the AUT1 station were evaluated for all three simulated mission and constellation scenarios. The results are shown in Fig. 4, where only one curve per scenario is displayed, as the predicted values are numerically identical across all accuracy cases. However, the forecast accuracy varies depending on the observation uncertainties assigned to each configuration. The prediction uncertainties are summarized in Table 1 as well as those when using the real time-series from GRACE/GRACE-FO for the entire period.

Figure 4 presents the predicted gravitational potential anomalies (ΔW) for the three simulated scenarios, shown alongside the original GRACE/GRACE-FO data, which serve as the validation reference. The black line corresponds to the GRACE/GRACE-FO observations, while the red, blue, and green lines represent the simulated predictions for Scenarios 1, 2, and 3, respectively. Figure 4 highlights the effect of mission duration on prediction performance as Scenario 1, based on only 3 years of NGGM and MAGIC data, shows the greatest deviation from the observed GRACE/GRACE-FO trend and signal. Scenario 2, which includes 5 years of NGGM and a 4-year overlap with MAGIC, shows improved agreement especially during the initial years of prediction. But the best alignment is observed in Scenario 3, which incorporates 7 years of NGGM data. In this case, the predicted anomalies closely follow the GRACE/GRACE-FO trend and remain stable almost throughout the forecast period, underscoring the value of longer mission durations for accurate and reliable gravitational potential estimates. The such-derived potential anomalies have been also compared to the "true" time-series for the entire period, with the residuals being shown in Fig. 4 (bottom) for all scenarios. Among these, Scenario 1 has a standard deviation (std) at the 0.015 m^2/s^2 level, which reduces to 0.012 m^2/s^2 and 0.011 m^2/s^2 for Scenarios 2 and 3, showing the benefits of having both a longer tandem mission and an NGGM duration.

In terms of the evaluation of the various cases for the uncertainties, the accuracies of the predicted gravitational potential anomalies at AUT1 are shown in Table 1 for both the true time-series and the simulated one. For Scenario 1, the uncertainty ranges between 0.00262 and 0.00379 m^2/s^2, in Scenario 2, the values increase slightly

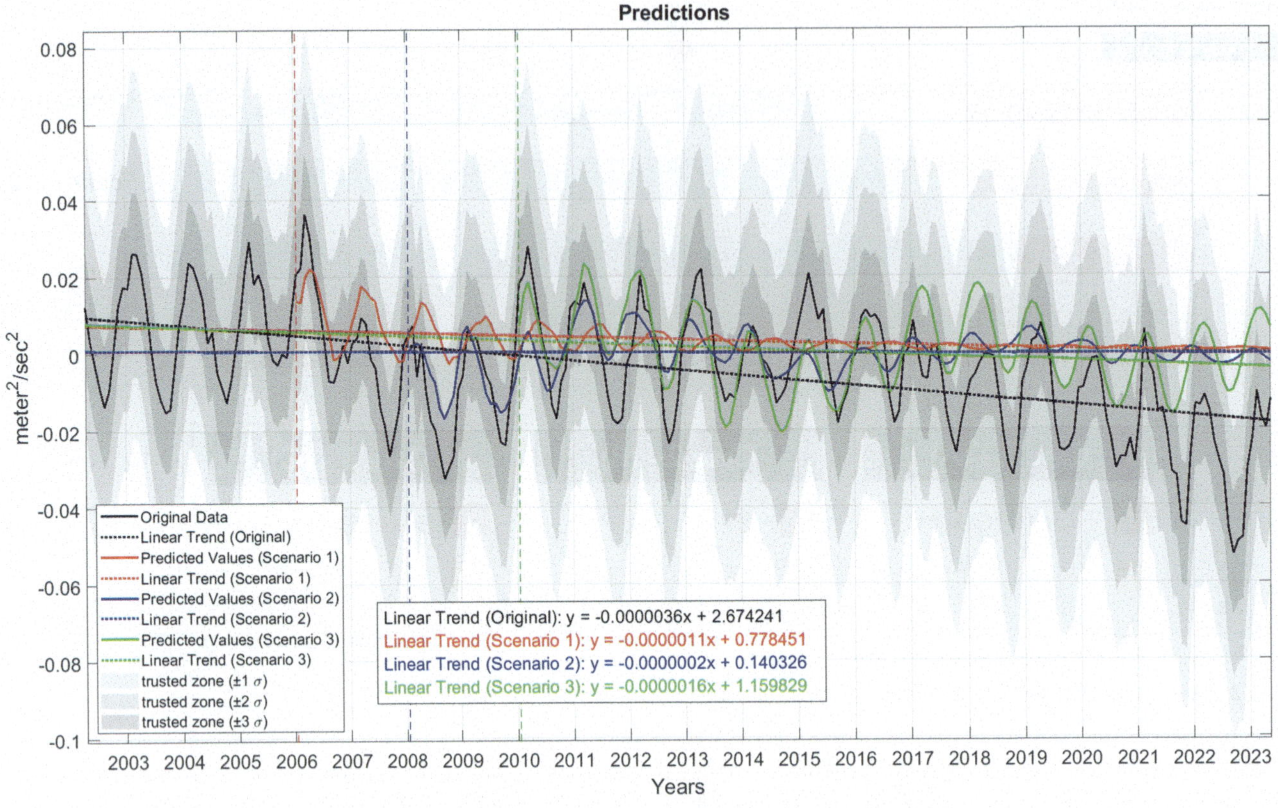

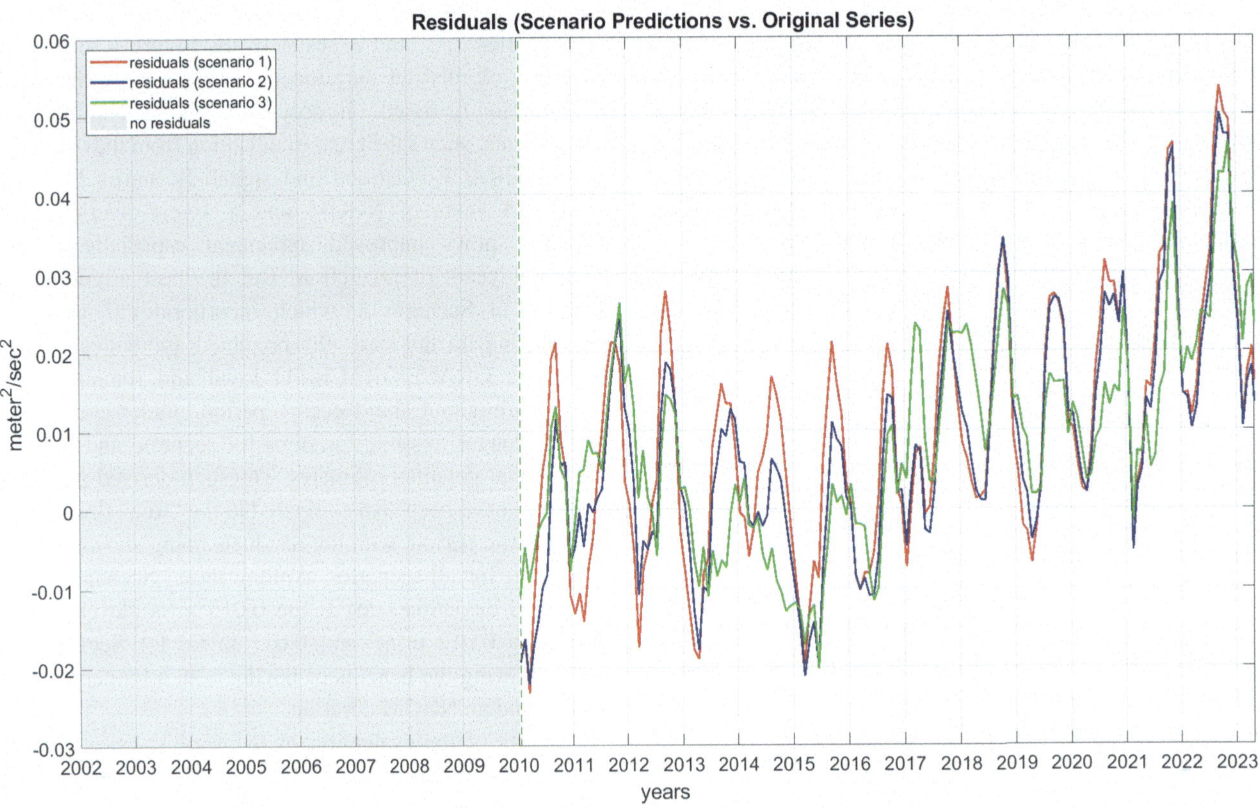

Fig. 4 Predicted gravitational potential anomalies (top) and linear trends at AUT1 for Scenarios 1 (red), 2 (blue), and 3 (green), compared to GRACE/GRACE-FO data (black) with 1σ–3σ uncertainty zones (Shaded areas). Potential residuals (bottom) at AUT1 relative to the real time-series for Scenarios 1 (red), 2 (blue), and 3 (green)

Table 1 Accuracies for GRACE/GRACE-FO and predicted gravitational potential anomalies at AUT1 under Scenarios 1, 2, and 3, for the three examined accuracy cases. Units: [m^2/s^2]

Accuracy	Real data	Case 1	Case 2	Case 3
Scenario 1	0.01300	0.00262	0.00314	0.00379
Scenario 2	0.01400	0.00303	0.00337	0.00438
Scenario 3	0.01500	0.00339	0.00478	0.00569

from 0.00303 m^2/s^2 to 0.00438 m^2/s^2, while Scenario 3 shows a range of 0.00339–0.00569 m^2/s^2. Although Scenario 3 presents slightly higher absolute uncertainties, due to the longer prediction period, it offers the best trend stability and signal prediction. This difference arises because the longer extrapolation period in Scenario 3 increases the formal uncertainty values shown in Table 1; however, as illustrated in Fig. 4, Scenario 3 reproduces the observed GRACE/GRACE-FO signal much more faithfully and remains stable throughout the entire 2010–2023 period, demonstrating superior temporal consistency and realism compared to the other scenarios. Notably, in Scenario 3, only Case 1 with the nominal error budget achieves an accuracy of 0.339 mm, which is statistically equivalent to the IHRF-required threshold of ± 0.3 mm when considering the standard deviation of the residuals and the modeled uncertainty. This level of agreement indicates that the result effectively meets the target accuracy set for future IHRF developments (Sánchez et al. 2021, 2024b). Scenario 1, Case 1 provides a slightly better prediction error by 0.0007 m^2/s^2 ($\sim$0.07 mm), which is not statistically significant compared the large std. of the residuals reported previously.

These findings clearly identify Scenario 3 as the most robust and reliable configuration. It not only offers the best alignment with the GRACE/GRACE-FO trend but also stands out as the only scenario that meets the IHRF-required accuracy threshold. This makes Scenario 3 the most appropriate choice for estimating gravitational potential change rates within the IHRF framework. Based on its performance, the linear trend derived from Scenario 3, shown in Fig. 4, has been estimated as 0.0157 cm^2/s^2, which can be applied as a predictive model to estimate future gravitational potential anomalies at the AUT1 station, based on the following equation:

$$\Delta W_{YY} = \Delta W_{XXXX.X} + \text{Trend}_{NGGM,MAGIC} \times YY, \quad (2)$$

where ΔW_{YY} is the predicted anomaly for the desired year YY, $\Delta W_{XXXX.X}$ is the reference value at a chosen epoch (e.g., 2017.0), and $\text{Trend}_{NGGM,MAGIC}$ is the linear trend derived from Scenario 3. Note that the gravitational potential anomaly at AUT1 has been independently estimated as: $\Delta W_{2017.0}^{AUT1} = 62636853.2914 \pm 0.0017\ m^2/s^2$ referring to the GRS80 ellipsoid and the Zero-Tide (ZT) system (Vergos and Tziavos 2017).

4.2 Geoid Height Anomalies (ΔN)

In addition to potential anomalies, we examined in this study the performance of predicted geoid height anomalies (ΔN) at the same IHRF station, AUT1. Here, ΔN represents the temporal variations of the geoid height N from its long-term mean value (2004.0–2009.9) and quantifies time-dependent changes in the geoid relative to the IHRF conventional W_0 value. This evaluation was based on Scenario 3, using the nominal uncertainty levels for NGGM at 1.8 cm and MAGIC at 1.0 cm (Case 1), which had previously demonstrated the best performance in terms of trend stability, compliance with IHRF standards and std. of the residuals to the ground truth data series.

The predicted ΔN values are shown in Fig. 5, with the red line representing the forecast and the black line corresponding to the original GRACE/GRACE-FO observations. In the initial forecast period, a strong agreement between predicted and observed geoid heights is evident. For periods larger than 10 years after the last monthly model acquisition, a gradual divergence emerges, consistent with trends observed in the gravity potential forecasts. The 1σ uncertainty associated with the ΔN prediction is approximately 0.42 mm, which slightly exceeds the IHRF-specified threshold of 0.3 mm. Despite slightly exceeding the IHRF threshold, the prediction remains within acceptable limits for practical geodetic use. Figure 5 also depicts the residuals between the predicted and observed ΔN values (green line), which are plotted on a secondary axis an in cm rather than m. Residuals show a variation between −0.55 and 0.25 cm for the entire period under study with a mean of the order of −0.08 cm and a std. of 0.14 cm, which suggest that the planned missions will manage to meet the IHRF requirements. In the initial prediction period and especially for the 6–7 years (up to 2017), the residuals vary between −0.3 and 0.25 cm only with a mean close to zero and a std. of 0.05 cm. After 2017 and especially after 2021, the residuals to the ground truth increase, indicating reduced forecast accuracy as the extrapolation period extends, while a negative trend in the estimated residuals can be seen.

In terms of the determination of geoid height trends for the AUT1 station, both long-term trends per period and a linear trend for the entire period of study have been derived (see Fig. 6). The linear trend of the forecast follows the ground truth trend with good coherence and starts deviating after 2017 when the residuals are increasing as well. Studying the entire period, the linear trend of the true signal is 0.005 mm/year and the one simulated from the NGGM/MAGIC timeseries 0.003 mm/year, which shows that even with this minor difference, the linear trend can serve as a valuable operational tool for estimating ΔN values at AUT1 in the absence of continuous satellite data, at least up to 6 years after the mission end.

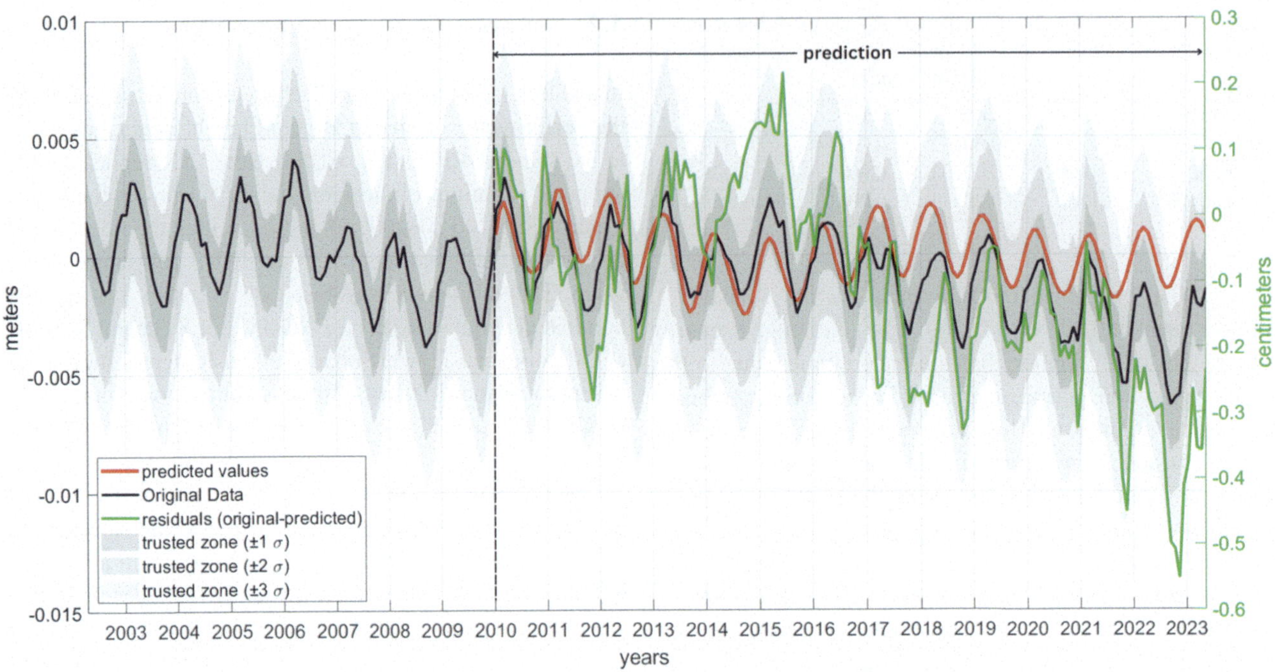

Fig. 5 Predicted geoid height anomalies at AUT1 (red line) based on Scenario 3, Case 1 and residuals (green line) relative to the ground truth timeseries

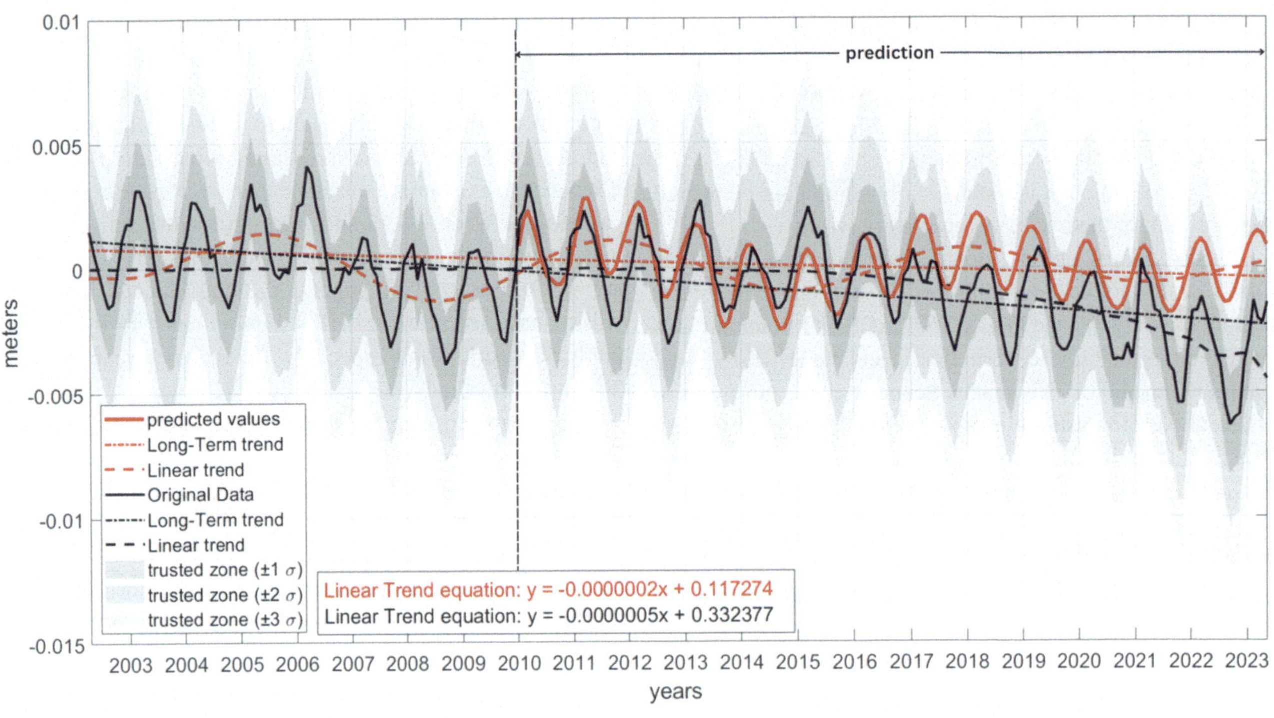

Fig. 6 Long-term and linear trends in observed (black) and predicted (red) ΔN values at AUT1

5 Conclusions and Outlook

This study assessed the potential contribution of the future NGGM satellite mission and MAGIC constellation to the dynamic realization of the International Height Reference Frame (IHRF), using the core station AUT1 as a case study. By processing GRACE/GRACE-FO data and simulating mission-specific configurations, gravitational potential anomalies (ΔW) and geoid height anomalies (ΔN) expected from these next-generation satellite missions have been forecasted.

Among the tested configurations, Scenario 3, which refers to 7 years of NGGM data with a 4-year overlap with MAGIC, demonstrated the most promising performance. For ΔW, this configuration achieved strong consistency with historical GRACE/GRACE-FO values and satisfied the IHRF-required accuracy threshold of ± 3 mm under Accuracy Case 1. Although ΔN predictions showed slightly higher uncertainty ($\sigma \approx 0.42$ mm), the results remained aligned with observed values and within acceptable ranges for geodetic applications.

These findings highlight the importance of mission duration and measurement precision in ensuring accurate forecasts of gravity-related parameters. Moreover, it is a valuable tool in order to derive long-term trends as a practical mechanism for extending IHRF support, particularly during periods without direct satellite observations.

Looking ahead, the forthcoming NGGM mission and the realization of the MAGIC constellation are expected to play a pivotal role not only in continuing the development of the IHRF but also in advancing the broader goal of Height System Unification (HSU) by linking local and national height systems into a consistent global frame. Based on the findings of this study, their use is expected to offer a promising way forward for maintaining a reliable and dynamic height reference that evolves over time.

References

Altamimi Z, Rebischung P, Collilieux X et al (2023) ITRF2020: an augmented reference frame refining the modeling of nonlinear station motions. J Geod 97:47. https://doi.org/10.1007/s00190-023-01738-w

Bíró P (1983) Time variation of height and gravity. Akadémiai Kiadó, Budapest

Daras I, March G, Joint Mass Change Mission Expert Group, Wiese D, Blackwood C, Forman F, Loomis B, Luthcke S, Peralta-Ferriz C, Sauber-Rosenberg J, Eicker A, Panet I, Visser P, Meyssignac B, Wouters B, Bamber J, Pail R, Flechtner F (2023) Next Generation Gravity Mission (NGGM) Mission Requirements Document, Issue 1.0. Earth and Mission Science Division, European Space Agency. https://doi.org/10.5270/ESA.NGGM-MRD.2023-09-V1.0

Daras I, March G, Pail R, Hughes CW, Braitenberg C, Güntner A, Eicker A, Wouters B, Heller-Kaikov B, Pivetta T, Pastorutti A (2024) Mass-change and geosciences international constellation (MAGIC) expected impact on science and applications. Geophys J Int 236(3):1288–1308. https://doi.org/10.1093/GJI/GGAD472

Drewes H, Kuglitsch F, Adám J, Rózsa S (2016) The geodesist's handbook 2016. J Geod 90(10):907–1205. https://doi.org/10.1007/s00190-016-0948-z

Drinkwater MR, Floberghagen R, Haagmans R, Muzi D, Popescu A (2003) GOCE: ESA'S first earth explorer core mission. In: Beutler G, Drinkwater MR, Rummel R, Von Steiger R (eds) Earth gravity field from space—from sensors to earth sciences. Space sciences series of ISSI, vol 17. Springer, Dordrecht. https://doi.org/10.1007/978-94-017-1333-7_36

Forootan E, Kusche J, Loth I et al (2014) Multivariate prediction of Total water storage changes over West Africa from multi-satellite data. Surv Geophys 35:913–940. https://doi.org/10.1007/s10712-014-9292-0

Forootan E, Schumacher M, Mehrnegar N, Bezděk A, Talpe MJ, Farzaneh S, Zhang C, Zhang Y, Shum CK (2020) An iterative ICA-based reconstruction method to produce consistent time-variable Total water storage fields using GRACE and Swarm satellite data. Remote Sens 12(10):1639. https://doi.org/10.3390/rs12101639

Gyawali B, Ahmed M, Murgulet D, Wiese DN (2022) Filling temporal gaps within and between GRACE and GRACE-FO terrestrial water storage records: an innovative approach. Remote Sens 14(7):1565. https://doi.org/10.3390/rs14071565

Haagmans R, Tsaoussi L (2020) Next generation gravity Mission as a mass-change and geosciences international constellation (MAGIC) mission requirements document. Earth and Mission Science Division, European Space Agency; NASA Earth Science Division. https://doi.org/10.5270/esa.nasa.magic-mrd.2020

Ihde J, Sánchez L, Barzaghi R, Drewes H, Foerste C, Gruber T, Liebsch G, Marti U, Pail R, Sideris M (2017) Definition and proposed realization of the international height reference system (IHRS). Surv Geophys 38(3):549–570. https://doi.org/10.1007/S10712-017-9409-3/FIGURES/4

Landerer FW, Flechtner FM, Save H, Webb FH, Bandikova T, Bertiger WI, Bettadpur SV, Byun SH, Dahle C, Dobslaw H, Fahnestock E, Harvey N, Kang Z, Kruizinga GLH, Loomis BD, McCullough C, Murböck M, Nagel P, Paik M et al (2020) Extending the global mass change data record: GRACE follow-on instrument and science data performance. Geophys Res Lett 47(12):e2020GL088306. https://doi.org/10.1029/2020GL088306

Loomis BD, Rachlin KE, Wiese DN, Landerer FW, Luthcke SB (2020) Replacing GRACE/GRACE-FO with satellite laser ranging: impacts on Antarctic ice sheet mass change. Geophys Res Lett 47(3):e2019GL085488. https://doi.org/10.1029/2019GL085488

Mandal N, Das P, Chanda K (2025) Machine-learning-based reconstruction of long-term global terrestrial water storage anomalies from observed, satellite and land-surface model data. Earth Syst Sci Data 17:2575–2604. https://doi.org/10.5194/essd-17-2575-2025

Pail R, Bingham R, Braitenberg C, Eicker A, Horwath M, Ivins E, Longuevergne L, Panet I, Wouters B, Balsamo G, Becker M, Bertrand D, Bolten JD, Boy J-P, van den Broeke M, Cazenave A, Chambers D, van Dam T, Diament M, van Dijk A, Dobslaw H, Döll P, Ebbing J, Famiglietti J, Feng W, Forsberg R, van de Giesen N, Greff M, Güntner A, Guo JY, Han S-C, Hanna E, Heki K, Hetényi G, Jayne S, Jiang W, Jin S, Kaser G, King M, Köhl A, Kunstmann H, Kusche J, Lay T, Löcher A, Luthcke S, Marcos M, van der Meijde M, Mikhailov V, Ohlwein C, Pollitz F, Pokhrel Y, Ponte R, Rodell M, Rolstad-Denby C, Save H, Scanlon B, Seneviratne S, Seyler F, Shepherd A, Song T, Spakman W, Shum CK, Steffen H, Sun W, Tang Q, Tiwari V, Velicogna I, Wahr J, van der Wal W, Wang L, Xie H, Yeh H-C, Yeh P, Zaitchik B, Zlotnicki V (2015) Observing mass transport to understand global change and to benefit society: science and user needs—an international multi-disciplinary initiative for IUGG. Deutsche Geodätische Kommission der Bayerischen Akademie der Wissenschaften, Heft 320, München

Richter HMP, Lück C, Klos A et al (2021) Reconstructing GRACE-type time-variable gravity from the swarm satellites. Sci Rep 11:1117. https://doi.org/10.1038/s41598-020-80752-w

Sánchez L, Ågren J, Huang J, Wang YM, Mäkinen J, Pail R, Barzaghi R, Vergos GS, Ahlgren K, Liu Q (2021) Strategy for the realisation of the international height reference system (IHRS). J Geod 95(3):1–33. https://doi.org/10.1007/S00190-021-01481-0/FIGURES/11

Sánchez L, Barzaghi R, Vergos G, Sánchez L (2024a) Operational infrastructure to ensure the long-term sustainability of the international height reference system and frame (IHRS/IHRF). In: Freymueller JT, Sánchez L (eds) Together again for geodesy. IUGG 2023.

International Association of Geodesy Symposia, vol 157. Springer, Cham, pp 1–10. https://doi.org/10.1007/1345_2024_250

Sánchez L, Vergos GS, Barzaghi R (2024b) Establishment of a geopotential-based world height system—the international height reference system (IHRS) and its realisation, the international height reference frame (IHRF). AVN Allgemeine Vermessungs-Nachrichten 2024(5–6):246–255. https://doi.org/10.14627/AVN.2024.5-6.3

Tapley BD, Bettadpur S, Watkins M, Reigber C (2004) The gravity recovery and climate experiment: mission overview and early results. Geophys Res Lett 31(9):9607. https://doi.org/10.1029/2004GL019920

Vergos GS, Tziavos IN (2017) Establishing an IHRS reference station AUT1 as an IHRS station (IAG IASPEI 2017). https://www.researchgate.net/publication/319351703_Establishing_an_IHRS_reference_station_AUT1_as_an_IHRS_station?channel=doi&linkId=59a68233a6fdcc61fcfb9ac2&showFulltext=true

Yi S, Sneeuw N (2021) Filling the data gaps within GRACE missions using singular Spectrum analysis. J Geophys Res Solid Earth 126(5):1–22. https://doi.org/10.1029/2020JB021227

Consistent Model-Based Determination of New Geopotential Numbers in a Levelling-Assisted Regional IHRS Realisation

A. Alfredsson and J. Ågren

Abstract

A common global vertical reference frame for physical heights is essential for many applications, such as studying and monitoring climate-related changes within the Earth system. The International Height Reference System (IHRS), defined by the International Association of Geodesy (IAG) in 2015, provides a global reference for physical heights. On a regional or local level, a combination of GNSS, a gravimetric (quasi)geoid model and precise levelling observations is an option to realise the IHRS. A previous study has shown that this so-called levelling-assisted realisation has the potential to significantly improve the standard (GNSS- and) model-based method in case a properly weighted least squares adjustment is made. This can result in a dense IHRF solution with reduced uncertainty. However, model-based geopotential numbers determined without levelling at a later stage will then not agree with the already existing levelling-assisted realisation. This disagreement is due to the smoothing effect obtained in the network adjustment process. The smoothing effect is caused by the low relative uncertainty of levelling observations, especially over short distances. In this contribution, we study this potentially systematic bias and how to construct a correction model to be applied to model-based geopotential numbers to improve their agreement with a weighted levelling-assisted IHRS realisation. This is investigated in a case study over Sweden. It is found that the proposed correction model does not significantly improve the systematic agreement between a model-based realisation made at a later stage with the levelling-assisted realisation in Sweden. However, it is concluded that it is important to be aware of that this type of correction might be needed in other regions. This should be investigated for any new levelling-assisted realisation.

Keywords

Correction model · IHRF · IHRS · Levelling-assisted realisation · Precise levelling · Vertical reference system

1 Introduction

1.1 Background

Monitoring climate-related changes within the Earth system and disaster management with flood risk assessments and hazard predictions are a few examples of the numerous applications that demand a common global vertical reference frame for physical heights (Ihde et al. 2017). The International Height Reference System (IHRS) and its realisation,

A. Alfredsson (✉) · J. Ågren
Faculty of Engineering and Sustainable Development, University of Gävle, Gävle, Sweden

Department of Geodetic Infrastructure, Geodata Division, Lantmäteriet, Gävle, Sweden
e-mail: anders.alfredsson@hig.se

J. T. Freymueller, L. Sànchez (eds.), *International Symposium on Gravity, Geoid and Height Systems 2024 (GGHS2024)*, International Association of Geodesy Symposia 158, https://doi.org/10.1007/1345_2025_294

the International Height Reference Frame (IHRF), provides such a common reference (Sánchez et al. 2021b). The definition of IHRS was adopted by the International Association of Geodesy (IAG) in 2015 (Drewes et al. 2016) and the work with IHRF is on-going task of the geodetic community (Sánchez et al. 2021b). A global IHRS realisation will soon be established based on different types of gravity field models and a global core network of IHRF stations. The core network is per definition sparse. Therefore, Sánchez et al. (2021b) recommends that regional model-based IHRS realisations (or densifications) are established to provide access to the global network. Regional gravimetric (quasi-)geoid models of highest quality together with accurate coordinates in the International Terrestrial Reference Frame (ITRF) (Altamimi et al. 2016) are utilised to establish a regional or national IHRS realisation (or densification). Compared to the global IHRS realisation, a regional IHRF solution offers significantly better accessibility for local users and applications. According to Sánchez et al. (2021b), precise levelling can be utilised in conjunction with a model-based IHRS realisation to support applications requiring enhanced resolution or reduced relative standard uncertainty.

For many practical purposes, high local availability and low relative standard uncertainty, the uncertainty between neighbouring height benchmarks, especially at shorter distances are desirable properties of a vertical reference frame. To fulfil these requirements, a regional IHRF solution realised by dense benchmarks on the ground is required. To compute geopotential numbers for these benchmarks using a model-based strategy, a gravity field model with the desired standard uncertainty is required, along with extensive GNSS-based ITRF coordinate determination to obtain accurate ellipsoid heights (cf. Filmer and Featherstone 2012). A good alternative is to include levelling data to reduce the relative uncertainty, which is particularly advantageous if a levelling network is already available. By combining a model-based realisation with levelling in a network adjustment, it is possible to obtain a regional IHRS realisation that meets the practical requirements. Depending on the adjustment method, the relative standard uncertainty may also be significantly reduced. Notice here that the term *model-based realisation* is used for a realisation that is made by combining ITRF coordinates with a gravity field model. *Levelling-assisted realisation* means combining ITRF coordinates in a few starting points, a gravity field model and a dense precise levelling network.

Previous research (Alfredsson and Ågren 2025) has shown significant improvements to the standard uncertainty of regional IHRF stations when precise levelling observations are included in the IHRS realisation process. Alfredsson and Ågren (2025) shows that the levelling-assisted solution that yields the lowest standard uncertainty is a weighted adjustment network with respect to the initial model-based IHRF solution. For instance, variance component estimation is utilised in the study to tune the a priori uncertainties of the geopotential numbers from the initial model-based IHRF solution and the levelling observations. As a result, the estimated standard uncertainties of the geopotential numbers in the IHRF solution are reduced by approximately 40% compared to those obtained using a constrained network adjustment. The study also illustrated that the number of height benchmarks could easily be significantly increased by the levelling-assisted IHRS realisation, providing a denser solution with improved local accessibility compared to for instance a simple model-based IHRF solution in a sparse GNSS-network. It is especially beneficial to include precise levelling observations in a regional or national IHRS realisation process when an existing levelling network is already in place. The extra effort required to include them is a minor issue.

One problem when model-based geopotential numbers are determined using GNSS after the establishment of the levelling-assisted IHRS realisation, is that the geopotential numbers might then not agree with the already existing realisation. The discrepancy is due to the adjustment of the levelling network and is related to the low relative uncertainty of levelling over short distances (Alfredsson and Ågren 2025). Newly determined model-based geopotential numbers might need to be corrected to be aligned to the levelling-assisted IHRS realisation.

In this study, the potentially systematic discrepancies between a model-based IHRF solution and a levelling-assisted IHRS realisation are investigated. To address these differences, a residual surface is constructed and evaluated. This residual surface is then applied to correct geopotential numbers from the model-based IHRF solution, aligning them more closely with those from the levelling-assisted IHRF solution.

1.2 Purpose and Delimitations

The main purpose of this study is to investigate how systematic discrepancies between a model-based IHRS realisation and a levelling-assisted IHRS realisation can be handled and corrected. The specific question examined is formulated as follows: How can a residual surface, derived from the differences between a model-based IHRF solution and a levelling-assisted IHRF solution, be used to align geopotential numbers determined by the standard model-based method with a weighted levelling-assisted IHRS realisation in Sweden, and how well does this method perform?

This article presents a case study over Sweden that is based on the regional levelling-assisted IHRF results from a previous study (Alfredsson and Ågren 2025). The results are therefore limited to the Swedish example. The proposed correction model needs to be investigated also in different geospatial contexts in the future.

1.3 Organisation of the Article

The introduction is presented in Sect. 1. As this article is based on the results from a previous article (Alfredsson and Ågren 2025), the used method is only briefly described in Sect. 2. Full details can be found in the aforementioned article. Section 2 also describes the creation and evaluation of the correction model. The results of the study are presented in Sect. 3. The results are finally discussed in Sect. 4, and a few recommendations are also provided.

2 Method

The correction model introduced in this case study might be used to align model-based geopotential numbers to a levelling-assisted IHRF solution in Sweden, which is materialised by geopotential numbers at the benchmarks in the Swedish levelling network. The correction model is computed as the difference obtained by adjusting the levelling network with the initial model-based geopotential numbers in the sparse GNSS starting points completely fixed (constrained adjustment) and by adjusting the network with the sparse initial GNSS starting points properly weighted (weighted adjustment). Since the former solution will agree exactly with the model-based solution in the GNSS starting points, this approach will provide us with the correction model; see further below. This means that the dense levelling network is used to first interpolate the discrepancies from the sparse GNSS starting points. In the next stage the discrepancies are interpolated to a regular grid. The right part of Fig. 1 shows how the correction model should be applied to model-based determined geopotential numbers. In the following subsections the included data sets and computation steps are described.

It should be mentioned that in this study all standard uncertainties associated with geopotential numbers are expressed in the milli geopotential unit (mGPU), where 1 mGPU = 1 gal m. The magnitude of mGPU is approximately equivalent to mm, as the gravity near sea level is approximately 0.98 kgal; cf. Hofmann-Wellenhof and Moritz (2005).

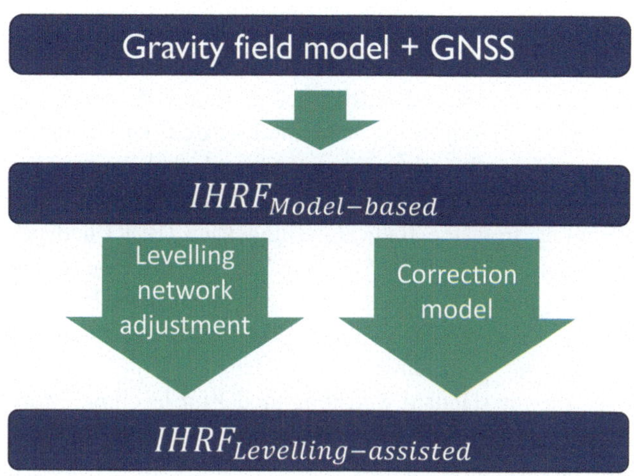

Fig. 1 Basic approach for determining a levelling-assisted IHRF solution. In the first stage, an initial model-based IHRF solution is computed. This is followed by the determination of a levelling-assisted IHRF solution, using the initial solution and a network adjustment, as illustrated on the left. The right-hand part of the figure shows the implementation of the proposed correction model

2.1 Initial Model-Based IHRF Solution in the GNSS Starting Points

The official Nordic regional gravimetric quasigeoid model, NKG2015 (Ågren et al. 2016), is utilised to compute IHRF geopotential numbers at 186 high-quality GNSS reference points. These reference points are used as starting points in the two adjustments of the levelling network in the study. The coordinates of the reference points are obtained by GNSS by repeated observations every sixth years with Dorne-Margolin antennas with 48 h of observations each time. The coordinates of the reference points are determined primarily in the Swedish reference system SWEREF 99 (Jivall and Lidberg 2000), and ITRF2014 (Altamimi et al. 2016) coordinates are obtained by utilising the Nordic official NKG transformation parameters (Häkli et al. 2016). This method includes both a 7-parameter Helmert transformation and the velocity field model NKG_RF17vel (Lantmäteriet 2023) for epoch conversion. The geopotential numbers of the 186 model-based IHRF reference points are calculated according to the guidelines in the strategy for regional IHRS realisations (Sánchez et al. 2021a, b). The reference epoch for IHRF is taken as 2021.04 (Huang and Ågren 2024).

The standard uncertainty of the obtained IHRF geopotential numbers of the initial model-based solution is estimated to be in the range of 12–15 mGPU depending on how the stochastic model is constructed; see Alfredsson and Ågren (2025) for further explanations of the stochastic models used.

2.2 Levelling-Assisted IHRF Solutions

The levelling-assisted IHRF solution from Alfredsson and Ågren (2025) utilises the initial model-based IHRF solution at the 186 GNSS reference points described above together with a height network adjustment of levelling observations stemming from the Baltic Levelling Ring (BLR) network (Mäkinen et al. 2006). Standard least squares adjustment using the Gauss-Markov model is used to adjust the levelling network. The weights of the levelling observations are set proportional to the inverse of the lengths of the levelling lines, i.e. the standard weighting model for levelling. More details on the height network adjustments can be found in Alfredsson and Ågren (2025).

The solutions from two adjustments of the levelling network are then used to derive the correction model.

A. A constrained adjustment with the geopotential numbers from the initial model-based IHRF solution in the GNSS starting points held fixed. The levelling observations are here used to densify the initial IHRF solution to the network of height benchmarks.
B. A weighted adjustment where the a priori uncertainties for the two groups of observations (initial model-based solution in GNSS-starting points and levelling observations) are computed using variance component estimation. This will result in a levelling-assisted IHRF solution.

The geopotential difference between the levelling-assisted (B) and the constrained adjustment (A) represent the discrepancy that occur when the levelling observations are allowed to smooth the initial model-based solution in the GNSS starting points for the weighted adjustment. As the two adjustments are performed with an identical levelling network, geopotential differences are computed for all height benchmarks. In total 3,536 height benchmarks are included in the levelling-assisted IHRF solution including the 186 starting points, which are also used as control stations in the cross validation below.

2.3 Gridded Residual Surface

The geopotential number differences between the constrained and weighted adjustments are interpolated into a regularly distributed correction grid. The grid can be used to correct newly determined geopotential numbers to agree better with the levelling-assisted IHRS realisation at a later stage. The resolution of the correction grid is $0.01° \times 0.02°$ in latitude and longitude, respectively. The Gravsoft software Geogrid (Forsberg and Tscherning 2014) is used to resample the differences in the discrete scattered benchmarks to a regular grid using the least squares collocation method. The correlation length $d_{1/2}$ is set to 50 km and the signal variance C_0 to 0.5 mGPU.

2.4 Cross-Validation

Cross-validation is used to evaluate the correction model. In this process, one station at the time from the initial model-based solution is removed temporary from the computations. The correction model is recalculated based on a new temporary constrained and weighted adjustments of the levelling network without the removed station. The removed station is then used as a control station to calculate a correction to the initial model-based determined geopotential number. The station is then included again, and another station removed, and the procedure is repeated until all stations in the initial solution have been used as control station. The cross-validation differences between the corrected and the adjusted geopotential numbers are used for assessing the uncertainty of the correction model.

3 Results

The 3,536 geopotential differences obtained between the two levelling network adjustments, the constrained and the weighted, which are the basis for the correction model, are in the range from -40.8 to $+30.1$ mGPU with a mean and standard deviation of -0.2 mGPU and 6.2 mGPU, respectively. The differences are visualised in Fig. 2.

In the corresponding residual surface, Fig. 3, the systematics of the differences are clearly visualised. The initial 186 GNSS reference points are used to compute the model-based IHRF solution, presented as black dots, are accentuated in the residual surface. The reason for this is that these reference points are treated differently in the two levelling network adjustments, the starting points are either fixed or weighted. They thereby get the largest discrepancies between the two levelling-assisted IHRF solutions.

The cross-validation utilises the correction model to calculate a correction for each of the GNSS starting points in the initial model-based IHRF solution as described in Sect. 2.4. The results from the cross-validation can be seen in Fig. 4 and Table 1.

4 Discussion and Conclusions

The realisation of the IHRS through a levelling-assisted IHRF solution presents challenges in aligning newly determined model-based IHRF geopotential numbers with the current levelling assisted realisation. This study introduces a correction model to address these discrepancies, ensuring consistency between geopotential numbers in the model-based and levelling-assisted IHRF solutions. The applied correction is based on the discrepancy between the results from a constrained and a weighted levelling network adjustment,

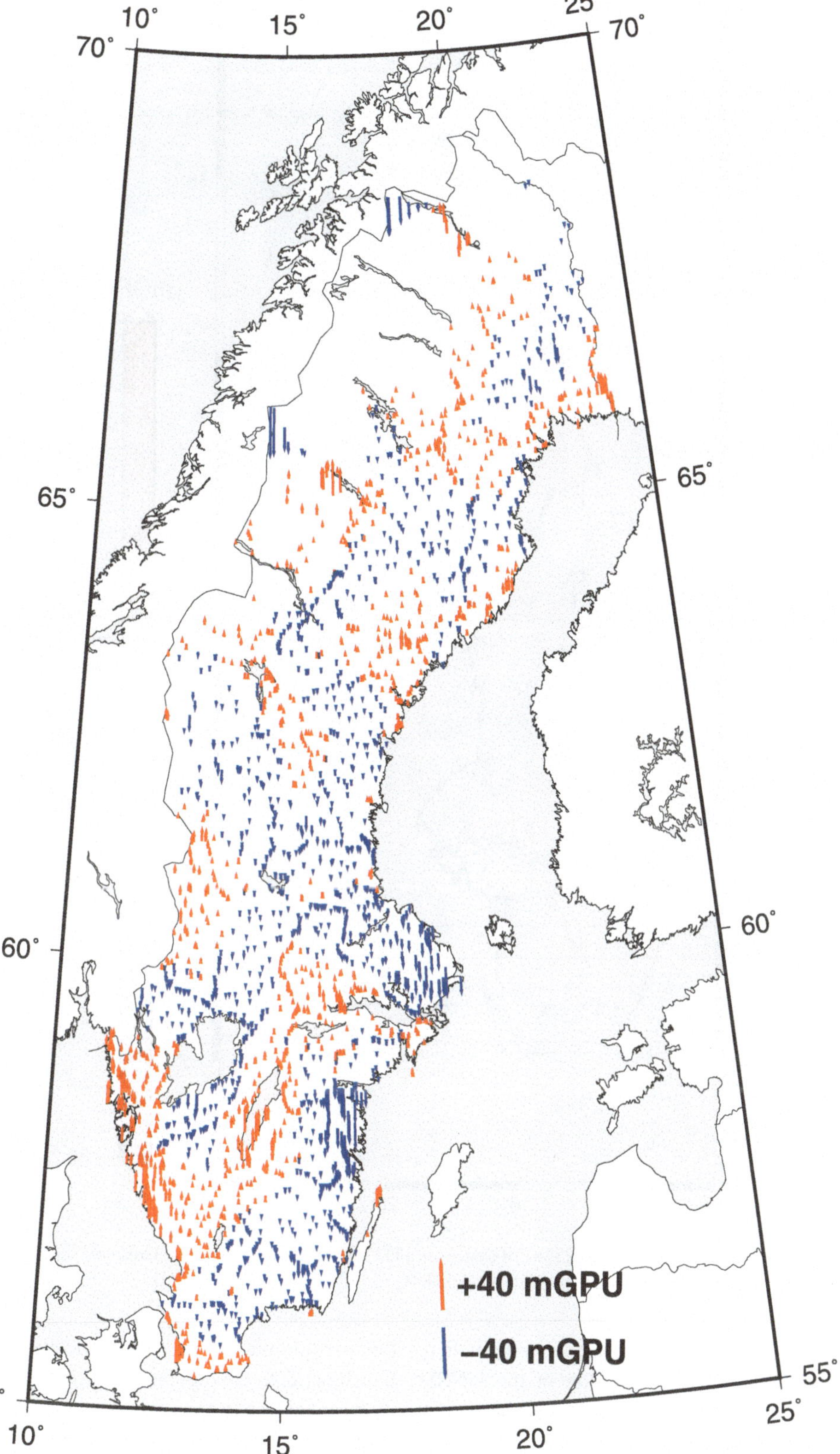

Fig. 2 Discrepancies between the weighted and constrained adjustments across the 3,536 evaluated benchmarks

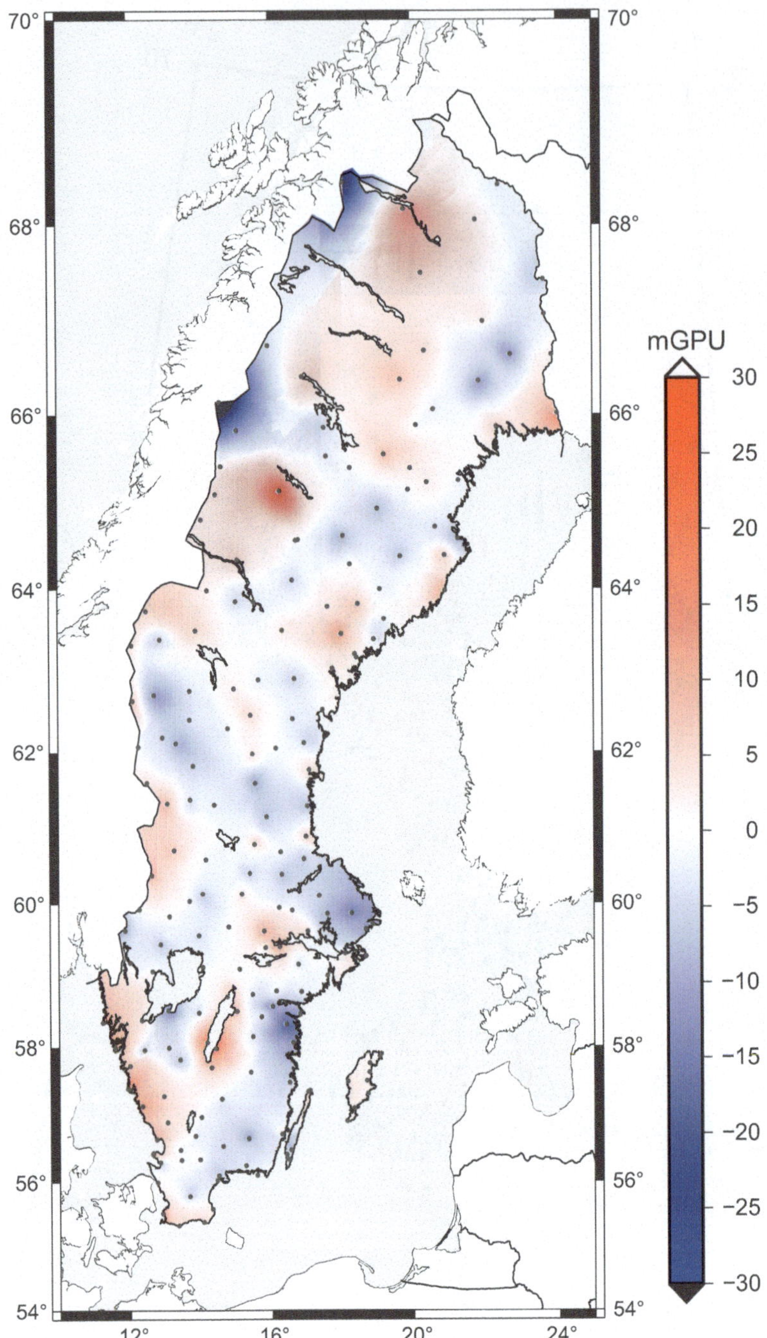

Fig. 3 Residual surface (correction model) based on the differences in Fig. 2, computed with least squares collocation. The 186 GNSS starting points used to compute the model-based solution are plotted as black dots

where the constrained adjustment represents a densified solution of the model-based IHRF solution in the GNSS reference points and the weighted adjustment is the actual levelling-assisted IHRS realisation.

In this case study over Sweden, the discrepancies between the two height network adjustments form the basis for a correction model. The discrepancies are computed for all adjusted benchmarks, see Fig. 2. The study utilises collocation to resample and interpolate the discrepancies given at the benchmarks to the regular correction model grid. The largest discrepancies between the weighted and the constrained solution are located at the benchmarks of the initial model-based IHRF solution, which are therefore accentuated in the correction model.

The correction model effectively addresses systematic discrepancies between different IHRS realisations. System-

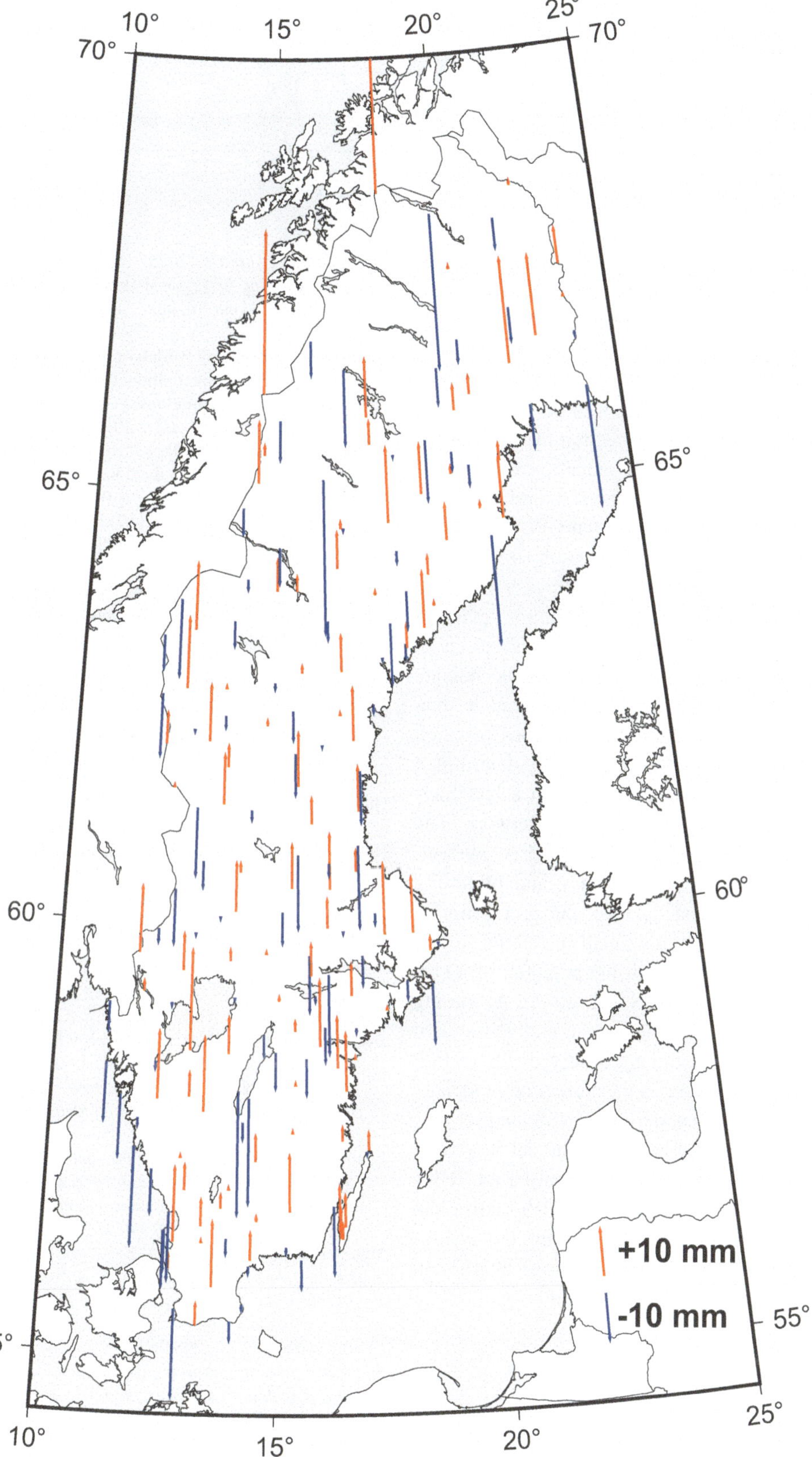

Fig. 4 Cross validation residuals at the 186 GNSS starting points

Table 1 Statistics for the discrepancies at the 186 benchmarks before the correction on the left, i.e. the geopotential differences between the constrained and the weighted adjustment. On the right, statistics from the cross-validation based on the 186 benchmarks (unit: mGPU)

	Geopotential differences before correction	Cross-validation of corrected geopotentials
# Points	186	186
Min	–40.8	–36.5
Max	30.1	31.3
Mean	–0.3	0.0
Std. dev.	10.9	9.9

atic errors, which may arise from the underlying geopotential model or levelling observations, can be corrected to align the model-based IHRF geopotential numbers to the levelling-assisted IHRS realisation. However, random errors, likely due to the GNSS observations, cannot be modelled or corrected by this approach.

However, the evaluation of the correction model through cross-validation shows only a small reduction of the discrepancies. As presented in Table 1, the standard deviation of the cross-validation residuals is 9.9 mGPU compared to the uncorrected geopotential differences of 10.9 mGPU. The improvement of the standard deviation is 9.4%. The total range between the maximum and minimum deviation is reduced from 70.9 to 67.8 mGPU, or by 4.4%.

The relatively small reduction in discrepancies indicate that the deviations are due to random errors rather than systematic effects. This study builds on previous research by providing a practical solution to align newly determined model-based IHRF geopotential numbers with a levelling-assisted IHRS realisation through a levelling network. The correction model offers a robust method to address systematic discrepancies, enhancing the accuracy and reliability of IHRS realisations. In the study area, the proportion of systematic deviations is clearly too small in relation to the random uncertainties to fully exploit the potential of a correction model between the two IHRS realisations. According to Alfredsson and Ågren (2025), the largest relative standard uncertainties are related to coordinates observed with GNSS. The random uncertainties present in the correction model are therefore to greatest extent related to the GNSS data set.

In conclusion, the proposed correction model is a valuable tool to align newly determined model-based IHRF geopotential numbers to levelling-assisted IHRS realisation. Future research should focus on further refining this type of correction model and investigating its application in different geographical contexts.

References

Ågren J, Strykowski G, Bilker-Koivula M, Omang O, Märdla S, Forsberg R, Ellmann A, Oja T, Liepins I, Parseliunas E, Kaminskis J, Sjöberg L, Valsson G (2016) The NKG2015 gravimetric geoid model for the Nordic-Baltic region. https://doi.org/10.13140/RG.2.2.20765.20969

Alfredsson A, Ågren J (2025) Investigations on the contribution of precise levelling for regional realization of IHRS – a case study over Sweden. J Geod 99(70). https://doi.org/10.1007/s00190-025-01992-0

Altamimi Z, Rebischung P, Métivier L, Collilieux X (2016) ITRF2014: a new release of the International Terrestrial Reference Frame modeling nonlinear station motions. J Geophys Res Solid Earth 121:6109–6131. https://doi.org/10.1002/2016JB013098

Drewes H, Kuglitsch F, Adám J, Rózsa S (2016) IAG Resolution No. 1, 2015 in the geodesist's handbook 2016. J Geod 90(10):907–1205. https://doi.org/10.1007/s00190-016-0948-z

Filmer MS, Featherstone WE (2012) Three viable options for a new Australian vertical datum. J Spat Sci 57(1):19–36. https://doi.org/10.1080/14498596.2012.679248

Forsberg R, Tscherning C (2014) An overview manual for the GRAVSOFT geodetic gravity field modelling programs, 3rd edn. https://ftp.space.dtu.dk/pub/RF/gravsoft_manual2014.pdf. Accessed 29 Apr 2025

Häkli P, Lidberg M, Jivall L, Nørbech T, Tangen O, Weber M, Pihlak P, Aleksejenko I, Paršeliunas E (2016) The NKG2008 GPS campaign – final transformation results and a new common Nordic reference frame. J Geod Sci 6(1):1–33. https://doi.org/10.1515/jogs-2016-0001

Hofmann-Wellenhof B, Moritz H (2005) Physical geodesy. Springer, Wien. https://doi.org/10.1007/b139113

Huang J, Ågren J (2024) IHRF convetions – simplified. https://ihrfcc.topo.auth.gr/wp-content/uploads/2024/08/IHRF_Conventions_Simplified_Jonas_v2.pdf. Accessed 29 Apr 2025

Ihde J, Sánchez L, Barzaghi R, Drewes H, Foerste C, Gruber T, Liebsch G, Marti U, Pail R, Sideris M (2017) Definition and proposed realization of the international height reference system (IHRS). Survey Geophys 38(3):549–570. https://doi.org/10.1007/s10712-017-9409-3

Jivall L, Lidberg M (2000) SWEREF 99 – an updated EUREF realisation for Sweden. EUREF Publication No. 9. EUREF Symposium, Tromsø, 22–24 Jun 2000. https://www.lantmateriet.se/globalassets/geodata/gps-och-geodetisk-matning/etrs89_proceed.pdf. Accessed 29 Jul 2025

Lantmäteriet (2023) Transformation between ITRF 2014/WGS 84 and SWEREF 99. https://www.lantmateriet.se/globalassets/geodata/gps-och-geodetisk-matning/transformations/transformation_itrf2014-sweref99.pdf. Accessed 29 Jul 2025

Mäkinen J, Lilje M, Ågren J, Engsager K, Eriksson P-O, Jepsen C, Olsson P-A, Saaranen V, Schmidt K, Svensson R, Takalo M, Vestøl O (2006) The baltic levelling ring. The Working Group for Height Determination of the Nordic Geodetic Commission. https://doi.org/10.13140/RG.2.2.33298.96961

Sánchez L, Huang J, Ågren J, Barzaghi R, Vergos GS (2021a) Recovering potential values from regional (quasi-)geoid models. Unpublished guidelines. IAG Joint Working Group 0.1.3

Sánchez L, Ågren J, Huang J, Wang YM, Mäkinen J, Pail R, Barzaghi R, Vergos GS, Ahlgren K, Liu Q (2021b) Strategy for the realisation of the International Height Reference System (IHRS). J Geod 95(3). https://doi.org/10.1007/s00190-021-01481-0

Impact of Different DEMs on High-Precision Geoid Modeling in the Region of CERN

Julia Azumi Koch, Benedikt Soja, and Markus Rothacher

Abstract

The Future Circular Collider (FCC), one of CERN's proposed next-generation particle accelerators, with a planned circumference of 91 km and an average depth of 300 m, requires an exceptionally precise geoid model to support its construction and the (pre-)alignment of its components. The target precision is sub-centimeter over 1 km and potentially sub-millimeter over shorter distances. Achieving such unprecedented accuracy requires a comprehensive assessment of all potential error sources. This study investigates the impact of different digital elevation models (DEMs) on gravimetric geoid determination.

Eleven global and regional DEMs were analyzed in terms of their height differences and their impact on the computation of gravity anomalies and geoid heights. While systematic differences between DEM heights were generally below 11 m, local discrepancies reached several hundred meters, leading to geoid height biases of up to 3.5 cm and standard deviations of 6 mm. Despite the magnitude of these effects, no single DEM produced a consistently superior geoid solution across all evaluation criteria. Comparisons with GNSS/leveling data were inconclusive, suggesting that the accuracy of the geoid solutions for the FCC may have reached or surpassed the accuracy of the reference datasets. These findings highlight the critical importance of DEM selection and preprocessing strategies in high-precision geoid modeling.

Keywords

CERN · Digital elevation model (DEM) · Gravity · Regional geoid modeling

1 Introduction

The Large Hadron Collider (LHC) at CERN (Conseil Européen pour la Recherche Nucléaire, the European Organization for Nuclear Research) is currently the largest particle accelerator of its kind, with a circumference of 27 km, located at an average depth of 100 m under the Earth's surface. In preparation for the period following its conclusion by the year 2040, CERN is planning a post-LHC particle accelerator. One of the proposed concepts is the Future Circular Collider (FCC), with a circumference of approximately 91 km and situated 300 m below the surface.

From a geodetic point of view, the FCC project poses several challenges. As illustrated in Fig. 1, the FCC extends the current area of CERN from around 90 km^2 to approximately 1000 km^2. Furthermore, the FCC spans widely across the borders of Switzerland and France. Due to differences in national height systems and administrative frameworks, unified geodetic models and reference systems must be specifically established for the extended CERN region. In addition, although the region of the FCC is topographically relatively flat, the planned trajectory encircles Mont Salève and passes beneath Lake Geneva, intersecting an area with several tectonic faults. The FCC area is also bordered by the Jura Mountains to the northwest and the (Pre-)Alps to the

J. A. Koch (✉) · B. Soja · M. Rothacher
Institute of Geodesy and Photogrammetry, ETH Zurich, Zurich, Switzerland
e-mail: jukoch@ethz.ch; sojab@ethz.ch; markus.rothacher@ethz.ch

J. T. Freymueller, L. Sànchez (eds.), *International Symposium on Gravity, Geoid and Height Systems 2024 (GGHS2024)*, International Association of Geodesy Symposia 158, https://doi.org/10.1007/1345_2025_292

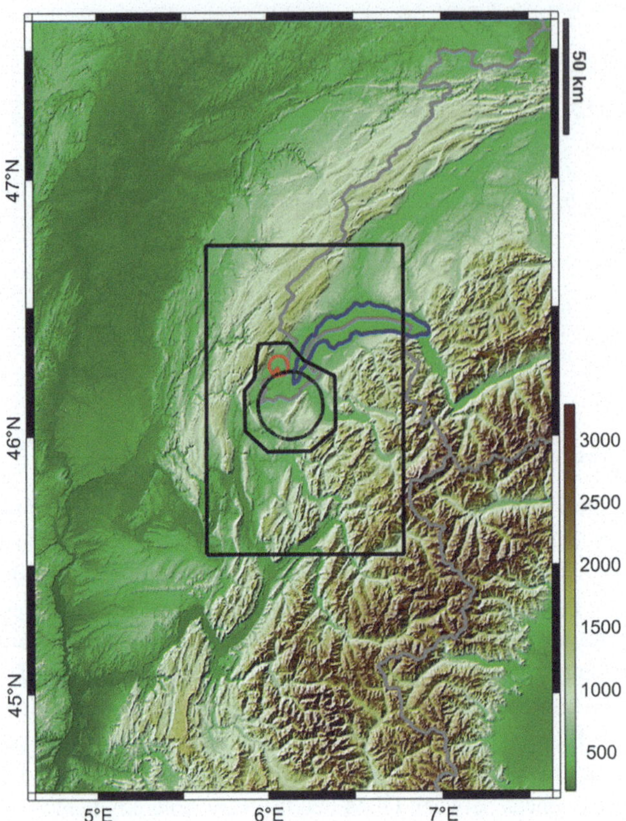

Fig. 1 The 3° × 3° section of the ASTER GDEM v3 in ellipsoidal heights [m], representing the full region used in this analysis. The black rectangle outlines the inner region where gravity data is available for this study and the geoid is computed. The black polygon indicates the area of interest (AoI) of the FCC Geodesy project and in which the (quasi)geoid solutions are validated, the black circle marks the approximate FCC location, and the red circle denotes the current LHC at CERN. The shoreline of Lake Geneva is shown as a blue polygon. All figures in this paper were generated using MATLAB R2025a (The MathWorks Inc., academic license, ETH Zurich)

southeast. In the region we investigated in this study, the elevation ranges from below 500 m in the Po and Alpine river valleys and the northwestern Jura region, up to around 4700 m in the Alps.

Within the FCC Geodesy project, we are developing concepts to establish the reference network on the surface and in the tunnel, as well as to realize new high-precision geoid models for this specific area. While the specific requirements on the geoid model are still under discussion, the ultimately targeted precision is in the lower millimeter to sub-millimeter range over a sliding window of 1 km to 200 m.

Meeting such unprecedented requirements on the geoid modeling necessitates a comprehensive evaluation of all potential error sources that could cause deviations from the true geoid. This includes the input data, simplifications and assumptions made during the calculation, and, eventually, even the numerical precision of the software used.

In this study, we focus on assessing the impact of the selected digital elevation model (DEM) on the geoid solution.

In geoid modeling, DEMs are used to compute topographic and downward continuation corrections for gravity observations collected on the Earth's surface. Therefore, any errors in the DEM directly influence the resulting geoid solution. Additionally, in the analysis for the calculation of a local geoid along the tunnel trajectory with a precision at the sub-mm level, a substantial amount of gravity measurements must be simulated by interpolation or prediction utilizing DEM information.

Uncertainty in geoid modeling is often attributed to limited data coverage, measurement accuracy, or poor geolocation of the observations. When DEM uncertainty is considered, the stated vertical accuracy of the DEM is typically used in error propagation. However, the actual DEM quality can vary substantially depending on factors such as acquisition method (e.g., national datasets, optical stereo reconstruction, LiDAR, InSAR), data age, processing schemes, and local surface characteristics.

In this study, firstly, differences between the DEMs are examined. Secondly, the effect of these differences on the terrain and downward continuation correction, and on the resulting geoid height, is assessed and compared to GNSS/leveling points.

This paper is structured as follows: Sect. 2 introduces the DEMs and gravity data used in this study. Section 3 presents the methods employed for data preprocessing and geoid computation. Section 4 contains the comparison between the available DEMs, their impact on the terrain correction, and the resulting geoid heights, which are compared against GNSS/leveling data. The study is concluded in Sect. 5.

2 Data

2.1 Digital Elevation Models

For this analysis, a total of 11 DEMs were used (see Table 1). Only DEMs that cover the entire 3° × 3° region centered on the FCC site were selected. Therefore, national DEMs of France and Switzerland are not included in this study. Most of the DEMs used in this study are available at a resolution of 1 arcsec, corresponding to approximately 30 m. EuroDEM (EO) is available in 2 arcsec resolution, while the MERIT DEM (ME) has a coarser resolution of 3 arcsec. EO is constructed using purely national data collected between 1999 and 2007. ASTER GDEM v3 (AS) and TanDEM-X DEM (TA) are derived from single satellite missions: Advanced Spaceborne Thermal Emission and Reflection Radiometer (ASTER) and TerraSAR-X ADD-on for Digital Elevation Measurement (TanDEM-X), respectively. SRTM GL1 v3.0 (SR) is primarily based on the Shutter Radar Topography Mission (SRTM) data, with data gaps filled with AS. ALOS World 3D v3.1 (AW) is based mainly on the satellite mission Advanced Land Observing Satellite (ALOS), complemented by publicly available DEMs. Other datasets combine multiple sources. EU-DEM (EU) and NASA DEM (NA) are merged products of ASTER and SRTM satellite mission data. The Copernicus DEM (GLO-30) (CO) is primarily based on TanDEM-X data, supplemented with satellite-based DEMs and national data from Spain and Norway. The Viewfinder Panorama DEM (VF) is mainly based on SRTM data, with data gaps filled from various sources. In Europe, VF utilizes freely available national elevation datasets. ME combines SR data up to 60 °N with AW above 60 °N, with additional gaps filled using VF. All of the above DEMs are, to varying degrees, composite digital surface models. In contrast, FABDEM V1-2 (FA) applies random forest machine learning algorithms to CO to remove vegetation and buildings, resulting in a bare-earth DEM that represents the ground surface only.

The year stated in Table 1 is the year of the published dataset for the CERN region. For example, VF underwent a major update in 2022 for the European Alps, therefore the table lists 2022 as the reference year, even though other parts of the dataset were processes earlier and have not been updated since. Some DEMs, such as SR, have clearly defined versions. This dataset is based on data collected in 2000 by SRTM on the Endeavour space shuttle. Its 3.0 version, published in 2015, includes gap-filling based on AS, which itself underwent multiple reprocessing cycles, with version 3 released in 2019. Due to the complexity of processing chains and data merges, it is often not possible to fully trace which datasets and versions were used in specific regions, making it difficult and impractical to quantify DEM quality based solely on input data and the published year. Furthermore, developing reliable and consistent quality metrics to identify the "best" DEM remains an open field of research (Bielski et al. 2024).

2.2 Gravity Data

In the region of the FCC, approximately 21,000 gravity measurements are available for geoid computation. Their spatial distribution is shown in Fig. 2 (center). The variety of data sources contributing to this dataset is described in detail in Koch et al. (2024). In this study, we do not distinguish between the different sources, even though the year of observation, the instruments used, and the reported precision vary across the dataset.

As shown in Fig. 2 (center), the distribution of gravity measurements across the region is highly heterogeneous. In particular, the western and southwestern parts of the area of interest (AoI), indicated by the black polygon, exhibit sparse

Table 1 Overview of the DEMs used in this comparison and their characteristics. The year refers to the publication date of the dataset for the CERN region

Abbr.	Name	Resolution	Year	Reference
AS	ASTER GDEM v3	1 arcsec	2019	Abrams et al. (2020)
AW	ALOS World 3D v3.1	1 arcsec	2016	Takaku et al. (2014, 2021)
CO	Copernicus DEM (GLO-30)	1 arcsec	2022	Strobl (2020)
EU	EU-DEM	1 arcsec	2016	Dufourmont et al. (2014)
EO	EuroDEM	2 arcsec	2008	Claus et al. (2023)
FA	FABDEM V1-2	1 arcsec	2022	Hawker et al. (2022)
ME	MERIT DEM	3 arcsec	2018	Yamazaki et al. (2019)
NA	NASA DEM	1 arcsec	2021	Crippen et al. (2016)
SR	SRTM GL1 v3.0	1 arcsec	2015	EROS Center (2018)
TA	TanDEM-X DEM	1 arcsec	2016	Earth Observation Center (2018)
VF	Viewfinder Panorama DEM	1 arcsec	2022	de Farranti (2022)

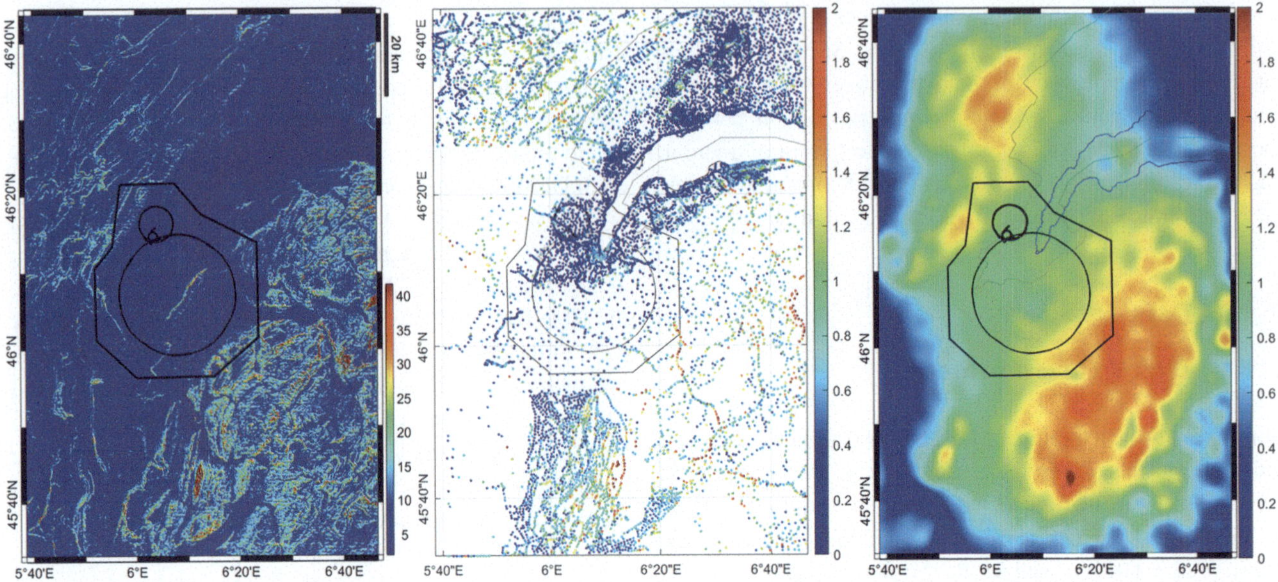

Fig. 2 Standard deviations (SD) computed at each grid cell or observation point using all 11 DEMs. Left: SD of DEM heights [m], center: SD of g modeled [mGal], right: SD of geoid heights [cm]

data coverage. This AoI was defined to include all current and planned facilities and infrastructure related to CERN and the FCC, which are expected to be located within a 5 km radius of eight vertical access shafts to the tunnel.

To address the data gaps and improve homogeneity of the data distribution, future plans for the FCC include additional gravity and deflection of the vertical measurements. One such gravity campaign was already conducted in 2022, targeting the data-sparse southwestern border region of the AoI.

3 Methods

3.1 Preprocessing of DEMs

Since the DEMs originated from different reference frames, some of which are associated with well-defined EPSG codes, which is a unique identifier for coordinate systems and geodetic properties like datums and ellipsoids, while others are only described in accompanying documentation, harmonization was necessary. Therefore, the horizontal coordinates of all DEMs were transformed to WGS84 using the `rastario` Python library. All datasets were clipped to the same $3^\circ \times 3^\circ$ region. In cases where data gaps existed within the region, a nearest-neighbor interpolation approach was applied to fill the missing values, in order to avoid the smoothing effect that could be introduced by other interpolation methods, such as linear interpolation, and to ensure that the highest variations between different DEMs are captured.

All DEM heights were then transformed to ellipsoidal heights using the relation between ellipsoidal h, orthometric H, and normal H^N heights: $h = H + N = H^N + \zeta$, where N and ζ are the geoid undulation/height and height anomaly, respectively. For datasets where the underlying geoid or quasigeoid was only available as gridded data (e.g., EGG08), the height transformation to ellipsoidal heights was performed. Bilinear interpolation was applied to the EGG08 grid (1 arcmin resolution) to match the WGS84 grid coordinates of the DEMs. For datasets with a reference surface represented via spherical harmonic coefficients, the geoid height at each DEM grid point was computed to the highest available degree and order (i.e. EGM96 until d/o 360, EGM2008 until d/o 2190) using the GROOPS software toolkit (Mayer-Gürr et al. 2021), and then added to the orthometric or normal height to obtain the ellipsoidal height. As the height difference between WGS84 and GRS80 is negligible compared to the absolute vertical accuracy of the DEMs, no additional adjustment was applied.

3.2 Geoid Computation with the GROOPS Software Toolkit

The geoid and quasigeoid computations were performed using the GROOPS software toolkit. GROOPS follows the classical remove-compute-restore (RCR) approach, employing a least-squares adjustment parametrization using radial basis functions (Pock 2017; Lercher 2023). Terrestrial gravimetric observations are combined with a global gravity field model up to degree and order 2160. The residual terrain correction is applied using the residual spherical-harmonic topographic potential (RSHTP) method, as proposed by Schwabe et al. (2024).

It has been shown that regional gravity field modeling benefits from the use of high-degree global gravity field models (Niedermaier et al. 2024). In this study, a combined global model was used, merging the GOCO06s and EGM08 models, with a smooth transition between d/o 180 and 300. This combination leverages the superior low-degree accuracy of GOCO06S and the high-resolution content of EGM08. However, any other global model with sufficient resolution could also be applied.

The residual gravity anomaly is computed as:

$$\Delta g = g - \gamma - \delta g_{\mathrm{GGM}} - \delta g_{\mathrm{DEM}} + \delta g_{\mathrm{SHRTC}} \tag{1}$$

where γ is normal gravity, δg_{GGM} is the gravity effect of the global model, δg_{DEM} accounts for the terrain correction based on the DEM, and $\delta g_{\mathrm{SHRTC}}$ is the residual terrain correction based on the spherical harmonics representation.

For this study, the only change applied to the GROOPS processing pipeline was the DEM input, which affects both δg_{DEM} and $\delta g_{\mathrm{SHRTC}}$. However, as the long-wavelength component of the DEM remains the same, $\delta g_{\mathrm{SHRTC}}$ remains largely unchanged. For δg_{DEM}, the terrain correction was computed using cuboids and cap radii of 15 km (maximum distance for the prism formula) and 100 km (for radial integration).

In the compute step, the observables are represented as a series expansion of radial basis functions:

$$\Delta g(P) = \sum_{i=1}^{n} a_i \, \phi(P, Q_i)$$

where a_i are the unknown coefficients to be determined with least squares, P is the evaluation point, and Q_i denotes the location of the spherical radial basis functions. Once the coefficients a_i are estimated from the reduced gravity observations, they are used with a corresponding set of kernels to compute other gravity field functionals beyond the original observables. The average distance between the spherical radial basis functions was set to 1.6 km to match the average distance between gravity observations in the study area.

4 Results and Discussion

4.1 Comparison of DEMs

Calculating the mean and standard deviation (SD) over all grid cells across the $3^\circ \times 3^\circ$ DEMs reveals no significant differences. Note that all statistical values reported in Table 2 refer to the full region as depicted in Fig. 1, while the following plots focus on the inner region to better visualize local variations. The mean differences range between 0 m (for DEM comparisons between AS-FA, AS-ME, FA-ME, CO-TA) and 11 m (for DEM comparisons between AW-EO, CO-EO, EO-TA), with EO exhibiting the highest offset overall. The SD is largest in all combinations involving ME (exceeding 20 m), while FA shows the smallest deviations, closely followed by CO and EU. The highest SD, at 29 m, is observed between ME and SR, whereas the smallest, at 9 m, is found in combinations of AS-FA, CO-FA, CO-NA, EU-SR. Examining specific DEM differences reveals much larger local variations. As an example, the difference in ellipsoidal heights between AS and VF is shown in Fig. 3. This comparison yields a near-zero bias and an SD of 17 m, which is within the mid-range of observed values. The three-sigma limit of all differences is within $\pm$ 50 m, but outliers reach up to 572 m in small localized areas. In the middle of Lake Geneva, due to the VF DEM, a distinct vertical line is visible, where the lake height east and west of it differ, indicating potential processing inconsistencies when data were merged. All DEM combinations show similar spatial variability patterns. To identify locations with the highest DEM disagreement over all DEMs, the SD of ellipsoidal heights across all DEMs was computed per grid cell (see Fig. 2, left). As expected, the lowest variability is found around Lake Geneva and in the relatively flat areas of the Molasse Basin in the northeast. The highest values appear

Table 2 Mean (top right triangle) and standard deviation (bottom left triangle) of the differences between DEM heights [m] for the full area of $3^\circ \times 3^\circ$ depicted in Fig. 1. DEM abbreviations are defined in Table 1

	AS	AW	CO	EU	EO	FA	ME	NA	SR	TA	VF
AS		−4	−3	6	8	0	0	−2	−1	−3	−0
AW	19		1	10	11	4	4	2	3	1	4
CO	11	17		9	11	3	3	1	2	−0	3
EU	13	13	14		1	−6	−6	−8	−7	−9	−6
EO	16	14	16	14		−7	−7	−10	−9	−11	−8
FA	9	18	9	12	13		−0	−3	−2	−4	−1
ME	23	27	21	25	20	22		−3	−2	−4	−0
NA	11	18	9	12	15	10	22		1	−1	2
SR	20	12	20	9	14	19	29	18		−2	1
TA	17	22	14	18	20	15	25	16	24		3
VF	17	16	18	15	15	15	22	17	15	21	

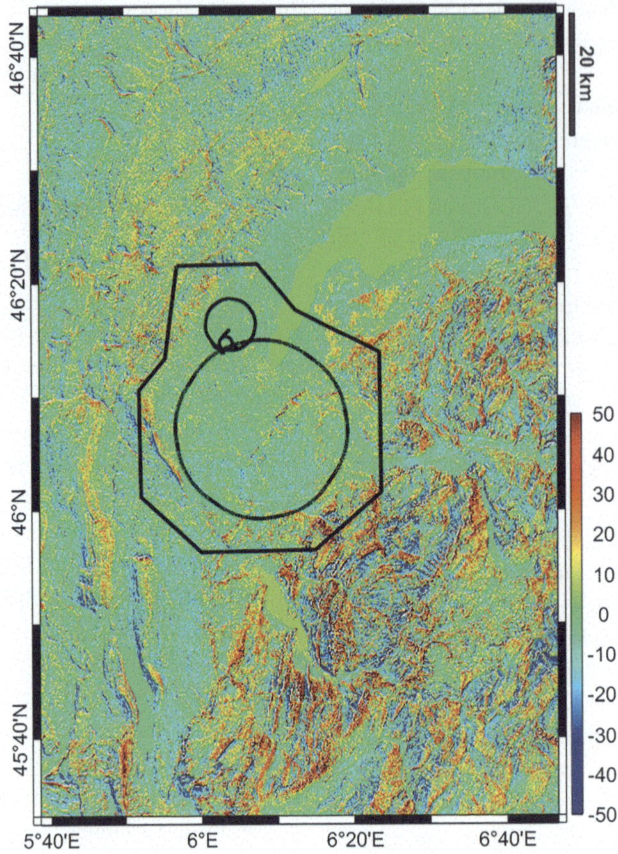

Fig. 3 Ellipsoidal height difference between ASTER GDEM v3 and Viewfinder Panorama DEM [m]

in the Prealps, where a ridge-like pattern suggests a possible horizontal misalignment between the DEMs. The differences could also result from the challenges in resolving steep terrain using satellite data, which may lead to shadowing effects, or from differing processing algorithms used to achieve the given resolution (e.g., 1 arcsec).

4.2 Comparison of Modeled Gravity Values

As noted in Sect. 2.2, the gravity data is heterogeneously distributed and the measurements vary in quality. Consequently, the impact of DEM differences on geoid solutions is not linearly propagated. Within the RCR approach, gravity residuals are computed using modeled gravity values at each observation point. Comparing these modeled gravity values at each measurement point using the different DEMs results in variations of up to 2 mGal in terms of SDs, closely matching the spatial patterns of DEM variability (see Fig. 2, center). However, some distinctions are notable: the northwestern area exhibits larger DEM impacts, and localized high-variability patches are found around Lake Geneva, despite only minor differences between the DEMs. The highest mean bias over all modeled gravity points occurs between AW and EO, and CO-EO (over 1 mGal), while the smallest differences are observed between NA-SR (2 μGal) and AW-TA (4 μGal). The largest SDs are found for AW-EO (1045 μGal), CO-EO (1060 μGal), EO-SR (1032 μGal), and EO-T (1167 μGal). The smallest are seen between AW-C (359 μGal), CO-NA (391 μGal), and CO-SR (402 μGal).

4.3 Comparison of Geoid Solutions

At the geoid level, Table 3 summarizes the differences between the computed solutions for the AoI. In almost all comparisons, the comparison to the EO-based geoid solution shows the largest bias (up to 3.5 cm relative to the TA geoid), while the smallest bias of only 1 mm is observed between ME and VF. The lowest SDs of 1 mm occur in comparisons of AW-NA, CO-NA, AW-SR, CO-SR, CO-TA, and NA-SR. The highest SDs are found between AS-ME, AW-VF, SR-VF, TA-VF (all around 5 mm), and ME-VF (6 mm). The spatial distribution of variability (see Fig. 2, right) reveals that the largest differences align with regions of high DEM variability. However, the geoid differences show patterns of longer wavelength. This includes areas around Lake Geneva, which display approximately 1 cm

Table 3 Mean (upper right triangle) and standard deviation (lower left triangle) of the differences between geoid solutions derived from different DEMs within the AoI [mm]. DEM abbreviations are defined in Table 1

	AS	AW	CO	EU	EO	FA	ME	NA	SR	TA	VF
AS		−6	−6	16	26	12	4	−2	−4	−8	5
AW	4		−0	22	32	18	10	4	2	−2	11
CO	3	2		22	32	18	10	4	3	−2	11
EU	3	4	3		10	−4	−12	−18	−20	−25	−11
EO	4	4	4	2		−14	−22	−28	−30	−35	−21
FA	4	2	2	3	3		−8	−14	−15	−20	−7
ME	5	2	3	4	4	3		−6	−8	−13	1
NA	3	1	1	3	4	2	2		−1	−6	7
SR	3	1	1	3	4	2	2	1		−5	8
TA	3	2	1	3	4	3	3	2	2		13
VF	3	5	5	2	3	4	6	4	5	5	

variations despite exhibiting relatively low DEM and gravity model variability. In peripheral regions with sparse gravity observations, lower variability is observed, likely due to the dominance of regularization in the geoid computation. This regularization was consistent across all DEMs, leading to similar solutions in these areas.

4.4 Comparison to GNSS/Leveling Points

To evaluate the quality of the geoid solutions, comparisons were made with GNSS/leveling datasets from the Federal Office of Topography swisstopo in Switzerland and the Institut national de l'information géographique et forestière (IGN) in France. The Swiss dataset includes only high-accuracy points, while the IGN dataset comprises points of varying age and quality. As individual uncertainties are not provided, both datasets are assumed error-free for this analysis.

Within the inner region displayed in Fig. 4, 15 Swiss and 213 French GNSS/leveling points are available. In the AoI, this reduces to 4 Swiss and 37 French points.

For the AoI, the mean offset of the 11 geoid solutions to the Swiss GNSS/leveling dataset is 305 mm, decreasing to 296 mm in the inner region. In the AoI, the SD in comparison with the Swiss GNSS/leveling points is less than 1 cm for all solutions, with all SDs between 5 mm and 7 mm. In the inner region, SD values increase to between 23 and 32 mm, but due to the limited number of points, these variations may not be statistically meaningful.

For the French dataset, the mean offset is −135 mm in the AoI and −134 mm in the inner region. SDs range from 13 to 15 mm in the AoI, and increase to between 13 mm (FA) and 32 mm (TA) in the full inner region.

The inner region is expected to show poorer agreement due to the scarcity of gravity measurements and the more complex mountainous terrain, where modeling is inherently more challenging.

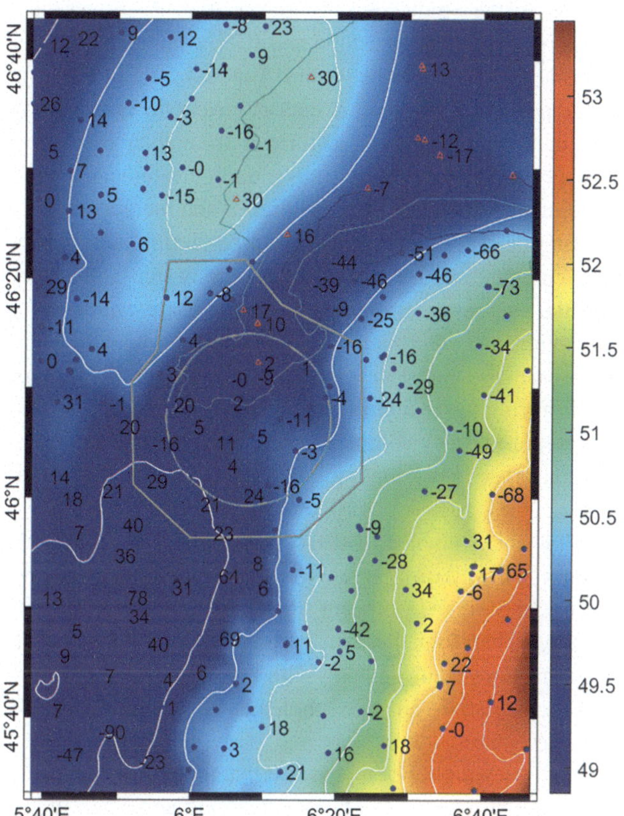

Fig. 4 Height anomalies ζ [m] derived from FABDEM V1-2. Blue dots and red triangles indicate the location of GNSS/leveling points from the French and Swiss datasets, respectively. Displayed values represent the differences from the GNSS/leveling dataset [mm]

5 Conclusion and Outlook

To summarize, this study shows that the use of different DEMs can result in geoid height biases of up to 3.5 cm and SDs of up to 6 mm in the study area. Therefore, the choice of the DEM has a significant impact on the geoid solution for the FCC. However, the intercomparison of solutions at the level of DEM heights, modeled gravity values, and final geoid heights did not reveal a clearly superior DEM. Similarly, the comparison against GNSS/leveling points was inconclusive. It is possible that the geoid solutions for the FCC region have reached the accuracy level of the GNSS/leveling dataset itself, highlighting the need for the development of new evaluation methods to assess potentially improved geoid models.

This study demonstrates that the accuracy of global and European DEMs is insufficient to produce a stable geoid solution in the study area. Therefore, the use of national DEMs published by swisstopo and IGN, which are publicly available at resolutions between 0.5 m and 5 m, is essential for the next steps. As neither dataset covers the entire study area, they must first be transformed from their respective national coordinate systems into a unified reference frame. Subsequently, the grids must be aligned and merged, and the resulting dataset down-sampled to a resolution suitable for geoid computation. Each of these processing steps introduces potential sources of error. Various methods are available for resampling and interpolation, ranging from standard techniques such as nearest-neighbor and bilinear interpolation or simple averaging, to more sophisticated approaches designed to preserve mass or volume, such as area-weighted averaging, conservative resampling, and mass-preserving aggregation.

While the influence of these choices may be negligible at the 1 cm precision level over a 1 km sliding window, they might become significant when targeting sub-millimeter precision. Therefore, these methods should be carefully evaluated in future studies, along with a thorough investigation of the calculation pipeline implemented in the GROOPS software toolkit.

Finally, it should be emphasized that the AoI of this study is in a region with relatively modest topographic variations. In more complex terrains such as the Alps, the Rocky Mountains, or the Himalayas, where elevation gradients are much steeper and small-scale topographic features are more pronounced, the influence of the DEM on geoid modeling is expected to be even more substantial. This highlights the need for region-specific assessments and careful DEM handling in geoid determination.

Acknowledgements This work was performed under the auspices and with support from the Swiss Accelerator Research and Technology (CHART) program (www.chart.ch).

We would like to express our sincere gratitude to the Federal Office of Topography swisstopo, Switzerland, and the Institut national de l'information géographique et forestière, France, for kindly providing access to their gravity and GNSS/leveling databases. We also thank Heiner Denker for providing the gridded EGG models.

References

Abrams M, Crippen R, Fujisada H (2020) ASTER Global Digital Elevation Model (GDEM) and ASTER Global Water Body Dataset (ASTWBD). Remote Sensing 12(7):1156. https://doi.org/10.3390/rs12071156

Bielski C, López-Vázquez C, Grohmann CH, et al (2024) Novel approach for ranking DEMs: Copernicus DEM improves one arc second open global topography. IEEE Trans Geosci Remote Sens 62:1–22. https://doi.org/10.1109/TGRS.2024.3368015

Claus S, Elling R, Persson V, et al (2023) EuroDEM - Pan-European height dataset at medium scale. Tech. rep., EuroGeographics AISBL. https://ome-download-data.s3.eu-west-1.amazonaws.com/euro-dem/documents/EuroDEM_2023_Specification.pdf

Crippen R, Buckley S, Agram P, et al (2016) Nasadem Global Elevation Model: Methods and progress. Int Arch Photogrammetry Remote Sens Spat Inf Sci XLI-B4:125–128. https://doi.org/10.5194/isprs-archives-XLI-B4-125-2016

de Farranti J (2022) Viewfinder Panorama Digital Elevation Models (VFPDEMs). https://viewfinderpanoramas.org/dem3.html

Dufourmont H, Gallego J., Reuter H, et al. (2014) EU-DEM statisticalValidation. Tech. rep., DHI GRAS c/o Geocenter Denmark, Copenhagen. https://ec.europa.eu/eurostat/documents/7116161/7172326/Report-EU-DEM-statistical-validation-August2014.pdf

Earth Observation Center (2018) TanDEM-X DEM products specification document. Tech. rep. https://geoservice.dlr.de/web/dataguide/tdm30/pdfs/TD-GS-PS-0021_DEM-Product-Specification_v3.2.pdf

EROS Center (2018) USGS EROS Archive - Digital Elevation - Shuttle Radar Topography Mission (SRTM) 1 Arc-Second Global. https://doi.org/10.5066/F7PR7TFT. https://www.usgs.gov/centers/eros/science/usgs-eros-archive-digital-elevation-shuttle-radar-topography-mission-srtm-1

Hawker L, Uhe P, Paulo L, et al. (2022) A 30 m global map of elevation with forests and buildings removed. Environ Res Lett 17(2):024016. https://doi.org/10.1088/1748-9326/ac4d4f

Koch JA, Marti U, Herrera Pinzón ID, et al (2024) Geoid computation for the future circular collider at CERN. https://doi.org/10.1007/1345_2024_275

Lercher T (2023) Contribution of gravity forward modeling in regional geoid computation. PhD thesis, TU Graz, Graz. https://doi.org/10.3217/bt663-7px76

Mayer-Gürr T, Behzadpour S, Eicker A, et al (2021) GROOPS: A software toolkit for gravity field recovery and GNSS processing. Comput Geosci 155:104864. https://doi.org/10.1016/j.cageo.2021.104864

Niedermaier J, Pail R, Gruber T (2024) Error tree analysis of geopotential field models. In: Gravity Geoid and Height Systems 2024 Symposium - GGHS2024

Pock C (2017) Consistent Combination of satellite and terrestrial gravity field observations in regional geoid modeling. PhD thesis, Graz University of Technology. https://online.tugraz.at/tug_online/wbAbs.showThesis?pThesisNr=61642&pOrgNr=34102

Schwabe J, Mayer-Gürr T, Hirt C, et al (2024) A new spherical harmonic approach to residual terrain modeling: a case study in the central European Alps. J Geodesy 98(7):65. https://doi.org/10.1007/s00190-024-01843-4

Strobl P (2020) The new Copernicus digital elevation model. GSICS Quart Newsl 14(1):1–20. https://doi.org/10.25923/enp8-6w06

Takaku J, Tadono T, Tsutsui K (2014) Generation of high resolution global DSM from ALOS PRISM. Int Arch Photogrammetry Remote Sens Spat Inf Sci XL-4:243–248. https://doi.org/10.5194/isprsarchives-XL-4-243-2014

Takaku J, Tadono T, Doutsu M, et al (2021) Updates of 'AW3D30' Alos Global Digital Surface Model in Antarctica with other open access datasets. Int Arch Photogrammetry Remote Sens Spat Inf Sci XLIII-B4-2021:401–408. https://doi.org/10.5194/isprs-archives-XLIII-B4-2021-401-2021

Yamazaki D, Ikeshima D, Sosa J, et al (2019) MERIT hydro: A high-resolution global hydrography map based on latest topography dataset. Water Resour Res 55(6):5053–5073. https://doi.org/10.1029/2019WR024873

Towards an Improved Gravimetric Geoid in Albania

G. S. Vergos, D. A. Natsiopoulos, Kristaq Qirko, Oltjon Balliu, Endri Qershija, Perparim Ndoj, Grigorios Tsokas, and Alexandros Stampolidis

Abstract

The determination of a high-accuracy and high-resolution geoid is of indispensable importance as it forms one of the cornerstones of a modern geodetic reference system. With that in mind, the State Authority for Geospatial Information (ASIG) of the Republic of Albania has launched an effort to modernize the gravity infrastructure of the country and as a consequence to make available a new gravimetric geoid model of enhanced resolution and accuracy for the country. In that respect, the necessary procedures to establish the primary and secondary leveling network have been detailed, collocated with gravity measurements, as well as the determination of a preliminary gravimetric geoid based on existing and new gravity data. In this work, we first focus on the gravity campaigns, data processing, reductions, and adjustments carried out as part of establishing the reference benchmarks of the second-order gravity network. Then the methodological steps followed for the determination of the geoid on a $1' \times 1'$ grid over Albania, in an area bounded between $38.5° \leq \varphi \leq 43.5°$ and $18.0° \leq \lambda \leq 22.0°$, are outlined. For the geoid determination, the remove-compute-restore (RCR) concept is employed using two global geopotential models (GGMs), namely EIGEN6c4 and XGM2019e, and two approaches for the topographic effects through a residual terrain model (RTM) correction with the spectral and classical computation. Then, the estimation of the geoid is carried out using least squares collocation (LSC) and fast Fourier transform (FFT) methods, and the validation is performed against both GNSS/leveling data over Albania and a high-accuracy leveling traverse connecting Durrës and Tirana. From the results achieved, the FFT-based geoid presented a mean value of the differences at the 3 cm level and an std of 14.1 cm. When compared to the new leveling traverse, the new geoid has shown differences at the $\sim$16 cm level which indicates the bias between the new gravimetric geoid from FFT and the Albanian vertical datum. The relative differences over the measured leveling BMs are at the 1–7 mm level. A first-guess hybrid geoid was also determined, after the parametric fit with the third-order polynomial model, reaching the 12.7 cm agreement with the old vertical datum, which is very satisfactory considering the fact that the acquired land gravity data manage to cover only a small portion of the gravity variability due to the long distances between them.

G. S. Vergos (✉)
Laboratory of Gravity Field Research and Applications (GravLab), Department of Geodesy and Surveying, Aristotle University of Thessaloniki, Thessaloniki, Greece
e-mail: vergos@topo.auth.gr

D. A. Natsiopoulos
Department of Surveying and Geoinformatics Engineering of the School of Engineering, International Hellenic University, Serres, Greece

K. Qirko · O. Balliu · E. Qershija
State Authority for Geospatial Information (ASIG), Tirana, Albania

P. Ndoj
Land & Co. Ltd., Tirana, Albania

G. Tsokas
Department of Mineralogy, Petrology, Economic Geology, Faculty of Geology, Aristotle University of Thessaloniki, Thessaloniki, Greece

A. Stampolidis
Exploration Geophysics Laboratory, Faculty of Geology, Aristotle University of Thessaloniki, Thessaloniki, Greece

J. T. Freymueller, L. Sànchez (eds.), *International Symposium on Gravity, Geoid and Height Systems 2024 (GGHS2024)*, International Association of Geodesy Symposia 158, https://doi.org/10.1007/1345_2025_289

Keywords

FFT · Geodetic infrastructure · Geoid modeling · Gravity densification · Least squares collocation

1 Introduction

During the last years, many countries worldwide have been actively engaged in developing new geoid models, utilizing advancements in satellite gravimetry, terrestrial gravity surveys, and geodetic techniques to enhance the accuracy of height reference systems (Vergos et al. 2023; Wang et al. 2023; Saari et al. 2021). Albania is located in the southeast part of Europe, and in the west part, it stretches along the Adriatic and Ionian Seas with a long shoreline. Bounded between four countries, it is one of the most mountainous European countries, presenting a varying topography, with Albanian Alps in the north, the Pindus Mountains in the southeast, and the Ceraunian Mountains in the southwest.

In countries with such various morphological features like Albania, the calculation of high-resolution and high-accuracy geoid models is crucial as global geopotential models cannot represent the high frequencies of gravity spectrum. This is especially important in view of modernizing the vertical datum of the country and developing a geoid model in support of GNSS/leveling.

The State Authority for Geospatial Information (ASIG) of Albania is the responsible organization for the establishment of the New Geodetic Reference Frame in Albania, including among others GNSS, leveling, gravimetric, tide-gauge station, and magnetometric networks. As a result, ASIG acknowledging the necessity for high-accuracy and high-resolution geoid, that will support accurate and efficient vertical positioning throughout Albania, has initiated the effort for the collection of new land gravity data and the determination of new nationwide geoid model. The current gravimetric network refers to the zeroth-order network with three absolute gravimetric points at Shkodra (northernmost BM), Tirana (BM in the central part of the country), and Saranda (southernmost BM), measured from September to October 2015 (see black stars in Fig. 1) relying on observations with the absolute gravimeter FG5-242 (Ullrich et al. 2016). The first network order consists of 42 points measured with CG-5 relative gravimeters in 2018 and an accuracy of 10 mGal, while part of the second- and third-order network includes 38 and 138 points, respectively. The aim of the current work was first to upgrade the national gravity database and then to determine a new geoid model of enhanced resolution and accuracy for the country.

This study aims to present the realization, analysis, and processing to collect new gravity measurements for the second-order gravity network in the territory of Albania and the methodologies, processing steps, estimation, and validation for the determination of the new geoid model on a regular 1 arcmin × 1 arcmin grid. After the relative gravity data collection and processing, the whole gravity database of free-air gravity anomalies spanning both land and marine regions was determined. The geoid modeling development was based on the well-known remove-compute-restore (RCR) concept, and its validation was performed against both GNSS/leveling data over Albania and a high-accuracy leveling traverse connecting Durrës (demoted with a red star in Fig. 1) and Tirana.

2 Gravity Database of Reductions

2.1 Relative Gravity Data Collection and Processing

In an effort to densify the currently available terrestrial gravity database over the Albanian territory, relative gravity campaigns have been carried out over 300 benchmarks (BMs), with a varying density of an average of 10 × 10 km in mountainous regions and 5 × 5 km in the western lowlands. Two Scintrex CG5 gravimeters were used during 35 days of observations in closed loops, one owned by Land&Co and the other owned by the Laboratory of Gravity Field Research and Applications (GravLab), of the Aristotle University of Thessaloniki (AUTH).

For the calculation of the calibration constant, the difference of the gravitational acceleration was measured four times in one day for the points AGP01-Tirana and AGP02-Shkodër (3 days, at the beginning, in the middle, and at the end of the campaign), while the stationary drift was estimated twice a week at one of the base stations. With all measurements based on zeroth- or first-order gravity BMs, the star method (A-B-A), the profile method (123454321), or a hybrid approach combining both was applied to acquire the observations (Torge 1989). Moreover, at each BM, the coordinates of the point were determined through 1 h of dedicated static GNSS observations, which were then processed with the Bernese software (Dach et al. 2015).

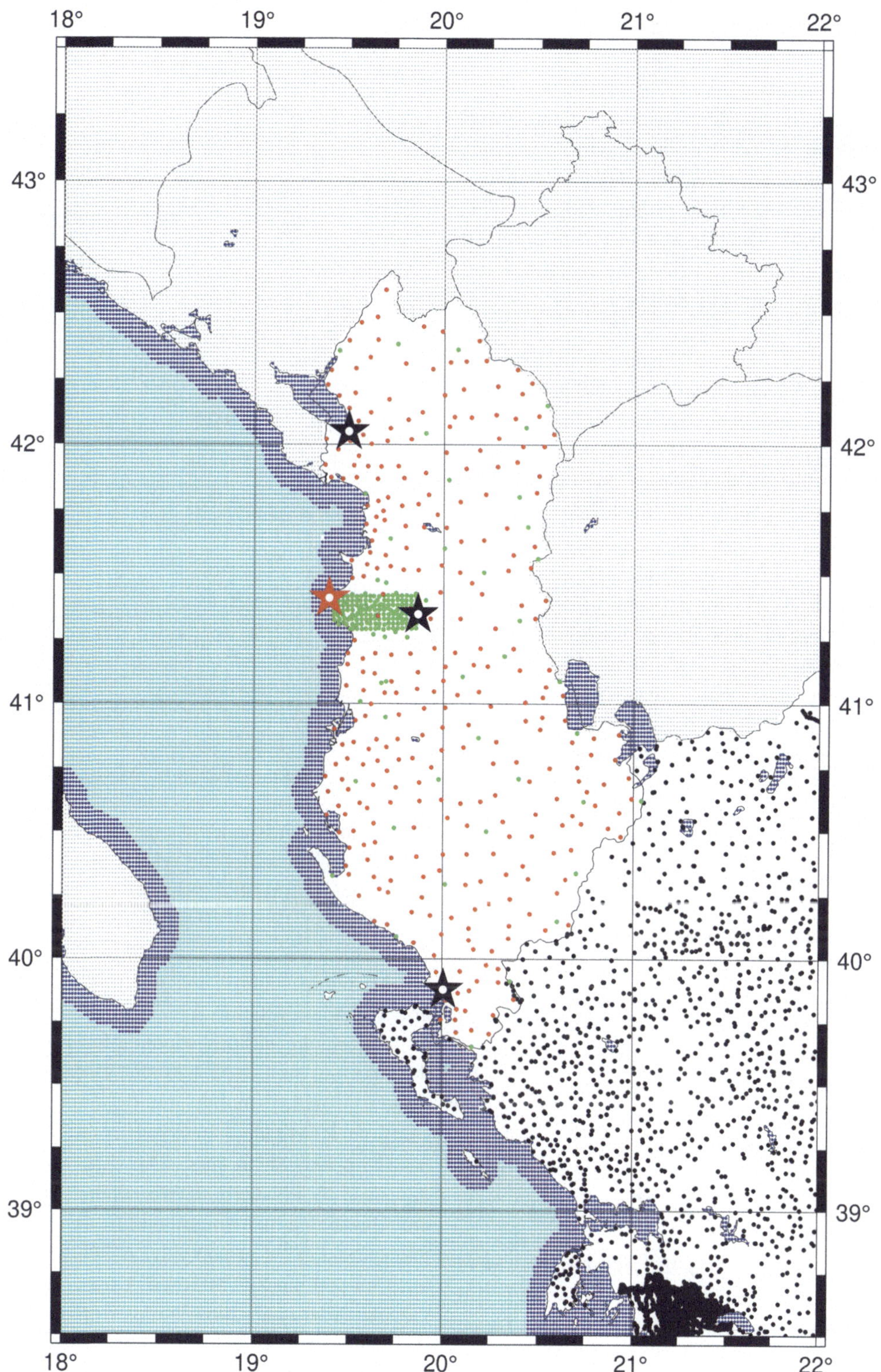

Fig. 1 The land and marine gravity database used for the gravimetric geoid determination over Albania. Land gravity from the current (red) and past (green) campaigns, DTU2021 (blue) and SIO29.1 (cyan) marine gravity data, land gravity data for Greek mainland (black), and EGM2008 fill-in data (gray) for neighboring countries. Black stars denote the zeroth-order absolute gravity stations (Shkodra in the north, Tirana in the center, and Saranda in the south) and the red star the Durrës tide-gauge station

2.2 Reduction of Gravity Data and Residual Modeling

The full dataset used for the determination of the gravimetric geoid refers to 52,517 free-air gravity anomalies spanning both land and marine regions, bounded between $38.5° \leq \varphi \leq 43.5°$ and $18.0° \leq \lambda \leq 22.0°$ (Fig. 1). This dataset includes (1) 25,013 satellite altimetry data obtained from DTU2021, used up to 10 km from the coastlines and within lakes (blue) (Andersen and Knudsen 2019), as well as SIO29.1, for areas located at greater distances offshore (cyan) (Sandwell et al. 2014); (2) 564 land gravity data collected across mainland Albania, from which 300 points were measured during the current campaign (red color) and 264 points (green color) refer to historical gravity data acquired in 2018; and (3) 2,246 point values provided by GravLab (black color) for the Greek mainland which refer to recent and historical, validated, discrete free-air gravity anomalies for the Greek territory (Grigoriadis et al. 2023; Tziavos et al. 2010; Vergos et al. 2025). Moreover, to minimize edge effects in the calculation of the Albanian gravimetric geoid model, the computation area has been extended by 1° in all directions using gravity anomalies derived from EGM2008 (Pavlis et al. 2012), complete to degree and order (d/o) 2,159, as fill-in for the neighboring countries (24,694 points with gray color).

The geoid development was based on the well-known remove-compute-restore (RCR) method (Sansò and Sideris 2013a). Given the availability of gravity field-related data, i.e., the collected free-air gravity anomalies as well as the historical data over the land and marine areas, the approach involves removing from the original data the influence of topography and that of the long and medium wavelengths from a global geopotential model (GGM). The short- and medium-frequency part was represented by EIGEN6c4 (Förste et al. 2014) and XGM2019e (Zingerle et al. 2020) in an effort to evaluate their performance over Albania. The so reduced gravity anomalies can be computed as

$$\Delta g_{\text{red}} = \Delta g_{\text{f}} - \Delta g^{\text{GGM}} \tag{1}$$

As evaluated by Barzaghi et al. (2018) and Grigoriadis et al. (2021), the GGMs can be either employed at their full degree of expansion or selected to a cutoff degree. Based on the results in the aforementioned studies, EIGEN6c4 performs best at $n_{\max} = 1{,}000$ and subsequently combined with higher-order residual terrain model (RTM) effects. In contrast, XGM2019e is more effective when utilized at its full d/o $n_{\max} = 2{,}190$. Based on this, the GGM contribution in Eq. (1) was computed for EIGEN6c4 to d/o 1,000 and 2,190 as well as XGM2019e to d/o 2,190, so that reduced data were estimated as

Table 1 Statistics of the original gravity data, GGM contribution, RTM effects for all approaches and degrees, reduced and residual gravity anomalies (units: mGal)

	Max	Min	Mean	Std
Δg	259.746	−181.654	1.145	63.884
$\Delta g_{\text{EIGEN6c4}}\vert_{n=2}^{1,000}$	178.540	−188.540	2.396	60.218
Δg_{red1}	131.552	−166.805	−1.246	21.804
$\Delta g_{\text{EIGEN6c4}}\vert_{n=2}^{2,190}$	258.469	−177.231	2.042	63.972
Δg_{red2}	52.538	−110.018	−0.892	6.120
$\Delta g_{\text{XGM2019e}}\vert_{n=2}^{2,190}$	226.589	−185.240	1.949	62.822
Δg_{red3}	58.539	−138.164	−0.798	9.467
$\delta g_{\text{RTM}}\vert_{n=1,001}^{2,190}$	110.622	−122.727	−0.354	18.951
$\delta g_{\text{RTM}}\vert_{n=2,191}^{90,000}$	86.740	−146.867	−3.767	18.258
$\delta g_{\text{RTM}}^{\text{C1}}$	109.638	−116.736	−0.843	7.880
$\delta g_{\text{RTM}}^{\text{C2}}$	65.068	−134.764	−0.264	3.382

$$\Delta g_{\text{red1}} = \Delta g_{\text{f}} - \Delta g_{\text{EIGEN6c4}}\Big|_{n=2}^{1,000} \tag{2}$$

$$\Delta g_{\text{red2}} = \Delta g_{\text{f}} - \Delta g_{\text{EIGEN6c4}}\Big|_{n=2}^{2,190} \tag{3}$$

$$\Delta g_{\text{red3}} = \Delta g_{\text{f}} - \Delta g_{\text{XGM2019e}}\Big|_{n=2}^{2,190} \tag{4}$$

From Table 1, it is evident that each model contributed to a reduction of the std of the gravity anomalies from 63.855 to 21.839 mGal for EIGEN6c4 to d/o 1,000, 6.184 mGal for EIGEN6c4 to d/o 2,190, and 9.500 mGal for XGM2019e to d/o 2,190.

The remove step aims to create a smoothed residual gravity anomaly field (Δg_{res}) suitable for effective prediction. From the various topographic reduction techniques, the one that was used was the so-called RTM reduction, since it provides smooth gravity anomaly residuals. In this study, the RTM reduction to gravity anomalies was computed based on two techniques. The first was the classical one described by Forsberg (1984), where the short wavelengths are represented by a combined DTM model, generated from the national DTM provided by ASIG for the entire Albanian territory and the SRTM-based 3 arcsec DTM for the rest of the area (Farr et al. 2007; Tziavos et al. 2010). The spectral RTM approach (Rexer et al. 2018) is based on a combination of a spherical harmonics expansion of the Earth's potential (Hirt et al. 2014) and ultrahigh-resolution RTM effects from a pre-computed global model.

From the so derived RTM reduction, δg_{RTM}, which represents the RTM reduction, the reduced fields derived from Eqs. (2)–(4) can lead to residual gravity anomalies as

$$\Delta g_{\text{res}} = \Delta g_{\text{red}} - \delta g_{\text{RTM}} \tag{5}$$

Table 2 Statistics of the residual gravity anomalies with the various reduction schemes (units: mGal)

		Max	Min	Mean	Std
EIGEN6c4 d/o 1,000	Δg_{res11}	131.552	−166.805	−0.284	21.232
	Δg_{res12}	122.871	−96.271	1.385	15.084
EIGEN6c4 d/o 2,190	Δg_{res21}	64.789	−83.884	−0.243	4.851
	Δg_{res22}	60.934	−84.407	−0.132	4.779
XGM2019e d/o 2,190	Δg_{res31}	64.993	−67.559	−0.150	8.733
	Δg_{res32}	66.511	−67.559	−0.038	8.597

and by combining the available reduced fields with the different approaches for the RTM effects, the residuals were estimated as

$$\Delta g_{res11} = \Delta g_{red1} - \delta g_{RTM}^{C1} \quad (6)$$

$$\Delta g_{res12} = \Delta g_{red1} - \delta g_{RTM}^{S1} \quad (7)$$

$$\Delta g_{res21} = \Delta g_{red2} - \delta g_{RTM}^{C2} \quad (8)$$

$$\Delta g_{res22} = \Delta g_{red2} - \delta g_{RTM}^{S2} \quad (9)$$

$$\Delta g_{res31} = \Delta g_{red3} - \delta g_{RTM}^{C2} \quad (10)$$

$$\Delta g_{res32} = \Delta g_{red3} - \delta g_{RTM}^{S2} \quad (11)$$

In Eqs. (6)–(11), the superscript C denotes the classic RTM approach, S denotes the spectral one, 1 denotes the RTM effects evaluated referenced to a GGM complete to d/o 1,000, and 2 denotes the RTM effects evaluated referenced to a GGM complete to d/o 2,190. The statistics of the gravity residuals (summarized in Table 2) show that the EIGEN6c4 to d/o 2,190 and spectral RTM combination provides the overall best results, being better by 0.1 mGal to the classic RTM effects. Comparing the residual fields from the two GGMs, the one referenced to XGM2019e presents a larger std value by ~4 mGal, while the mean is equivalent. As a further evaluation step of the derived residuals, the empirical covariance functions have been derived and presented in Fig. 2. It is evident that the residuals to EIGEN6c4 to d/o 1,000 have a much larger variance, even after the removal of the topographic effects, being at ~150 mGal^2 for the spectral and ~420 mGal^2 for the classic RTM effects. The fact that the classical RTM does not manage to reduce the data as the spectral one does is attributed to the limitations of the DTM/DBM and the method itself given a high d/o expansion of the GGM. The EIGEN6c4 to d/o 2,190 residuals provide the smaller variances, 17.37 and 17.79 mGal^2 for the spectral and classical RTM, compared to the ones relative to XGM2019e, being at the 69.99 and 68.32 mGal^2 level for the spectral and classical RTM, respectively. The same can be concluded from the respective correlation lengths, which are at the 8.29 and 7.11 km, for the spectral and classical RTM residuals to EIGEN6c4, and at the 8.45 and 7.98 km, for the corresponding ones to XGM2019e. Therefore, in both cases, the residuals with the spectral RTM effects show a larger correlation length, compared to the ones with the classic RTM effects, hence offering a smoother prediction field. It should be noted that in Fig. 2, the empirical covariance functions of the two residual fields to EIGEN6c4, with classical and spectral RTM effects, practically coincide; hence, the two lines are indistinguishable. The same holds for the empirical covariance functions of the residuals fields, relative to XGM2019e.

3 Geoid Determination

From the residuals calculated during the remove step, several gravimetric geoid models with both GGMs as reference and both approaches for the reduction of the topography have been examined. For geoid estimation in Albania, two approaches, namely least squares collocation (LSC) and fast Fourier transform (FFT) (Sideris 2013; Sansò and Sideris 2013b; Haagmans et al. 1993) with a Wong-Gore modification of the Stokes' kernel function (Wong and Gore 1969), have been selected. For the validation of the geoid models, 225 GNSS/leveing BMs provided by ASIG.

have been used, 113 of which come from past campaigns originating in the old ALB86 vertical datum and 112 from newly acquired GNSS/leveling measurements. Moreover, 50 BMs from the precise leveling lines Durrës–Tiranë and Vorë–Shëngjin, where ellipsoidal and orthometric heights are available, have been employed.

In the FFT approach, the fields of available residual gravity values have been gridded onto a $1' \times 1'$ grid employing kriging, while various intermediate geoid models were determined in order to find the optimal cutoff degrees for the Wong-Gore modification of Stokes' kernel. This was investigated for various cutoff degrees between 60 and 180, with the optimal performance, in terms of the standard deviation of the differences to the GNSS/leveling data, being found for the band 100–120 for all solutions. The model with the residuals to EIGEN6c4 d/o 2,190 and the spectral RTM approach provided the smallest std (±14.1 cm), with a mean value of 3.4 cm (see Table 3), while the one with the classical RTM effects gives a similar std and a mean at the few mm level. Some useful insights on the quality of the solutions, which are practically identical, can be gained by forming the absolute geoid height differences at the first-order BMs of the precise level traverse. For six BMs (BM1, TGBM, BM2, BM3, D17HaHpVG, and RII006) of this leveling line, both orthometric and ellipsoidal heights were available, being referenced to the mean sea level (MSL) of the Durrës tide gauge. Table 4 summarizes these results, where both gravimetric geoid models perform equally well, with relative

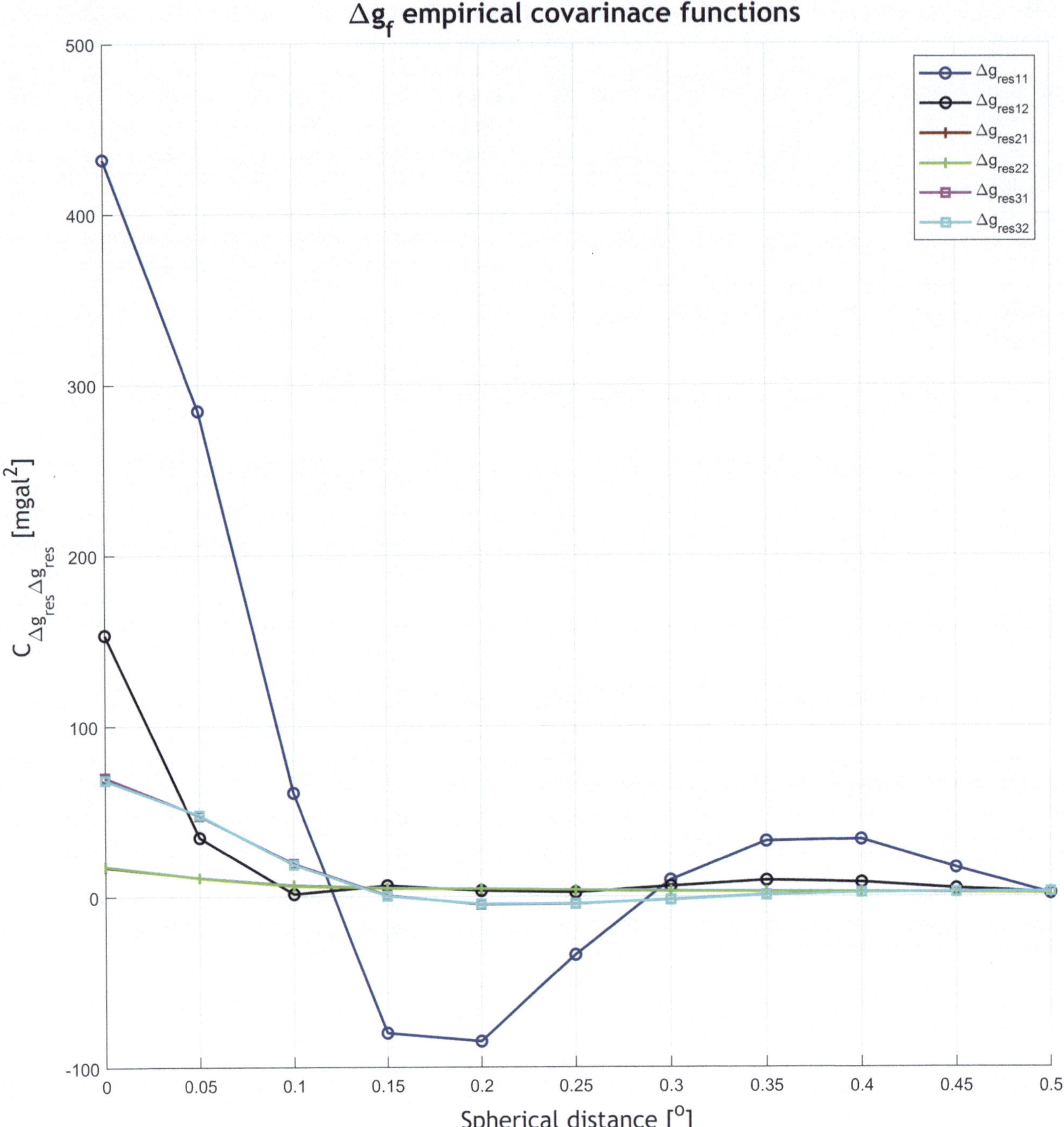

Fig. 2 Empirical covariance functions of the residual gravity anomaly fields after reduction to EIGEN6c4 (d/o 1,000 and 2,190) and XGM2019e (d/o 2,190) and removal of RTM effects with the classic (RF) and spectral (HK) approaches

differences at the few mm level, with the exception of BM D17-RII6 which is probably biased, while the absolute ones reach the 6 cm level for the classic RTM effects and 10 cm for the spectral ones. If the problematic BM is removed from the statistics, then the std is reduced by 5 cm for both solutions.

Finally, LSC-based geoid models have been determined for the two best-performing FFT-based models, specifically, EIGEN6c4 with classical RTM and spectral effects, as this configuration yielded the most accurate overall fit to the GNSS/leveling benchmarks. Table 5 tabulates both the differences of the LSC geoid models to the GNSS/leveling data as well as their differences to the respective FFT geoids. The differences of the LSC geoid models are worse than those of the 2D FFT solution, by ~2 cm in terms of the std, while the mean for the LSC geoid based and the classical RTM effects is significantly larger. These larger errors of the LSC solution

Table 3 Statistics of the geoid height differences between the various 2D FFT gravimetric geoid models and the GNSS/leveling data (units: m)

	Max	Min	Mean	Std
$N^{grav}_{EIGEN2190HK} - N^{GNSS/Lev}$	0.482	−0.349	0.034	0.141
$N^{grav}_{EIGEN2190RF} - N^{GNSS/Lev}$	0.348	−0.351	−0.005	0.146

Table 4 Statistics of the geoid height differences between the various 2D FFT gravimetric geoid models and the GNSS/leveling data (units: m)

	BM1–BM2	BM2–BM3	BM3–D17	D17–RII6	
EIGEN2190HK	−0.002	−0.005	0.007	−0.068	
EIGEN2190RF	−0.001	−0.005	0.008	−0.066	
Abs. dif	**BM1**	**BM2**	**BM3**	**D17**	**RII6**
$h_i - H_i - N^{grav}_{EIGEN2190HK}$	0.168	0.166	0.161	0.169	0.103
$h_i - H_i - N^{grav}_{EIGEN2190RF}$	0.130	0.127	0.123	0.130	0.062

Table 5 Statistics of LSC-based geoid heights with spectral and classic RTM effect differences with the GNSS/leveling data and over the entire area with the 2D FFT gravimetric geoid models (units: m)

	Max	Min	Mean	Std
$N^{LSC}_{EIGEN2190HK} - N^{GNSS/Lev}$	0.738	−0.759	−0.033	0.162
$N^{LSC}_{EIGEN2190RF} - N^{GNSS/Lev}$	0.592	−0.823	−0.130	0.161
$N^{2D-FFT}_{EIGEN2190HK} - N^{LSC}_{EIGEN2190HK}$	0.375	−0.328	0.013	0.035
$N^{2D-FFT}_{EIGEN2190RF} - N^{LSC}_{EIGEN2190RF}$	0.532	−0.155	0.050	0.066

Table 6 Statistics of the relative and absolute geoid height differences between the various LSC gravimetric geoid models and the leveling BMs along the precise leveling traverse

Abs. dif.	**BM1**	**BM2**	**BM3**	**D17**	**RII6**
$\delta N^{LSC}_{EIGEN2190HK}$	−0.069	−0.071	−0.078	−0.068	−0.199
$\delta N^{LSC}_{EIGEN2190RF}$	−0.090	−0.093	−0.100	−0.090	−0.217
Rel. dif.	**BM1–BM2**	**BM2–BM3**	**BM3–D17**	**D17–RII6**	
EIGEN2190HK	−0.002	−0.005	0.007	−0.068	
EIGEN2190RF	−0.001	−0.005	0.008	−0.066	

are probably due to the few land gravity data available over the country, which have a sparse distribution in order to predict the wanted final $1' \times 1'$ ($\sim$1.8 km) geoid resolution. This is also depicted in the bias between the LSC and FFT geoid models, ranging from 1 cm for the spectral RTM residuals to 5 cm for the classical ones. Similar to the FFT-based models, the LSC-based ones have been used for direct comparisons at the newly established leveling traverse (see Table 6). Once again, the results achieved, despite the fact that the LSC models show a larger std with the differences to the GNSS/leveling data of the old vertical network, are very encouraging. The relative differences are similar to those of the FFT geoid models, while the absolute ones are of an opposite sign and differ from the FFT-based estimates by several centimeters.

4 Geoid Validation, Parametric Fit to GNSS/Leveling Data and Relative Differences

To assess the degree of improvement offered by the new gravimetric geoid models, their differences to the available GNSS/leveling data have been used. The validation of the geoid models was performed through comparisons with the available GNSS/leveling observations on land, by estimating absolute and relative differences between all formed baselines. After this initial validation, the next step was to use a deterministic parametric model to estimate the final hybrid geoid model. Two models were validated, the 2D FFT geoid using EIGEN6c4 to d/o 2,190 with both approaches for the RTM effects. In the evaluation process, we assessed various parametric models, like four- and five-parameter similarity transformation (Heiskanen and Moritz 1967) and second- and third-order polynomial ones (Grebenitcharsky et al. 2023) and evaluated both absolute and relative differences of the geoid models.

The 2D FFT gravimetric geoid with the spectral RTM approach (see Table 7) before and after the parametric fit has a better std (12.7 cm) which is 0.7 cm better than the one with the classical RTM approach (see Table 8). This level of agreement, which is inferior than that along the first order leveling line, is due to the fact that in this comparison

Table 7 Statistics of the geoid height differences between the 2D FFT gravimetric geoid model (EIGEN6c4 to d/o 2,190 and spectral RTM effects) and the GNSS/leveling data (units: m)

	Max	Min	Mean	Std
$N^{grav}_{EIGEN2190HK} - N^{GNSS/Lev}$	0.482	−0.349	0.034	0.141
Parametric model				
Second-order polynomial	0.412	−0.400	0.000	0.131
Third-order polynomial	0.338	−0.374	0.000	0.127
Four-parameter	0.456	−0.383	0.000	0.134
Five-parameter	0.438	−0.376	0.000	0.132

Table 8 Statistics of the geoid height differences between the 2D FFT gravimetric geoid model (EIGEN6c4 to d/o 2,190 and classical RTM effects) and the GNSS/leveling data (units: m)

	Max	Min	Mean	Std
$N^{grav}_{EIGEN2190RF} - N^{GNSS/Lev}$	0.348	−0.351	−0.005	0.147
Parametric model				
Second-order polynomial	0.366	−0.383	0.000	0.135
Third-order polynomial	0.336	−0.361	0.000	0.134
Four-parameter	0.361	−0.379	0.000	0.144
Five-parameter	0.340	−0.365	0.000	0.136

also historical GNSS/leveling data have been used, which is necessary, since a uniform distribution over the country is needed for future hybrid geoid model development in support of leveling through GNSS.

Figure 3 (top) depicts the relative differences, as a function of the baseline length S_{ij}, of the geoid model with the spectral RTM effects, where the improvement of the fitted geoid model is once again visible. After the fit, the fitted geoid reaches relative accuracies lower than 10 ppm for baselines as short as 10–20 km. For longer baselines, the relative error decreases rapidly, dropping below 1 cm/km and reaching levels below 1 mm/km for baseline lengths exceeding 80 km. For the geoid model after the third-order parametric fit (bottom), 47.4% of the differences are below the 1 cm$\sqrt{S_{ij}}$ level and 73.4% below the 2 cm$\sqrt{S_{ij}}$ one. Finally, Fig. 4 depicts the final gravimetric geoid model for Albania, before any parametric fit, based on EIGEN6c4 and spectral RTM effects, giving a preliminary solution for the geoid model of the country, using the currently available land gravity data.

5 Conclusion

The scope of this work has been the determination of an analytical gravimetric geoid model for the area of Albania with a 1 arcmin resolution. In that respect, first the relative gravity campaigns, data processing, reductions, and adjustments carried out as part of establishing the reference benchmarks of the second-order gravity network have been described. Various geoid models have been determined, investigating different GGMs to model the long-wavelength part of the spectrum, and two approaches to treat the terrain through RTM effects, towards the modeling of the short-wavelength part of the spectrum using the RCR technique. Two main approaches with various variants have been used to estimate geoid heights, namely a spectral FFT-based and a stochastic LSC-based one.

From the results achieved, after evaluation relative to both the GNSS/leveling data from the old Albanian vertical datum and the newly acquired leveling traverse, the FFT-based geoid has a mean value of differences at the 3 cm level and an std of 14.1 cm. The relative differences over the measured leveling BMs are at the few millimeter level, with the exception of BM D17-RII6 which has probably biased orthometric and/or ellipsoidal height. After the fit with a third-order polynomial model, to remove any systematic differences in the formed residuals, the fit to the GNSS/leveling data reduces to 12.7 cm, with relative differences at the 10 ppm for baselines above 10 km and 73.4% being below the 2 cm$\sqrt{S_{ij}}$ standard error.

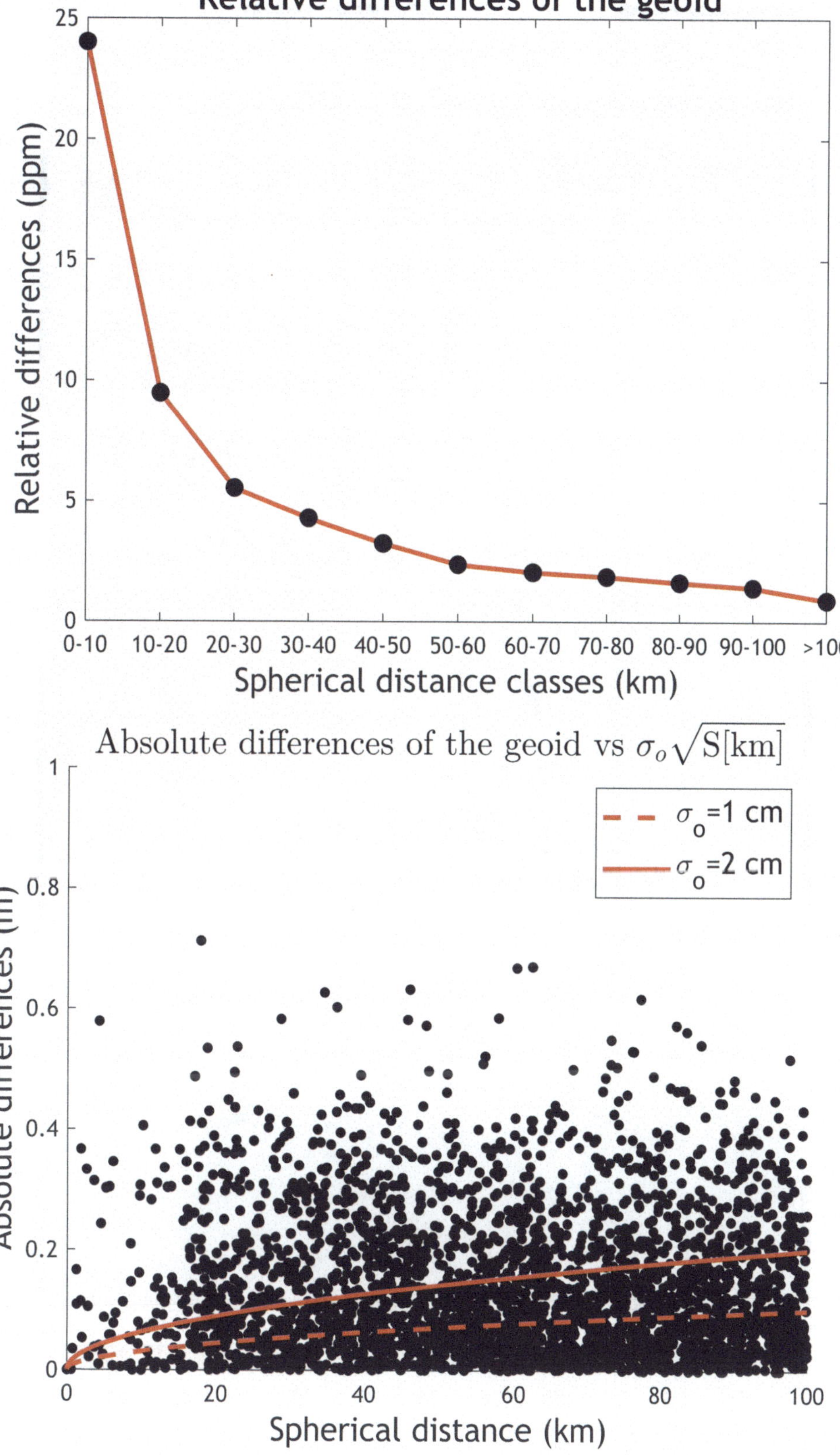

Fig. 3 Relative (top) and absolute (bottom) differences of the fitted with the third-order polynomial 2D FFT gravimetric geoid with the spectral RTM approach

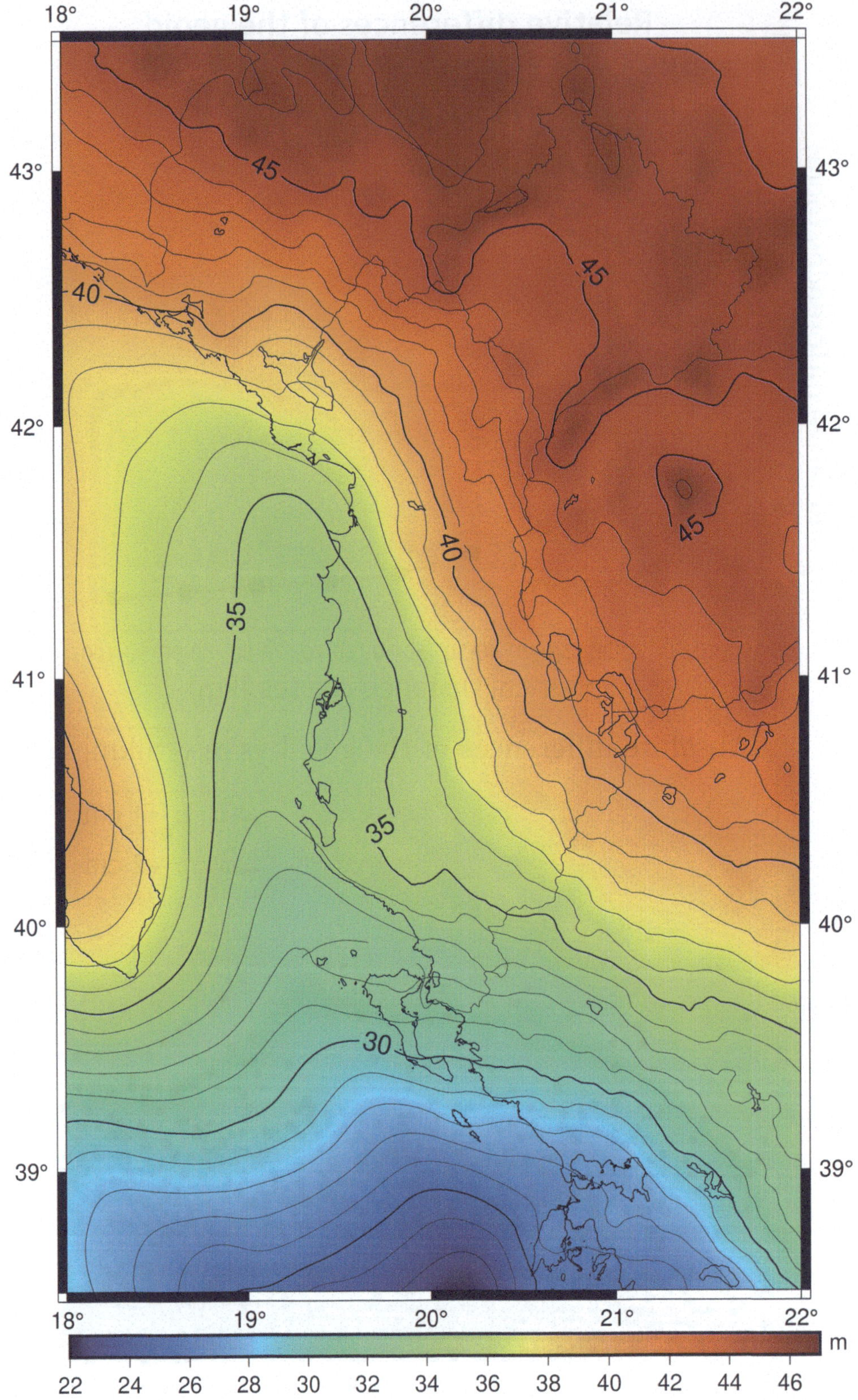

Fig. 4 The final Albanian gravimetric geoid model based on 2D FFT, using EIGEN6c4 to d/o 2,190 and spectral RTM residuals (units: m)

References

Andersen OB, Knudsen P (2019) The DTU17 global marine gravity field: first validation results. In: Mertikas S, Pail R (eds) Fiducial reference measurements for altimetry. International Association of Geodesy symposia, vol 150. Springer, Cham. https://doi.org/10.1007/1345_2019_65

Barzaghi R, Carrion D, Vergos GS, Tziavos IN, Grigoriadis VN, Natsiopoulos DA, Bruinsma S, Reinquin F, Seoane L, Bonvalot S, Lequentrec-Lalancette MF, Salaün C, Andersen O, Knudsen P, Abulaitijiang A, Rio MH (2018) GEOMED2: high-resolution geoid of the Mediterranean. In: International Association of Geodesy symposia. Springer, Berlin. https://doi.org/10.1007/1345_2018_33

Dach R, Lutz S, Walser P, Fridez P (2015) Bernese GNSS Software Version 5.2. University of Bern, Bern Open Publishing. https://doi.org/10.7892/boris.72297l

Farr TG et al (2007) The shuttle radar topography mission. Rev Geophys 45:RG2004. https://doi.org/10.1029/2005RG000183

Forsberg R (1984) A study of terrain reductions, density anomalies and geophysical inversion methods in gravity field modelling, Dept. of Geod. Sci. Rep. 355. Ohio State Univ., Columbus

Förste C, Bruinsma S, Abrikosov O, Jean-Michel L, Marty JC, Flechtner F, Balmino G, Barthelmes F, Biancale R (2014) EIGEN-6C4 the latest combined global gravity field model including GOCE data up to degree and order 2190 of GFZ Potsdam and GRGS Toulouse. GFZ Data Services. https://doi.org/10.5880/icgem.2015.1

Grebenitcharsky RS, Vergos GS, Al-Shahrani S, Al-Qahtani A, Iuri G, Othman A, Aljebreen S (2023) Hybrid geoid modeling for the Kingdom of Saudi Arabia. In: Freymueller JT, Sánchez L (eds) Gravity, positioning and reference frames. REFAG 2022. International Association of Geodesy symposia, vol 156. Springer, Cham. https://doi.org/10.1007/1345_2023_215

Grigoriadis VN, Vergos GS, Barzaghi R, Carrion D, Koç Ö (2021) Collocation and FFT-based geoid estimation within the Colorado 1 cm geoid experiment. J Geod 95:52. https://doi.org/10.1007/s00190-021-01507-7

Grigoriadis VN, Andritsanos VD, Natsiopoulos DA, Vergos GS, Tziavos IN (2023) Geoid studies in two test areas in Greece using different geopotential models towards the estimation of a reference geopotential value. Remote Sens 15(17):4282. https://doi.org/10.3390/rs15174282

Haagmans R, de Min E, von Gelderen M (1993) Fast evaluation of convolution integrals on the sphere using 1D FFT, and a comparison with existing methods for stokes' integral. Manuscr Geod 18:227–241. https://doi.org/10.1007/BF03655315

Heiskanen WA, Moritz H (1967) Physical geodesy. Freeman, San Francisco

Hirt C, Kuhn M, Claessens S, Pail R, Seitz K, Gruber T (2014) Study of the earth's short-scale gravity field using the ERTM2160 gravity model. Comput Geosci 73:71–80. https://doi.org/10.1016/j.cageo.2014.09.001

Pavlis NK, Holmes SA, Kenyon SC, Factor JK (2012) The development and evaluation of the Earth gravitational model 2008 (EGM2008). J Geophys Res Solid Earth 117:8916

Rexer M, Hirt C, Bucha B, Holmes S (2018) Solution to the spectral filter problem of residual terrain modelling (RTM). J Geod 92:675–690. https://doi.org/10.1007/s00190-017-1086-y

Saari T, Bilker-Koivula M, Koivula H, Nordman M, Häkli P, Lahtinen S (2021) Validating geoid models with marine GNSS measurements, sea surface models, and additional gravity observations in the Gulf of Finland. Mar Geod 44(3):196–214. https://doi.org/10.1080/01490419.2021.1889727

Sandwell DT, Müller RD, Smith WHF, Garcia ES, Francis R (2014) New global marine gravity model from Cryo-Sat-2 and jason-1 reveals buried tectonic structure. Science 346(6205):65–67. https://doi.org/10.1126/science.1258213

Sansò F, Sideris MG (2013a) Geoid determination: theory and methods. Springer-Verlag, Berlin. https://doi.org/10.1007/978-3-540-74700-0

Sansò F, Sideris MG (2013b) The local modelling of the gravity field by collocation. In: Sansò F, Sideris M (eds) Geoid determination, Lecture notes in Earth system sciences, vol 110. Springer, Berlin. https://doi.org/10.1007/978-3-540-74700-0_5

Sideris MG (2013) Geoid determination by FFT techniques. In: Sansò F, Sideris M (eds) Geoid determination, Lecture notes in Earth system sciences, vol 110. Springer, Berlin. https://doi.org/10.1007/978-3-540-74700-0_10

Torge W (1989) Gravimetry. De Gruyter

Tziavos IN, Vergos GS, Grigoriadis VN (2010) Investigation of topographic reductions and aliasing effects on gravity and the geoid over Greece based on various digital terrain models. Surv Geophys 31. https://doi.org/10.1007/s10712-009-9085-z

Ullrich C, Ruess D, Butta H, Qirko K, Pavicevic B, Murat M (2016) A highly accurate absolute gravimetric network for Albania, Kosovo and Montenegro. Geophys Res Abstr Vol. 18, EGU2016-5953-1.

Vergos GS et al (2023) Development of the national gravimetric geoid model for the Kingdom of Saudi Arabia. In: Freymueller JT, Sánchez L (eds) Gravity, positioning and reference frames. REFAG 2022. International Association of Geodesy symposia, vol 156. Springer, Cham. https://doi.org/10.1007/1345_2023_214

Vergos GS, Natsiopoulos DA, Mamagiannou EG, Tzanou EA, Triantafyllou AI, Tziavos IN, Ramnalis D, Polychronos V (2025) A regional gravimetric and hybrid geoid model in Northern Greece from dedicated gravity campaigns. Remote Sens 17(2):197. https://doi.org/10.3390/rs17020197

Wang YM, Veronneau M, Huang J, Ahlgren K, Krcmaric J, Li X, Avalos-Naranjo D (2023) Accurate computation of geoid-quasigeoid separation in mountainous region – a case study in Colorado with full extension to the experimental geoid region. Journal of Geodetic Science, 13(1). https://doi.org/10.1515/jogs-2022-0128

Wong L, Gore R (1969) Accuracy of geoid heights from modified Stokes kernels. Geophys J Int 18:81–91

Zingerle P, Pail R, Gruber T, Oikonomidou X (2020) The combined global gravity field model XGM2019e. J Geod 94:66. https://doi.org/10.1007/s00190-020-01398-0

Experimental Geoid Model for Greece and Evaluation Using a Velocity Field Model

Vassilios N. Grigoriadis, Stylianos Bitharis, Vassilios D. Andritsanos, and Dimitrios N. Natsiopoulos

Abstract

In the frame of the ModernGravNet project, a new experimental gravimetric quasi-geoid model for the broader Greek region has been developed. The computations were based on the remove-compute-restore approach, 1D spherical fast Fourier transform method with the Wong-Gore modification, XGM2019 as the reference geopotential model, and residual terrain modeling (RTM) scheme. The model's evaluation was conducted using GNSS/leveling data across the Greek mainland, taking into account the geoid separation term. Furthermore, a velocity field model, derived from 16 years of continuous observations at 227 GNSS permanent stations in Greece and the adjacent areas, was incorporated into the evaluation process to assess temporal vertical changes. This model reflects the dynamic geometrical properties associated with the crust's geophysical processes. As the gravimetric geoid was computed using gravity data from various time periods and the leveling data refer to an adjustment carried out four decades ago, the velocity field has been utilized to examine temporal changes in the GNSS/leveling data relative to the gravimetric geoid heights, aiming for more precise comparisons. The results obtained show an improvement in the comparisons up to 3 cm in terms of standard deviation. The study highlights the importance of adopting a dynamic geoid model and emphasizes the need for its thorough evaluation.

Keywords

Dynamic field model · Geoid model · Greece · Heights

1 Introduction

In the so-called Colorado Experiment (Wang et al. 2021), various quasi-geoid/geoid modeling methodologies were employed by different research groups to examine discrepancies and finally compare the results provided by each method. The comparisons revealed that most of the methodologies applied led to similar solutions with differences up to 2 cm in terms of standard deviation. In that study though, as in most geoid modeling computations, changes over time were not considered. Regarding gravity data, its time-dependent part causes only small variations in the quasi-geoid/geoid model (Filmer and Featherstone 2012), while temporal/seasonal variations may be of the order of a

V. N. Grigoriadis (✉)
Laboratory of Gravity Field Research and Applications – GravLab, Department of Geodesy and Surveying, Aristotle University of Thessaloniki, Thessaloniki, Greece
e-mail: nezos@topo.auth.gr

S. Bitharis
Laboratory of Topography, Department of Geodesy and Surveying, Aristotle University of Thessaloniki, Thessaloniki, Greece

V. D. Andritsanos
Geospatial Technologies Lab, Department of Surveying and Geoinformatics Engineering, University of West Attica, Athens, Greece

D. N. Natsiopoulos
Department of Surveying and Geoinformatics Engineering, School of Engineering, International Hellenic University, Serres, Greece

J. T. Freymueller, L. Sànchez (eds.), *International Symposium on Gravity, Geoid and Height Systems 2024 (GGHS2024)*, International Association of Geodesy Symposia 158, https://doi.org/10.1007/1345_2025_287

few mm (see, e.g., Godah et al. 2017). On the other hand, the traditional vertical datums incorporating a leveling network established using benchmarks (BMs) are more susceptible to changes due to land deformations attributed to geophysical activities.

These changes over time were evaluated by Rangelova et al. (2012), among others, in the frame of a transition to a geoid-based vertical datum in North America, including changes due to post-glacial rebound, water-mass changes, ice-mass changes, and vertical crustal displacements. Their conclusion was that the geoid needs to be reevaluated every 10 years but benchmark ellipsoidal and orthometric heights require shorter timeframes. Similarly, Szelachowska et al. (2022) focused on the correction of orthometric/normal heights for temporal variations, where in their study case over Poland, differences reached some centimeters. In China, Guo et al. (2024) investigated orthometric variations using GRACE satellite models, hydrological models, and vertical displacements at GNSS stations, acknowledging a significant annual change. Furthermore, changes in orthometric/normal heights over time are still a topic of investigation (e.g., Yadeta et al. 2023) in the frame of establishing the International Height Reference System (IHRF).

Apart from continuous variations over time, there are also abrupt changes that are caused by seismic activity. These changes may range from some millimeters (especially for geoid heights) up to several centimeters (for orthometric and ellipsoidal heights), but also cumulative events may have even more significant impact (Montecino et al. 2017; Castro and de Freitas 2021). In the same category, volcanic activity may also be included. Volcanic eruptions may also trigger other events, e.g., landslides, and especially for cataclysmic events, their impact may reach several centimeters for heights and some millimeters for geoid models (see, e.g., Jacob et al. 2012).

By analyzing GNSS-derived velocities, we can gain valuable insights into the deformation and movement of the Earth's crust. The geodetic velocity field serves on plate kinematics by elucidating the motion of the Earth's surface and its relationship to tectonic processes along plate boundaries. However, it is important to note that the velocity field differs between tectonic plates, as the interactions between these plates determine the patterns of crustal movement and deformation, leading to variations in the observed velocities. For example, in Greece, a region with high tectonic and seismic activity, these velocity variations are particularly pronounced, while significant crustal deformation is observed.

Considering the above, this study aims to examine the temporal changes from a different point of view, i.e., the validation of geoid models using GNSS/leveling data by considering changes over time in the validation dataset. Therefore, in this study, an experimental geoid model over Greece is first computed. Then, the latter is validated using GNSS/leveling data with and without considering a velocity field model for changes over time, with the aim to examine possible improvements by making the comparisons more consistent and hence leading to a more robust scheme for geoid model validation.

2 Data

For the computation of the gravity field effect of the topography, a digital terrain and bathymetry model (DTBM) was used. It is a combined model that was computed from the fusion of the Copernicus DEM (DLR e.V., Airbus Defense and Space GmbH 2018), the GreekSeas DTM of the Hellenic Military Navy, and EMODnet DTM (EMODnet 2020). The selection of aforementioned models was made based on the results provided by Grigoriadis et al. (2023a), who validated different global and regional models at various test areas in Greece.

Regarding the reduction of gravity field anomalies, the XGM2019 (Zingerle et al. 2020) geopotential model was used up to degree and order 2,190. This choice was based on the results of Grigoriadis et al. (2023b), who for two pilot areas tested 12 geopotential models for the ones that when combined with an RTM scheme produced the smoothest gravity field.

The gravity data (see Fig. 1) used in this study originate from different sources, and they were collected during a time span of more than 50 years. The data were preprocessed for

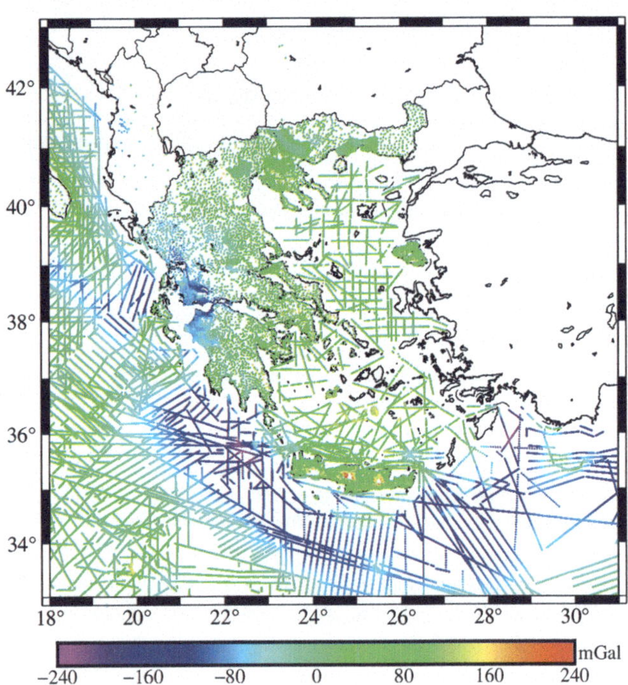

Fig. 1 Available free-air anomaly point value distribution

outlier and blunder detection. More information regarding the preprocessing scheme followed, and the various sources are given by Grigoriadis (2009). For the surrounding countries, the EGM2008 geopotential model (Pavlis et al. 2012) was used to fill in the blank areas since XGM2019 would be used in the reduction procedure. The final total number of point gravity anomaly values was more than 68,000.

For the validation of the computed geoid model, two GNSS/leveling traverses of 12.0 and 14.5 km each were used (Grigoriadis et al. 2023a), one located in Central (ModernGravNet Central) and the other in Northern Greece (ModernGravNet North), that were measured in the frame of the ModernGravNet project in 2020. Moreover, GNSS/leveling-derived geoid heights at BMs of the national trigonometric network obtained from the Hellenic Cadaster and Katsampalos et al. (2010) were also used during the validation procedure as it will be described next. The GNSS coordinates refer to the year 2016. Regarding the orthometric heights, these were derived from a network adjustment carried out in 1989 (see Takos 1990) and are based on trigonometric and leveling measurements carried out since 1962. The orthometric heights are tied to the tide-gauge station at Piraeus Port, in Attica, which is the starting point of the Greek Vertical Datum. Therefore, the orthometric heights are considered to follow the mean-tide approach (Mäkinen 2021).

The velocity field model computed by Bitharis et al. (2024) (see Fig. 2 and Table 1) was used to derive vertical temporal changes for geometrical heights at selected BMs in the Hellenic area. The estimation of the vertical geodetic velocities and their uncertainties was carried out from the daily position time series analysis, spanning 16 years from 2001 till 2016, using a Median Interannual Difference Adjusted for Skewness (MIDAS) robust trend estimator as described by Blewitt et al. (2016). GNSS time series analysis is sensitive to periodic or geophysical signals as well as to discontinuities caused by equipment changes or coseismic displacements. The MIDAS algorithm was selected because, as a robust regression estimator, it is not affected by such discontinuities. Additionally, it provides more realistic uncertainty estimates for geodetic velocity compared to the ordinary least squares (OLS) regression model, which can yield overoptimistic results in long-term time series. The vertical geodetic velocity model is defined strictly within the Greek territory with a cell dimension of 0.3° × 0.3° (Lon. × Lat.), using bi-cubic spline interpolation. According to the model, the magnitude of vertical velocities in Greece varies between areas, with values that can range from −1 to 0.9 mm/year. The highest values were recorded on the island of Santorini during the period of volcanic unrest from 2011 to 2012 (Bitharis 2021). It should be noted that no specific pattern can be observed in Greece, although in some parts of the mainland, there is a subsidence tendency.

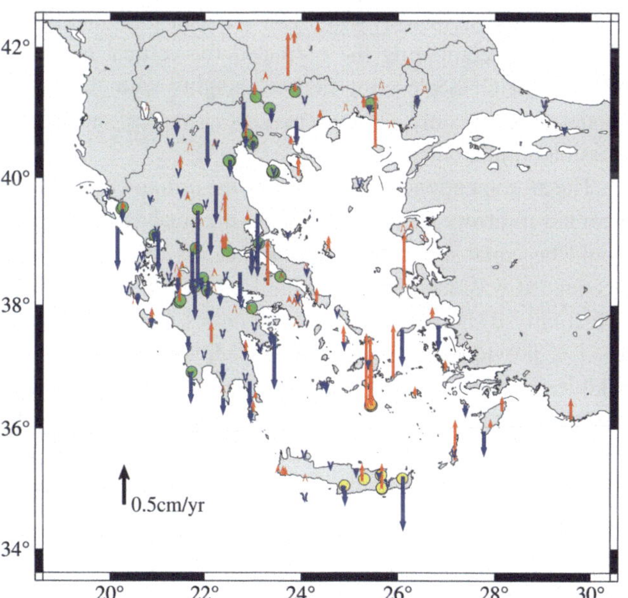

Fig. 2 The vertical velocity field and selected BMs of the national trigonometric network (circles – yellow: at Crete Island, orange: at Santorini Island, green: at mainland)

Table 1 Statistics of velocity field model (up) (unit: mm/year)

	Min	Max	Mean	Std
Up	−10.0	9.2	0.0	2.6

3 Methodology and Results

3.1 Geoid Computation

The methodology for computing the geoid for the wider Hellenic area is the one described by Grigoriadis et al. (2023b). More specifically, XGM2019 up to d/o 2,190 was removed from the available gravity data to produce the reduced field; the RTM effects were computed, using the combined DTBM and the mean DTBM, and were removed to obtain the residual field (see Table 2). Among the two approaches, the classical and spectral RTM scheme, the classical one (Forsberg 1984) was used with the aid of a mean DTBM after low-pass filtering the combined one using a moving average filter. As was previously mentioned, the decision was based on the results provided by Grigoriadis et al. (2023b).

Table 2 Statistics of original (free-air anomalies), reduced, and residual gravity field (unit: mGal)

	Min	Max	Mean	Std
Original	−233.49	236.87	−21.13	71.48
Reduced	−121.34	72.38	4.89	10.16
Residual	−31.10	59.59	5.85	8.55

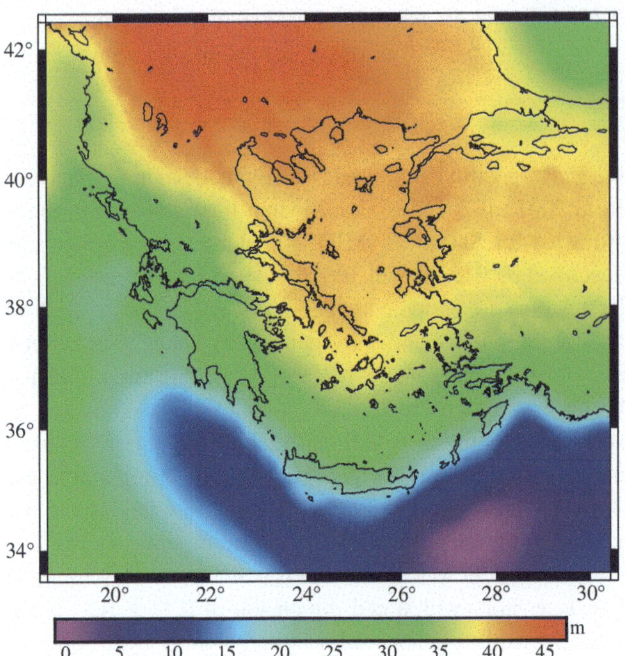

Fig. 3 The experimental geoid model XHGM2023

Table 3 Statistics of XHGM2023 (unit: m)

	Min	Max	Mean	Std
XHGM2023	−0.071	45.945	28.958	11.851

After that and as the original data had different spatial resolutions (area dependent), mean values were computed and gridded using spline interpolation. From the gridded mean residual field and using the 1D spherical FFT method, the residual quasi-geoid heights were obtained. It should be noted that the Wong-Gore kernel modification was used with a cutoff degree equal to 200 following Grigoriadis et al. (2023b). After restoring the contribution of the RTM and that of the geopotential model, the geoid-quasigeoid separation term was also added following the approach described by Flury and Rummel (2009). The final geoid model, named XHGM2023, is depicted in Fig. 3, while its statistical results are provided in Table 3. All the aforementioned computations were carried out following the zero-tide approach.

The RTM effects and geoid were computed using the GRAVSOFT software package (Tscherning et al. 1992), with all other computations performed using custom in-house software.

3.2 Validation

For the validation of XHGM2023, all quantities were converted to the mean-tide approach, i.e., geoid heights were converted from zero-tide to mean-tide according to Eq. (15) from Ekman (1989) and GNSS geometric heights from tide-free to mean-tide by combining Eqs. (23) and (25) from Mäkinen (2021). Hence, the statistical results of the validation against the available GNSS/leveling-derived geoid heights are presented in Table 4. It should be noted initially that the validation was not carried out for the Greek Islands but only for the mainland. This was decided because the vertical datum of the Islands is not connected to the mainland one. The minimum and maximum values for the BMs measured by the Hellenic Cadaster depict that there are problematic points that need further consideration, but this can be done only by remeasuring them.

Table 4 Validation statistics of XHGM2023 against GNSS/leveling data over the Greek mainland (unit: m)

	Points	Min	Max	Mean	Std
Hellenic Cadaster	1,691	−0.536	0.775	−0.045	0.137
ModernGravNet North	175	−0.179	0.143	−0.050	0.066
ModernGravNet Central	138	−0.233	−0.004	−0.130	0.040

The validation of the geoid model has one unaddressed issue. All quantities (orthometric, geometric, and geoid heights) refer to different time periods. For the geoid heights, using the described solution, it is impossible to define a specific reference year. Consequently, it is assumed that the long-term changes of the Earth's gravity field are minimal, i.e., it will be considered as a static field. For the orthometric heights, it will be assumed that their year of reference is 1989, the year the network adjustment took place. Therefore, the geometric heights, stemming from GNSS data that have a known reference epoch, will be converted backwards in time to make them more "compatible" to the orthometric heights. The velocity field model spans a period of only 16 years, and so there is still a gap between 1989 and 2001. Thus, when comparing the corrected for vertical temporal changes in GNSS/leveling-derived heights with the geoid model heights, only a trend may be revealed showing a possible improvement.

The aforementioned adjustment of the orthometric heights resulted to errors up to 2 cm, while the final BM accuracy could be equal up to several centimeters. Taking this into account, it was deemed preferable for the next validation to select only BMs with the highest accuracy (less than 1 cm), as it is provided from the network adjustment. Moreover, additional criteria were placed to avoid incorporating additional errors in the validation. Therefore, it was decided to only include BMs that are located within a distance less than 5 km from the stations used for deriving the velocity field model, excluding those stations whose GNSS solution uncertainty was higher than the computed velocity. This was made to avoid interpolation errors, i.e., errors due to the low resolution of the model, and include only the most accurate velocity predictions. Last, BMs that were determined nowadays as problematic after field inspection were also removed.

Table 5 Validation statistics of XHGM2023 against selected GNSS/leveling data before and after correcting with the velocity field model for the Greek mainland, Santorini island, and Crete island (unit: m)

		Min	Max	Mean	Std
Mainland	Before	−0.145	0.239	0.023	0.099
	After	−0.114	0.219	0.018	0.087
Santorini	Before	−0.053	0.007	−0.016	0.033
	After	0.079	0.126	0.101	0.024
Crete	Before	−0.006	0.286	0.106	0.112
	After	−0.013	0.188	0.107	0.082

Following the established criteria, a total of 30 BMs were selected throughout the Greek mainland. For the selected BMs (see Fig. 2), their geometric height was converted to refer to the year 2001, i.e., 15 years back using the velocity field model data. The derived GNSS/leveling geoid heights were compared to those of the geoid model, and the statistics are provided in Table 5. The improvement achieved is 1.2 cm in terms of standard deviation and 5.1 cm in terms of range. Although the average value also decreased by 5 mm, it should not be considered as it is mainly attributed to biases.

As it was previously mentioned, the Greek islands have their own vertical datum defined by a local tide gauge. Therefore, it was decided for the validation using the velocity field model to examine them separately. By following the same criteria as before, only two islands matched them and had an adequate number of BMs, i.e., Santorini (3 BMs) and Crete (6 BMs). It should be noticed that both islands are very tectonically active. Table 5 lists the results obtained before and after applying the corrections from the velocity field model. As in the mainland, an improvement is noticed. The standard deviation improves by 0.9 cm and 3 cm and ranges by 1.3 cm and 9.1 cm for Santorini and Crete, respectively.

4 Discussion and Conclusion

The validation of XHGM2023 showed that there are significant problems in the trigonometric BMs, whose comparison against the geoid model revealed differences up to 0.7 m. When compared to recent measurements (ModernGravNet North and Central), these differences dropped to about 12 cm.

The application of the velocity field model led to improved results for the selected BMs. This shows that geoid validation results might not be as expected when not considering temporal variations. Although the study makes many assumptions, i.e., static gravity field, orthometric heights referring to a specific epoch, and no horizontal variations over time, the reduction shown in the statistics is an indication that geoid model validation is more properly carried out when referring all quantities (geoid and GNSS/leveling heights) to a common epoch. Additionally, this study is in line with previous studies that stress the necessity to monitor changes over time, both in the Earth's gravity field and on its surface.

The velocity field vertical component, in comparison to the horizontal one, is quite more sensitive to the impacts of equipment changes and periodic signals in Continuously Operating Reference Stations (CORS). It has been observed that at some stations, the change of equipment might affect the time series, while special incidents, e.g., earthquakes, require special attention. Significantly more years of data are needed for a reasonable estimate of vertical velocities, and more intricate mathematical models are needed to predict the trend. The resolution of the model also plays an important role. When selecting BMs at distances of more than 5 km from the CORS stations, the velocity field prediction led in many cases to unreasonable results. As the resolution of the model cannot get better, as more CORS stations are required, it is necessary to combine information for vertical crust movement from other sources, like InSAR data, although at present further improvements in terms of measurement stability and accuracy are required.

In conclusion, geoid validation should be carried out after considering variations of heights and the Earth's gravity field over time, while improvements in the validation could reach up to 3 cm in terms of standard deviation and even several centimeters in terms of range, when considering the areas examined in this study.

Acknowledgements Part of this work was carried out in the frame of the ModernGravNet research project that was supported by the Hellenic Foundation for Research and Innovation (H.F.R.I.) under the "1st Call for H.F.R.I. Research Projects to Support Faculty Members & Researchers and the Procurement of High-Cost Research Equipment Grant" (Project Number: 1550).

References

Bitharis S (2021) Study of the geodynamic field in Greece using modern satellite geodetic methods. Ph.D. dissertation, Department of Geodesy & Surveying, School of Rural and Surveying Engineering, Aristotle University of Thessaloniki, Thessaloniki, Greece (in Greek). https://doi.org/10.12681/eadd/50310

Bitharis S, Pikridas C, Fotiou A, Rossikopoulos D (2024) GPS data analysis and geodetic velocity field investigation in Greece, 2001–2016. GPS Solut 28(16). https://doi.org/10.1007/s10291-023-01549-8

Blewitt G, Kreemer C, Hammond WC, Gazeaux J (2016) MIDAS robust trend estimator for accurate GPS station velocities without step detection. J Geophys Res Solid Earth 121(3):2054–2068. https://doi.org/10.1002/2015JB012552

Castro HDM, de Freitas SRC (2021) Implications of the temporal variations of geoid heights over South American vertical reference frames. Arab J Geosci 14:185. https://doi.org/10.1007/s12517-021-06562-0

DLR e.V., Airbus Defense and Space GmbH (2018) Copernicus DEM GLO-30. Provided under COPERNICUS by the European Union and ESA; all rights reserved. https://spacedata.copernicus.eu. Accessed 20 May 2021

Ekman M (1989) Impacts of geodynamic phenomena on systems for height and gravity. Bull Geod 63:281–296. https://doi.org/10.1007/BF02520477

EMODnet Bathymetry Consortium (2020) EMODnet Digital Bathymetry (DTM)

Filmer MS, Featherstone WE (2012) Three viable options for a new Australian vertical datum. J Spat Sci 57(1):19–36. https://doi.org/10.1080/14498596.2012.679248

Flury J, Rummel R (2009) On the geoid–quasigeoid separation in mountain areas. J Geod 83:829–847. https://doi.org/10.1007/s00190-009-0302-9

Forsberg R (1984) A study of terrain reductions, density anomalies and geophysical inversion methods in gravity field modelling. Technical Report, OSU Report No. 355, Department of Geodetic Science, The Ohio State University, Columbus, Ohio, USA

Godah W, Szelachowska M, Krynski J (2017) Investigation of geoid height variations and vertical displacements of the Earth surface in the context of the realization of a modern vertical reference system: a case study for Poland. In: Vergos G, Pail R, Barzaghi R (eds) International symposium on gravity, geoid and height systems 2016. International Association of Geodesy symposia, vol 148. Springer, Cham. https://doi.org/10.1007/1345_2017_15

Grigoriadis VN (2009) Geodetic and geophysical approach of the Earth's gravity field and applications in the Hellenic area. Ph.D. dissertation, Department of Geodesy & Surveying, School of Rural and Surveying Engineering, Aristotle University of Thessaloniki, Thessaloniki, Greece (in Greek). https://doi.org/10.12681/eadd/30653

Grigoriadis VN, Andritsanos VD, Natsiopoulos DA (2023a) Validation of recent DSM/DEM/DBMs in test areas in Greece using spirit leveling, GNSS, gravity and echo sounding measurements. ISPRS Int J Geo Inf 12(3):99. https://doi.org/10.3390/ijgi12030099

Grigoriadis VN, Andritsanos VD, Natsiopoulos DA, Vergos GS, Tziavos IN (2023b) Geoid studies in two test areas in Greece using different geopotential models towards the estimation of a reference geopotential value. Remote Sens 15(17):4282. https://doi.org/10.3390/rs15174282

Guo J, Shi T, Jin X, Liu X, Zhao B, Qiao X (2024) Geodetic analysis of orthometric height variations in mainland China using GRACE, hydrological models, and GPS data. IEEE Trans Geosci Remote Sens 62(1–15):4504815. https://doi.org/10.1109/TGRS.2024.3386879

Jacob T, Wahr J, Gross R et al (2012) Estimating geoid height change in North America: past, present and future. J Geod 86:337–358. https://doi.org/10.1007/s00190-011-0522-7

Katsampalos K, Kotsakis C, Gianniou M (2010) Hellenic Terrestrial Reference System 2007 (HTRS07): a regional realization of ETRS89 over Greece in support of HEPOS. Bollettino Geod Sci Affini 69:329–347

Mäkinen J (2021) The permanent tide and the International Height Reference Frame IHRF. J Geod 95:106. https://doi.org/10.1007/s00190-021-01541-5

Montecino HD, de Freitas SRC, Báez JC, Ferreira VG (2017) Effects on Chilean Vertical Reference Frame due to the Maule Earthquake co-seismic and post-seismic effects. J Geodyn 112:22–30. https://doi.org/10.1016/j.jog.2017.07.006

Pavlis NK, Holmes SA, Kenyon SC, Factor JK (2012) The development and evaluation of the Earth Gravitational Model 2008 (EGM2008). J Geophys Res 117(B04406). https://doi.org/10.1029/2011JB008916

Rangelova E, Wal W, Sideris M (2012) How significant is the dynamic component of the North American vertical datum? J Geod Sci 2(4):281–289. https://doi.org/10.2478/v10156-012-0005-7

Szelachowska M, Godah W, Krynski J (2022) Contribution of GRACE satellite mission to the determination of orthometric/normal heights corrected for their dynamics – a case study of Poland. Remote Sens 14(17):4271. https://doi.org/10.3390/rs14174271

Takos IK (1990) The new adjustment of Greek Trigonometric Networks. Tech Chron A 10(3):305–331 (in Greek)

Tscherning CC, Forsberg R, Knudsen P (1992) The GRAVSOFT package for geoid determination. In: Holota P, Vermeer M (eds) 1st Continental workshop on the geoid in Europe, Prague, 7–9 Jun 1993, pp 327–334

Wang YM, Sánchez L, Ågren J et al (2021) Colorado geoid computation experiment: overview and summary. J Geod 95:127. https://doi.org/10.1007/s00190-021-01567-9

Yadeta SM, Godah W, Szelachowska M, Fotopoulos G (2023) Assessment of temporal variations of orthometric/normal heights at proposed International Height Reference Frame sites using GRACE/GRACE-FO. Surv Rev 56(398):489–499. https://doi.org/10.1080/00396265.2023.2293367

Zingerle P, Pail R, Gruber T, Oikonomidou X (2020) The combined global gravity field model XGM2019e. J Geod 94:66. https://doi.org/10.1007/s00190-020-01398-0

Part III

Gravity for Climate and Geohazards

MAGIC's Ability to Estimate the Long-Term Trend in Climate-Related Mass Transport Signals

M. Schlaak and R. Pail

Abstract

GRACE and its successor, GRACE-FO, have successfully observed Earth's gravity field changes for over two decades. For the continuous data collection, GRACE-C (NASA/DLR) is planned to be launched in a polar orbit as its predecessor in 2028, followed by ESA's Next Generation Gravity Mission (NGGM) with a similar architecture but with an inclined orbit. Together, these two missions form the Mass change And Geosciences International Constellation (MAGIC), expected to deliver the anticipated higher temporal and spatial resolution for temporal variable gravity solutions. At present, the data recording of satellite gravity missions has improved our understanding of large-scale processes of the water cycle. Additional value will arise through the continuation of satellite gravity observations. A longer observation record will lead to significance concerning climate-related mass transport signals such as the essential climate variable terrestrial water storage (TWS) as well as the mass component of the sea-level rise. This contribution evaluates current and future satellite gravity missions in multidecadal numerical closed-loop simulations, showing their potential to resolve long-term trends. Individual error sources are investigated to show their contribution to the uncertainty in trend estimates, considering atmosphere and ocean de-aliasing errors (AOerr), ocean tide error (ΔOT), signal aliasing (ΔH_{al}), erroneous trends in the de-aliasing model (AO-DEAL), and instrument noise (ϵ). The results show that AOerr is the dominating error in the trend estimation for higher degrees. Investigation of several decades has shown that AOerr reduces while ΔOT converges after 20 years. The single-pair trend estimate is mainly affected by AOerr over the period of the Earth System Model (ESM). For low degrees, the erroneous trend of AO-DEAL becomes the limiting factor, especially as the double pair, with its intrinsic de-aliasing capabilities, reduces all other error contributions.

Keywords

ESA ESM · Long-term trend · MAGIC · Satellite geodesy · Satellite gravimetry

1 Introduction

In 2002, GRACE was launched to observe changes in Earth's gravity field until 2017. Since 2018, GRACE-FO has taken over, delivering high-quality observations of monthly mass changes. Currently, two complementary missions are in preparation to continue the time series of more than 20 years of observations. GRACE-C (NASA/DLR) is expected to be launched in 2028 after the solar maximum to ensure continuity of mass transport observations in a similar orbital plane with a polar inclination. In 2032, ESA's Next Generation Gravity Mission (NGGM) is expected to be launched. NGGM follows the same measurement

M. Schlaak (✉) · R. Pail
Astronomical and Physical Geodesy, TUM School of Engineering and Design, Technical University of Munich, Munich, Germany
e-mail: Marius.schlaak@tum.de

J. T. Freymueller, L. Sànchez (eds.), *International Symposium on Gravity, Geoid and Height Systems 2024 (GGHS2024)*, International Association of Geodesy Symposia 158, https://doi.org/10.1007/1345_2025_286

principle but is located on a more inclined ($i = 70$) orbital plain (Heller-Kaikov et al. 2023). Both missions are planned to overlap for at least 4 years (Daras et al. 2023), forming the MAGIC constellation. Due to the improved observation geometry and the additional number of available observations, MAGIC is expected to deliver higher temporal and spatial resolution.

This study compares the resolution capabilities of a GRACE-like single-pair mission and a MAGIC double-pair mission, focusing on the long-term trend retrieval and the impact of different error sources in the observation system on long-term trend estimates. After illustrating the two satellite constellations, the applied simulation environment is shortly introduced, followed by a description of the typical error sources in satellite gravimetry. The results section presents the individual contribution of the error sources to the respective missions.

2 Methodology

In this study, a reduced-scale closed-loop simulation environment (Hauk 2020; Murböck 2015), adapted for long-term trend estimation (Schlaak et al. 2022), is used to evaluate the performance of a single-pair constellation as well as of a double-pair constellation, in retrieving long-term trend estimates. The error sources considered in this study are instrument noise (ϵ), background model errors (AOerr, ΔOT, AO-DEAL), and signal aliasing effects (ΔH_{al}).

The established Earth System Model (ESM) of the European Space Agency (ESA) (Dobslaw et al. 2015) is considered for simulating the mass change signals. The ESM spans 12 years and considers mass change signals in atmosphere (A), ocean (O), hydrology (H), glaciers and ice sheets (I), as well as solid earth (S).

For the instrument error (ϵ), the key payloads, namely accelerometer and laser ranging instruments, are considered. The background model error consists of the ocean tide model error (ΔOT) and the de-aliasing product error (DEAL + AOerr) from the ESM (Dobslaw et al. 2016). To evaluate the effect of the chosen input model, a second input model is introduced, which consists of components of the terrestrial water storage extracted from a climate model (GFDL-CM4) (Guo et al. 2018) combined with trends over the ice sheets of Greenland and Antarctica from GRACE.

2.1 Satellite Mission Characteristics

The single- and double-pair constellations investigated in this study are based on the same LL-SST observing system (Tapley et al. 2004). Each pair consists of two satellites in the same near-circular orbital plane, orbiting with a distance of ca. 220 km and a ground repeat of 30 days.

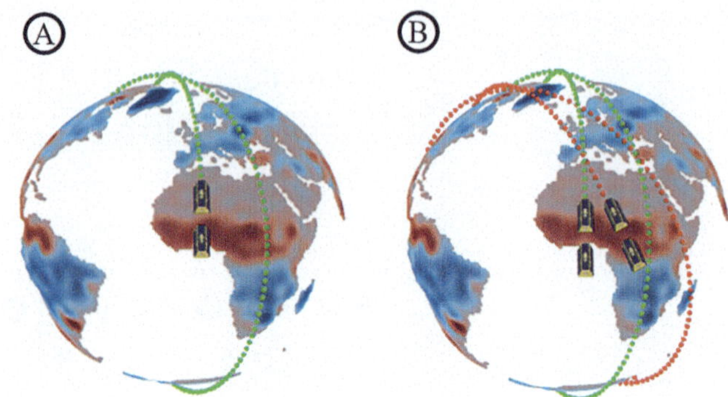

Fig. 1 Mission constellations are shown for the GRACE-like single-pair case (**a**) and the MAGIC double-pair case (**b**)

The orbits follow the MAGIC mission design study by Heller-Kaikov et al. (2023). For the single pair (Fig. 1a), a GRACE-like orbit is assumed with an inclination of 89° and a mean altitude of 488 km. The double pair (Fig. 1b) combines the observations of the single-pair constellation with a second inclined pair. The inclined pair orbits with an inclination of 70° at 397 km.

2.2 Simulation Environment for Climate Estimations

To ensure a realistic representation of the GRACE-type observation system, an adapted version of the reduced-scale simulator (RSS) is used (Murböck 2015; Schlaak et al. 2022). Synthetic observations are computed using the acceleration approach for a given input signal (Rummel 1979). The observations (l_{SST}) can be directly derived from the difference in the gradient of the disturbing potential (∇T) acting on the two satellites along the line of sight

$$l_{\mathrm{SST}} = \nabla T_2 \cdot e_{2,1} - \nabla T_1 \cdot e_{1,2} \quad (1)$$

with $e_{2,1}$ and $e_{1,2}$ being the unit vectors of the respected satellite positions. The gravitational accelerations are computed from the temporal variable gravity field of interest (Sect. 2.3) and the noise components expected for the constellations (Sect. 3). The gravity field is retrieved up to spherical harmonic degree 90 in a least squares adjustment using a standard Gauss-Markov model (Niemeier 2008).

2.3 Input Models

In the case of satellite gravity simulations, typically, the signals of interest are mass changes in continental hydrology (H), ice sheets (I), and solid earth (S). AO signals are usually taken as a priori knowledge from background models

(Shihora et al. 2022); therefore, only their error is introduced in this study (Sect. 3.3). A global time series of representative mass changes considering HIS components over the time period 1995–2006 is provided by ESA's ESM (Dobslaw et al. 2015). The ESM is available up to spherical harmonic degree and order 180 with a temporal resolution of 6 h.

According to Dobslaw et al. (2015), the individual components of the ESM consider ice mass changes over Greenland and Antarctica based on regional climate models (Ettema et al. 2009). The hydrology is driven by the land surface discharge model (Dill 2008), which considers soil moisture, snow storage, wetlands, rivers, and lakes. The solid Earth component is represented by glacial isostatic adjustment as well as co- and post-seismic deformation of the Sumatra-Andaman earthquake (magnitude 9.1).

As an additional model available for several decades, climate model outputs (GFDL-CM4) (Guo et al. 2018) in combination with current ice sheet trends from mascon solutions are used in this study. The climate model considers components of terrestrial water storage (TWS), namely soil moisture and snow. The GFDL-CM4 model was selected out of the CMIP6 ensemble because its long-term behavior agrees with the observational record of the GRACE mission (Jensen et al. 2019). The climate model represents H and I components with a temporal resolution of 1 month. The second model is included to evaluate whether climate models can give similar realistic error behavior to the ESM.

Overall, depending on the assumed input model, the target signal is HIS, represented either in the ESM or in the climate signal, as described above. Uncertainties of the observations accumulate through different error sources, which are described in the following section.

3 Error Sources

Typical error sources in satellite gravimetry can be categorized as instrumental errors (Sect. 3.1) of the payload instruments, such as accelerometer and laser ranging instruments, that contribute uncertainties to the range measurements. The second error source is temporal aliasing, which has tidal (Sect. 3.2) and nontidal (Sect. 3.3) contributions. Temporal aliasing is the main error source in satellite gravimetry products and is caused by high-frequency mass transport signals that are over- or underestimated due to insufficient sampling.

3.1 Instrument Errors

Instrument errors are represented as the uncertainty in non-gravitational acceleration measurements from Heller-Kaikov et al. (2023), but projected in the line of sight (LOS). The accelerometer uncertainties are expressed in amplitude spectral densities (ASD) for the polar pair (P1) and the inclined pair (P2) in $[\mathrm{m}/\left(\mathrm{s}^2\sqrt{\mathrm{Hz}}\right)]$:

$$\mathrm{AS\,D_{accP1}} = \sqrt{2}\cdot 10^{-10}\sqrt{1+\frac{0.005\,\mathrm{Hz}}{f}} \tag{2}$$

$$\mathrm{AS\,D_{accP2}} = 10^{-11}\cdot\sqrt{\frac{\left(\frac{10^{-3}\mathrm{Hz}}{f}\right)^2}{\left(\left(\frac{10^{-5}\mathrm{Hz}}{f}\right)^2+1\right)}+1+\left(\frac{f}{10^{-1}\mathrm{Hz}}\right)^4} \tag{3}$$

For the ranging instruments, updated noise assumptions are considering the inter-satellite distance ($L = 220$ km) in $[\mathrm{m}/\sqrt{\mathrm{Hz}}]$ (Heller-Kaikov et al., personal communication, March 2025):

$$\mathrm{AS\,D_{lriP1}} = \sqrt{\left(L\cdot\frac{10^{-15}}{\sqrt{f}}\right)^2+\left(\frac{10^{-12}}{f^2}\right)^2} \tag{4}$$

$$\mathrm{AS\,D_{lriP2}} = L\cdot 2\cdot 10^{-13}\cdot\sqrt{1+\left(\frac{0.01\mathrm{Hz}}{f}\right)^2} + \sqrt{1+\left(\frac{0.001\mathrm{Hz}}{f}\right)^2} \tag{5}$$

The combined instrument noise (ϵ) is computed into accelerations and added to the observation time series of acceleration differences between the satellites.

3.2 Tidal Aliasing Errors

In gravity field retrieval, the effect of ocean tides (OT) is modeled by estimating the potential changes due to the major frequencies at which OT occur. To represent the model error (ΔOT), the differences between OT models EOT11a (Savcenko et al. 2012) and GOT4.7 (Ray 1999) are used in this study. As OT appear on distinct excitation frequencies, their effect is expected to stay significant over long timescales. Imperfections in the OT models are propagated to the gravity solutions as temporal aliasing. These effects contribute to the long-term trend uncertainties in single-pair and double-pair missions.

3.3 Nontidal Aliasing Errors

Nontidal aliasing is mainly caused by high-frequency mass transport events that are not resolved by the observing system. Typically, atmospheric (A) and oceanic (O) components are compensated by de-aliasing products (AOD1B, Shihora et al. 2022). The background model errors on small and large spatial scale events lead to significant striping effects in the gravity solution. To assess their impact, the de-aliasing background model (DEAL) and its errors (AOerr) provided by the ESM (Dobslaw et al. 2015, 2016) are considered in this study. In addition, to evaluate the effects of AOerr over longer periods, the updated AO error, namely AOe07, is investigated (Shihora et al. 2024). While AOerr is a dominant error source in monthly gravity field data, the error behavior of long-term trend estimates over large time spans has not been investigated.

DEAL consists of an AO model neglecting small-scale events and barystatic sea-level variability (Dobslaw et al. 2016), leading to a residual erroneous trend in the de-aliased observation, which is believed to be too pessimistic considering the latest background model release (Shihora et al. 2022). In addition to the de-aliasing errors, additional uncertainties are incorporated by unmodeled high-frequency events in the target signal that are not removed by a priori models. In the case of the ESM and the GFDL model as inputs, the main contribution will come from hydrology temporal aliasing (ΔH_{al}).

Consolidating all error sources, the simulated observations contain all error components in addition to the target signal:

$$\mathrm{HIS} + \epsilon + \Delta\mathrm{OT} + \mathrm{AO} - \mathrm{DEAL} + \mathrm{AOerr} + \Delta H_{\mathrm{al}}$$

In the following, the simulation results are evaluated for retrieving long-term trends considering single-pair and double-pair constellations as observing techniques. In addition, the contribution of the individual error sources is illustrated for each constellation.

4 Results

The results show the error contribution of individual error sources, their combination, and signal aliasing effects on long-term trend estimates in mm/year equivalent water height (EWH). By looking at observations spanning three decades, the long-term behavior of the main error contributors and their implications are discussed.

4.1 Contributions of Individual Error Sources

The results presented in this subsection consider direct long-term trend estimations of 12 years of observation if not explicitly specified different.

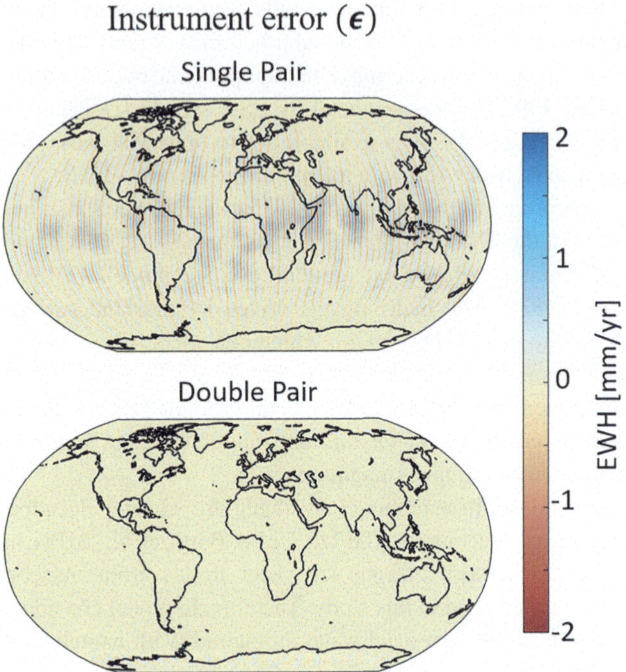

Fig. 2 Instrument errors propagated for single-pair (top) and double-pair (bottom) mission in mm/year EWH

4.1.1 Instrument Errors

Instrument errors (ϵ) consider acceleration and ranging instrument. Their contribution to the error budget is small (see Fig. 2, max. 1.67 mm/year for a single pair with a global rms of 0.28 mm/year; max 0.50 mm/year for a double pair with a global rms of 0.07 mm/year) and further reduced with increased number of observations.

4.1.2 Tidal Background Model Errors

Ocean tide errors (ΔOT) show higher magnitude effects in satellite observations.

As can be seen in Fig. 3, ΔOT leads to global striping of the single-pair solution, with higher amplitudes at the equatorial areas, where the sampling is sparser in time and space compared to the polar areas. The striping of the double-pair solution is smaller and is confined to areas of larger OT signals, such as Oceania and Southeast Asia. Moreover, polar areas covered only by observations of the polar pair show increased error levels.

4.1.3 Nontidal Background Model Errors

Nontidal background model errors (AOerr) lead to intense striping patterns. The effects of temporal aliasing occur globally for both satellite constellations, as can be seen in Fig. 4 (left column). The global rms for a single-pair mission is 28.12 mm/year, while the rms of the double-pair mission is 0.83 mm/year. For the double pair, the error patterns are elevated in the tropics, where higher magnitude

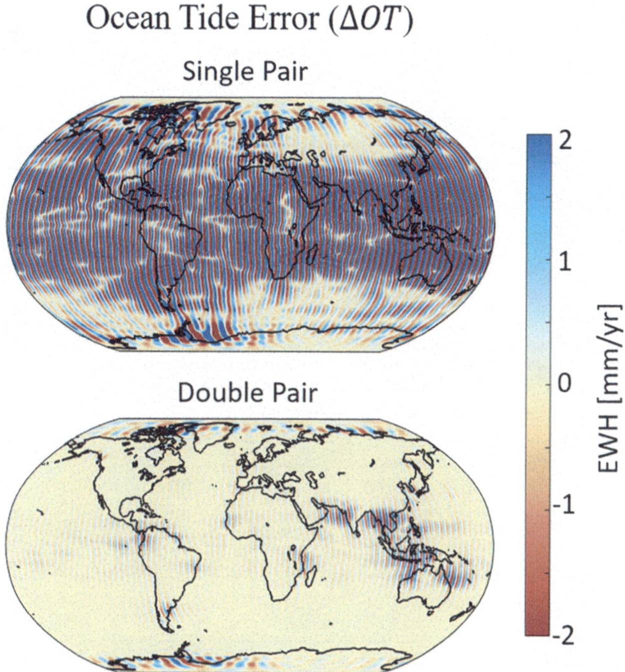

Fig. 3 EWH map of ΔOT. For a single-pair mission (top), the global rms is 6.66 mm/year with a max of 101.65 mm/year. For a double-pair mission (bottom), the rms constitutes 0.42 mm/year, with a max of 6.85 mm/year

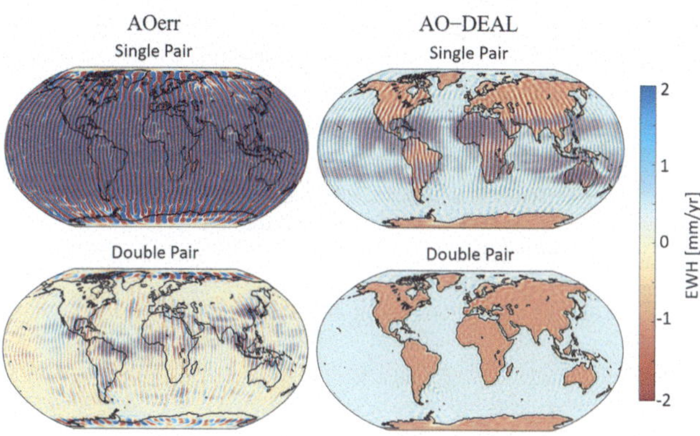

Fig. 4 Nontidal temporal aliasing error contributions to single-pair mission (top rows) and double-pair mission (bottom rows). The contributions of AOerr are presented in the left column, while the de-aliasing product difference to AO (AO-DEAL) is illustrated in the right column

atmospheric events occur regularly, and the transition zone between the double-pair and single-pair coverage occurs at latitudes ±70°.

Focusing on the right column of Fig. 4, the effects of the DEAL model on satellite gravimetry missions are shown. For a single-pair mission, the model differences cause some temporal aliasing, but the magnitude is negligible compared to the temporal aliasing effects caused by AOerr (Fig. 4, top left). More significant and visible for both satellite constellations, the model differences cause a trend signal

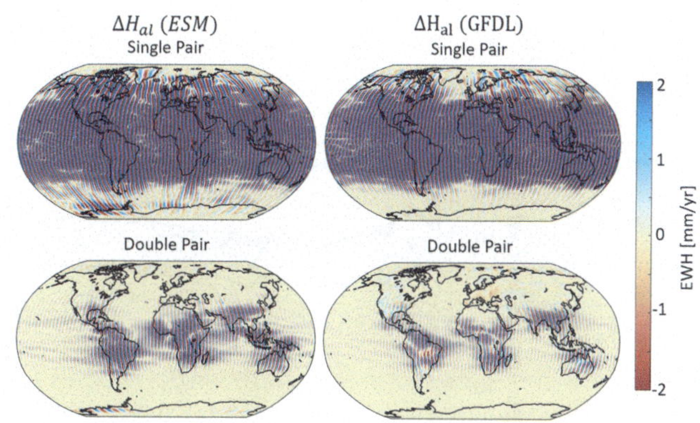

Fig. 5 Signal aliasing effects are displayed for the ESM (left column) and the GFDL (right column) for single-pair (top row) and double-pair constellation (bottom row)

that manifests as a residual trend signal in the solution independently of the mission performance.

4.1.4 Signal Aliasing

In addition to background model errors, target signal aliasing can introduce errors caused by temporal undersampling of the target signal. As described in Sect. 2.3, this study compares the signal aliasing (ΔH_{al}) effects of the target signal of the ESM with the climate model GFDL.

As can be seen in Fig. 5, both input models cause aliasing effects of similar magnitude to the respective observing systems. For a single pair, the global rms constitutes 55.23 mm/year for the ESM and 59.10 mm/year for the GFDL. The double-pair observing system leads to rms of 1.25 mm/year (ESM) and 0.64 mm/year (GFDL). Here, the signal aliasing effects are especially visible in areas with stronger hydrological signals (South America, Africa, and Southeast Asia).

4.2 Sum of Error Contributions

To evaluate the global effect of the individual error contributions to long-term trend estimates, the degree amplitudes are shown in Fig. 6 for a single-pair mission and in Fig. 7 for a double-pair mission.

In each constellation, the individual error contributors play different roles. For the single-pair constellation, the dominant error source is AOerr over most of the spectrum. In the low degrees (do < 15), the induced trend of the model difference (AO-DEAL) is dominating. As can be seen in Fig. 6, the full noise simulation scenario overlaps with the sum of all error contributors in the single-pair case.

For the double-pair constellation, the error behavior differs from the single pair. The overall error magnitude is strongly reduced, especially in high degrees. However, the

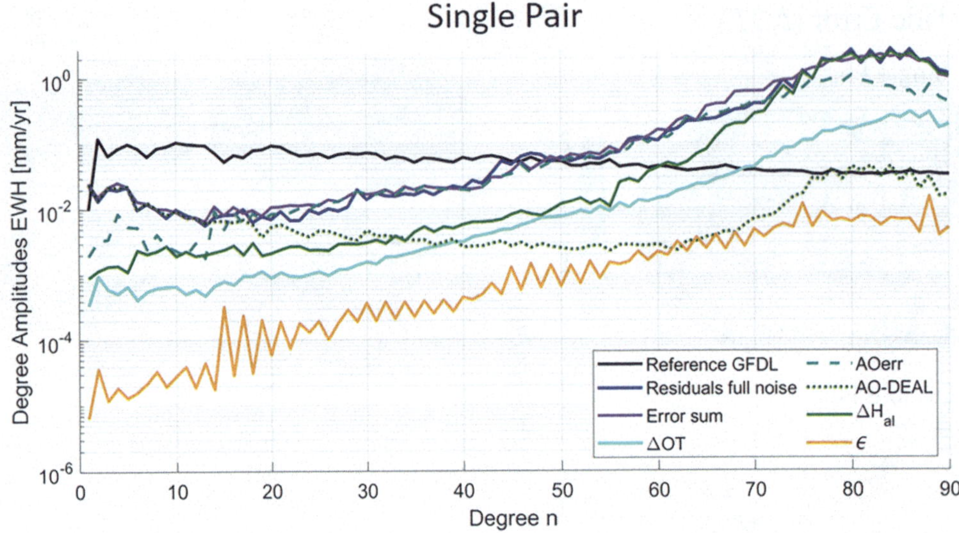

Fig. 6 Degree amplitudes for a single-pair scenario. Displayed are the individual components for ϵ, AOerr, AO-DEAL, ΔOT, and ΔH_{al}, as well as their sum and the residuals considering a full noise scenario

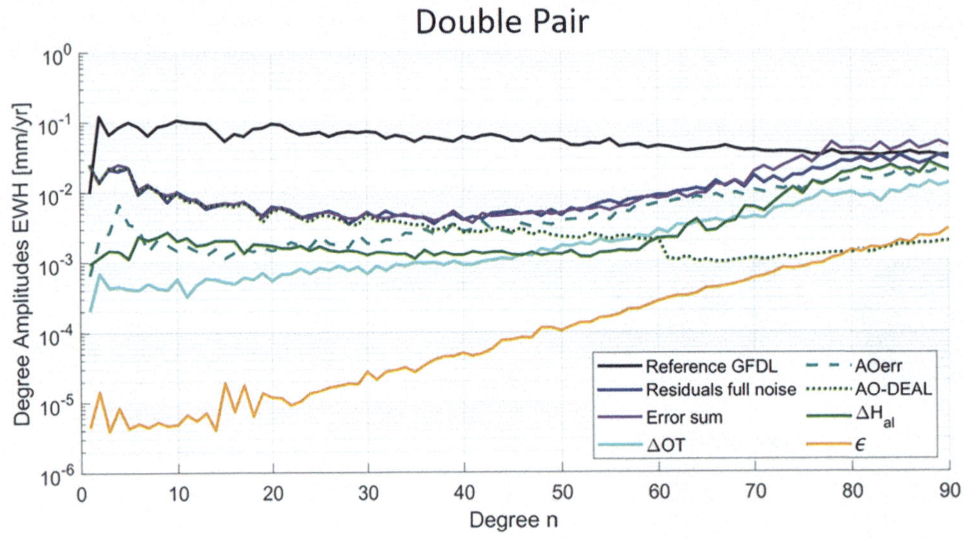

Fig. 7 Degree amplitudes for a double-pair scenario. Displayed are the individual components for ϵ, AOerr, AO-DEAL, ΔOT, and ΔH_{al}, as well as their sum and the residuals considering a full noise scenario

role of the individual error contributors also changes. AOerr is still dominant, but only in high degrees; in lower degrees (do < 40), the erroneous trend of AO-DEAL is dominating. At the same time, the magnitudes of ΔH_{al} and ΔOT are of a similar order as that of AOerr.

4.3 Error Contribution at Multidecadal Scale

The previous sections showed that over a time period of 12 years, the dominating error sources are AOerr, ΔOT, and long-term trend deficits in the DEAL model. In this section, these three error sources are discussed in relation to their behavior over three decades.

On the one hand, the error introduced through the model difference of AO-DEAL is a geophysical signal. This means it is a real trend that the gravity missions resolve and is therefore included in the model. Because this signal is almost completely resolved by a single pair and completely resolved by a double pair, it will not improve with longer observation periods. It is a background model error that is independent of the observation system, as they are sensitive to the signal.

AOerr and ΔOT, on the other hand, introduce effects based on errors over time that influence the trend estimation.

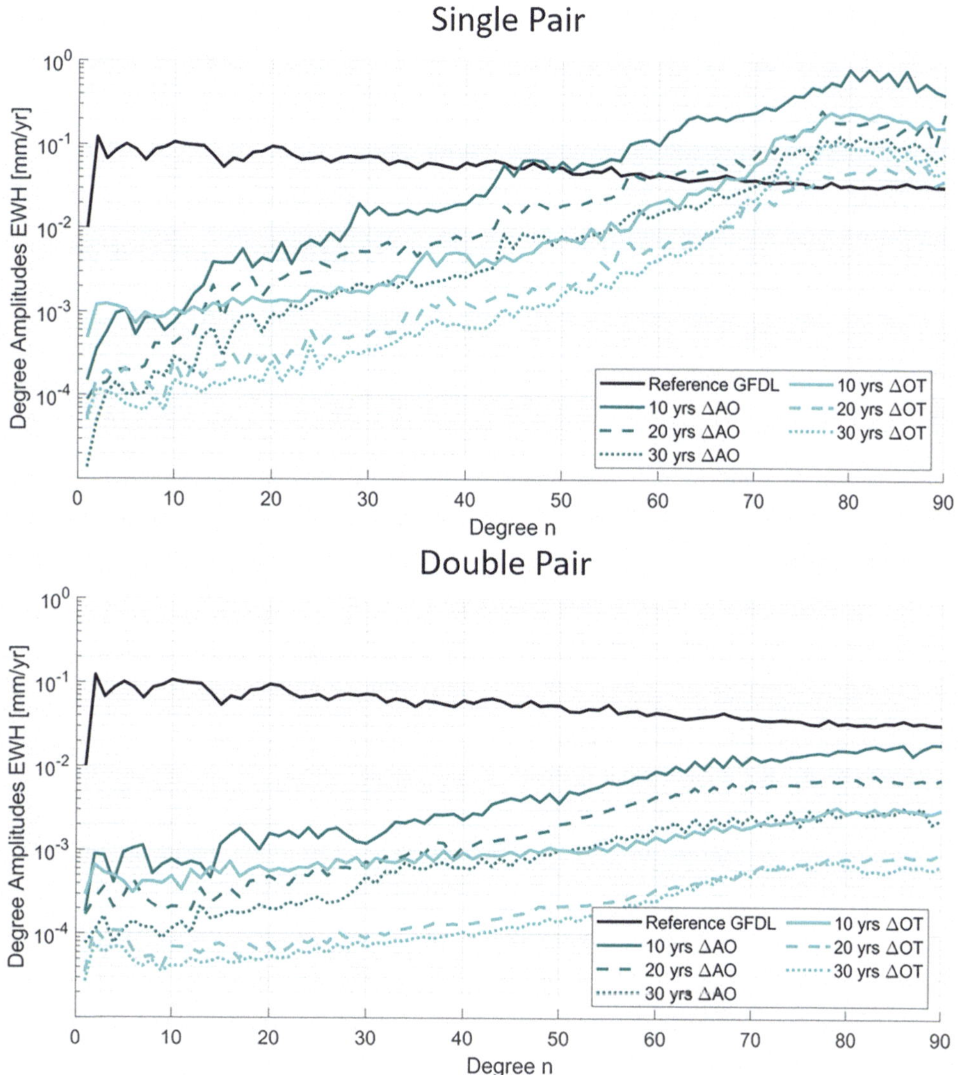

Fig. 8 Long-term trend behavior of AOerr and ΔOT over 10, 20, and 30 years for a single-pair (top) and double-pair mission (bottom)

The errors introduced do not correspond to a geophysical signal, but they are temporal aliasing effects. Because AOerr is only available for 12 years (Dobslaw et al. 2016), the updated AOerr, available for 25 years (AOe07) (Shihora et al. 2024), is used in this study. The time series of AOe07 has been artificially extended by repeating the first 17 years of AOe07 twice. In Fig. 8, long-term behavior of AOe07 together with ΔOT has been evaluated for over 30 years.

As can be seen in Fig. 8, both error sources contribute less for longer periods of observation. Looking at the ΔOT difference between 20 and 30 years of observations, they show a convergence, indicating that the distinct excitation frequencies of ΔOT errors limit the improvement by averaging. The AOerr is still the dominant error source after 30 years of observation, but as the range of frequencies is much wider compared to ΔOT, the improvement does not saturate as fast, which indicates that ocean tide background model errors will become more significant for even longer time series.

5 Discussion

The separation of the individual error sources shows clearly that some sources are currently more significant than others. For instance, instrument noise (ϵ) is orders of magnitude lower than current background model errors. Continuously improving the instrument performance is still essential, as it defines the maximum resolution of the observing system. Background model errors could be improved in the future, which would lead to an improved resolution of satellite gravity products.

Moreover, background model errors are a sampling issue in satellite gravimetry, not depending on the instrument

Table 1 Summary table of the global rms of the individual error sources in mm/year EWH

Error source	Single pair (global rms [mm/year])	Double pair (global rms [mm/year])
ϵ	0.28	0.07
ΔH_{al} (ESM)	55.27	1.25
ΔH_{al} (GFDL)	59.10	0.64
ΔOT	6.66	0.42
AO-DEAL	1.22	0.63
AOerr	28.12	0.83

performance on board the satellites (only because the instruments are performing orders of magnitude better than the background models). This becomes apparent when comparing temporal aliasing effects of single- and double-pair missions. Due to the improved observation geometry of the MAGIC mission, the impact of background model errors is significantly reduced, as can be seen in Table 1. Furthermore, only the sources of strong events causing temporal aliasing, e.g., Oceania for ΔOT or equatorial region in case of AOerr, are significant in the double-pair constellation, while the single-pair constellation suffers globally from these errors.

The error introduced through long-term instability of de-aliasing models (AO-DEAL) is different from temporal aliasing. AO-DEAL introduces an artificial trend through the de-aliasing model, which is (almost) independent of the mission performance. For the single pair, it only dominates in low degrees (d/o 15), but for the double-pair constellation, it is the dominating source of error in the trend until d/o 40. Because satellite gravimetry is sensitive to this erroneous trend and resolves it, there is no improvement possible through better mission performance, more efficient sampling, or longer observation periods. According to Shihora et al. (2022), the erroneous trend introduced through DEAL is pessimistic and current AOD1B products do have better long-term stability. Updated versions of Earth System Models for satellite gravimetry simulation should consider updating the long-term stability of de-aliasing models to the state of the art, as it is already a limiting factor for mission performances, planned to be launched in the near future.

The largest rms values in Table 1 are for ΔH_{al} and are caused by estimation instabilities up to d/o 90 in the case of the single pair and are limited to high degrees (Fig. 6). Within the 12 years of the ESM, AOerr is the most significant error source in both constellations and dominates most of the spectrum of the single-pair retrieval (Fig. 6). The picture changes in the double-pair constellation (Fig. 7). Here, ΔOT and ΔH_{al} are on a similar order of magnitude because the intrinsic de-aliasing due to the sampling of the inclined pair reduced the overall magnitude of errors.

Looking at the ΔOT and AOerr over multiple decades shows that only parts of the error are averaged out over the years. Overall, it indicates that AOerr will continue to dominate the error spectra over 30 years but that the improvements over time for ΔOT are attenuating after 20 years, while AOerr is averaged more efficiently. For long-term applications (>30 years), ΔOT will become the dominant error source.

6 Conclusions and Outlook

In conclusion, this study considers three types of error sources on long-term trend retrievals with single-pair and double-pair satellite gravimetry missions. These are instrument errors (ϵ), erroneous trends (AO-DEAL), and temporal aliasing errors (AOerr, ΔOT, and ΔH_{al}).

Currently, instrument noises are not limiting the observation capabilities, but they are the physical limit of the observation system. The low degrees are dominated by the erroneous trend introduced by a pessimistic de-aliasing model (DEAL), which will be a limiting factor independent of the observation system. Current de-aliasing products show better long-term stability than those represented in the ESA ESM.

The dominant error source in the single-pair spectrum is AOerr. For a double-pair scenario, AOerr is still dominating, but the error level is reduced overall, and background model errors are of a similar order of magnitude. Over longer periods of observation, the trend estimates improve as the dominating error sources are reduced.

While improving the observation system is indispensable, improved background models and additional processing strategies, e.g., stochastic modeling of background model errors (Abrykosov et al. 2021; Kvas and Mayer-Gürr 2019), can further improve the gravity retrieval.

In the case of a double-pair mission, an additional improvement could be gained by optimizing the relative weight between the two different satellite pairs in the gravity field retrieval process. As the temporal sampling of the inclined pair is more efficient, its observations are dominating within its coverage ($\pm70°$ latitude). As there are no observations of the inclined pair above 70°, the polar pair observations dominate there. An optimal weighting of the two pairs might reduce the temporal aliasing errors currently observed in the transition band between the two satellite pairs (e.g., Fig. 5, bottom row).

References

Abrykosov P, Sulzbach R, Pail R, Dobslaw H, Thomas M (2021) Treatment of ocean tide background model errors in the context of GRACE/GRACE-FO data processing. Geophys J Int 228(3):1850–1865. https://doi.org/10.1093/gji/ggab421

Daras I, March G, Joint Mass Change Mission Expert Group, Wiese D, Blackwood C, Forman F et al (2023) Next Generation Gravity Mission (NGGM) mission requirements document. ESA

Dill R (2008) Hydrological model LSDM for operational Earth rotation and gravity field variations. Scientific Technical Report STR 08/09. https://doi.org/10.2312/GFZ.b103-08095

Dobslaw H, Bergmann-Wolf I, Dill R, Forootan E, Klemann V, Kusche J, Sasgen I (2015) The updated ESA Earth System Model for future gravity mission simulation studies. J Geodesy 89(5):505–513. https://doi.org/10.1007/s00190-014-0787-8

Dobslaw H, Bergmann-Wolf I, Forootan E, Dahle C, Mayer-Gürr T, Kusche J, Flechtner F (2016) Modeling of present-day atmosphere and ocean non-tidal de-aliasing errors for future gravity mission simulations. J Geodesy 90(5):423–436. https://doi.org/10.1007/s00190-015-0884-3

Ettema J, van den Broeke MR, van Meijgaard E, van de Berg WJ, Bamber JL, Box JE, Bales RC (2009) Higher surface mass balance of the Greenland ice sheet revealed by high-resolution climate modeling. Geophys Res Lett 36(12):2009GL038110. https://doi.org/10.1029/2009GL038110

Guo H, John JG, Blanton C, McHugh C, Nikonov S, Radhakrishnan A et al (2018) NOAA-GFDL GFDL-CM4 model output. With assistance of contact, GFDL Climate Model Info

Hauk M (2020) Simulation studies for gravity field retrieval in the context of a next generation satellite gravity mission. Dissertation. Technische Universität München, München

Heller-Kaikov B, Pail R, Daras I (2023) Mission design aspects for the mass change and geoscience international constellation (MAGIC). Geophys J Int 235(1):718–735. https://doi.org/10.1093/gji/ggad266

Jensen L, Eicker A, Dobslaw H, Stacke T, Humphrey V (2019) Long-term wetting and drying trends in land water storage derived from GRACE and CMIP5 models. J Geophys Res Atmos 124(17–18):9808–9823. https://doi.org/10.1029/2018JD029989

Kvas A, Mayer-Gürr T (2019) GRACE gravity field recovery with background model uncertainties. J Geodesy 93(12):2543–2552. https://doi.org/10.1007/s00190-019-01314-1

Murböck M (2015) Virtual constellations of next generation gravity missions. Dissertation. Technische Universität München, München

Niemeier W (2008) Ausgleichungsrechnung. Statistische Auswertemethoden, 2. überarb. und erw. Aufl. De Gruyter (de Gruyter Lehrbuch), Berlin

Ray RD (1999) A global ocean tide model from TOPEX/POSEIDON altimetry: GOT99.2. National Aeronautics and Space Administration, Goddard Space Flight Center

Rummel R (1979) Determination of short-wavelength components of the gravity field from satellite-to-satellite tracking or satellite gradiometry. An attempt to an identification of problem areas. Manuscripta Geodaetica 4(2):107–148

Savcenko R, Bosch W, Dettmering D, Seitz F (2012) EOT11a – Global empirical ocean tide model from multi-mission satellite altimetry, with links to model results, supplement to: Savcenko R, Bosch W (2012) EOT11a – Empirical ocean tide model from multi-mission satellite altimetry, vol 89. Deutsches Geodätisches Forschungsinstitut (DGFI), München, 49 pp

Schlaak M, Pail R, Jensen L, Eicker A (2022) Closed loop simulations on recoverability of climate trends in next generation gravity missions. Geophys J Int 232(2):1083–1098. https://doi.org/10.1093/gji/ggac373

Shihora L, Balidakis K, Dill R, Dahle C, Ghobadi-Far K, Bonin J, Dobslaw H (2022) Non-tidal background modeling for satellite gravimetry based on operational ECWMF and ERA5 reanalysis data: AOD1B RL07. J Geophys Res Solid Earth 127(8):e2022JB024360. https://doi.org/10.1029/2022JB024360

Shihora L, Liu Z, Balidakis K, Wilms J, Dahle C, Flechtner F et al (2024) Accounting for residual errors in atmosphere–ocean background models applied in satellite gravimetry. J Geodesy 98(4). https://doi.org/10.1007/s00190-024-01832-7

Tapley BD, Bettadpur S, Watkins MM, Reigber C (2004) The gravity recovery and climate experiment: Mission overview and early results. Geophys Res Lett 31(9). https://doi.org/10.1029/2004GL019920

Intercomparison of Spherical Harmonics and Mascons for GRACE-based Mass Change Estimates

Huiyi Wu, Marius Schlaak, and Roland Pail

Abstract

This study evaluates two common methods for estimating mass changes from GRACE inter-satellite observations: spherical harmonics (SH) and mass concentration (mascon) solutions. Using a numerical simulation framework, we compare their performance for single-pair GRACE-like and double-pair Bender-type missions based on global RMS of residuals and degree-dependent spectral power. The analysis focuses on gravity field recovery using global 3° equal-area mascons and SH coefficients up to degree and order (d/o) 60. Mascon parameters are converted into equivalent SH coefficients for consistent spatial and spectral comparison. Results show that mascon solutions achieve lower global RMS residuals due to spatial averaging and geophysical-based regularization, whereas SH solutions preserve high-degree spectral power but are more affected by aliasing. These findings highlight the complementary strengths of the two parameterizations in gravity field recovery.

Keywords

GRACE · Mascon · Satellite gravimetry · Temporal gravity field variations

1 Introduction

Changes in the Earth's gravity field reflect the dynamic redistribution of mass driven by processes in the atmosphere, oceans, hydrology, cryosphere, and solid Earth. Over the last two decades, GRACE (Tapley et al. 2004) and GRACE-FO missions (Landerer et al. 2020) have provided crucial data on temporal variations in mass, revealing insights into phenomena such as ice loss in Greenland and Antarctica. The upcoming satellite gravity constellation Mass-Change and Geosciences International Constellation (MAGIC) (Massotti et al. 2021) aims to enhance this understanding by improving the spatial and temporal resolution of gravity field measurements. Currently, optimum mission designs for this future mission are explored (Heller-Kaikov et al. 2023).

H. Wu (✉) · M. Schlaak · R. Pail
Institute for Astronomical and Physical Geodesy, Technical University of Munich, Munich, Germany
e-mail: huiyi.wu@tum.de; marius.schlaak@tum.de; roland.pail@tum.de

Traditionally, GRACE mass change estimation relies on changes in the Earth's gravitational potential represented by spherical harmonics. SHs are widely used in satellite geodesy for historical and practical reasons, as they provide a compact representation of the global gravity field, enable efficient computation of the potential and its derivatives, and allow the separation of spectral contributions by spatial wavelength (Heiskanen and Moritz 1967). High-degree coefficients have a relatively minor influence on satellite orbits compared to low-degree harmonics, with their effects decreasing rapidly with altitude. Unconstrained SH solutions often produce unrealistic north-south striping patterns due to temporal aliasing (Swenson and Wahr 2006), typically mitigated by smoothing or de-striping techniques (Wahr et al. 1998; Swenson and Wahr 2006), which, however, can introduce signal leakage and attenuation.

Mascon solutions offer an alternative approach by dividing the Earth's surface into localized blocks, allowing direct estimation of mass anomalies. This method uses locally defined basis functions rather than global ones, enabling better incorporation of a-priori information to mitigate aliasing

J. T. Freymueller, L. Sànchez (eds.), *International Symposium on Gravity, Geoid and Height Systems 2024 (GGHS2024)*, International Association of Geodesy Symposia 158, https://doi.org/10.1007/1345_2025_306

effect and modeling errors. Research groups have applied mascon techniques both for specific regions of interest (Rowlands et al. 2005; Luthcke et al. 2013) and on a global parameterization (Sabaka et al. 2010; Watkins et al. 2015; Save et al. 2016; Tregoning et al. 2022). According to Watkins et al. (2015) and Save et al. (2016), three main mascon approaches exist:

(a) Analytical mascons: Directly relate inter-satellite range-rate or range-acceleration measurements to individual mascons through an analytical gravitational potential.
(b) Lumped spherical harmonic mascons: Utilize spherical harmonic functions and their partial derivatives to connect measurements to mascons via the chain rule.
(c) Post-processing mascons: Serve as a post-processing tool to derive regional mass changes from existing spherical harmonic solutions without direct access to GRACE observations.

This study employs a point-mass mascon approach, explicitly linking inter-satellite measurements to an analytical mascon formulation. The objective of this paper is to conduct a comparative analysis of mascon solutions and spherical harmonic solutions, focusing on their performance in gravity retrieval under varied spatial basis functions and satellite configurations in a simulated environment. Following Save et al. (2012), Tikhonov regularization is used to reduce striping in gravity field solutions, eliminating the need for post-processing in our simulations.

2 Theory

2.1 Spherical Harmonics Gravity Model

The gravitational potential of the Earth V is usually parameterized as an infinite spherical harmonic series. This parameterization results from the assumption of a source-free external space and the solution of *Laplace*' s equation $\Delta V = 0$ in spherical coordinates

$$V(r, \varphi, \lambda) = \frac{GM}{R} \cdot \sum_{n=0}^{\infty} \sum_{m=0}^{n} \left(\frac{R}{r}\right)^{n+1} \overline{P}_{nm}(\sin\varphi) \cdot \left(\overline{C}_{nm} \cos m\lambda + \overline{S}_{nm} \sin m\lambda\right) \quad (1)$$

where G is the gravitational constant, M is the Earth mass, and R is Earth's reference radius. The $\overline{P}_{nm}$ terms are fully normalized Legendre polynomials, and $\overline{C}_{nm}, \overline{S}_{nm}$ are fully normalized spherical harmonic coefficients of degree n and order m. r, φ and λ are radial distance, latitude and longitude of the spacecraft. Practical applications limit the series to a maximum degree n_{max}, with spatial resolution determined $\kappa_{min}\,[km] \approx \frac{20000km}{n_{max}}$. As the altitude increases, the degree in the exponent leads to a faster attenuation of the contribution of high-degree harmonics to the Earth's potential as compared to low-degree harmonics.

2.2 Surface Point-Mass Mass Concentration Model

The mascon approach divides the Earth's surface into localized blocks, allowing direct estimation of mass anomalies over specified regions and discrete time intervals. This approach models time-variable gravity signals by defining regular or irregular blocks where mass changes are due to equivalent water load on the Earth's surface.

For a point P at position $\boldsymbol{r}_P$, the gravitational potential from mass over volume v_Q with density ρ_Q at source point Q at position $\boldsymbol{r}_Q$ is given by Newton's integral

$$V_P = \iiint_{v_Q} \frac{G\rho_Q}{\left\|\boldsymbol{r}_P - \boldsymbol{r}_Q\right\|_2} dv_Q \quad (2)$$

In case of involving surface mass, where mascons represent uniform surface mass density σ_Q, with the unit *kg/m*^{2} or as millimeters of Equivalent Water Height (EWH) over area S_Q, the gravitational potential at point P is approximated by

$$V_P = \oiint_{S_Q} \frac{G\sigma_Q}{\left\|\boldsymbol{r}_P - \boldsymbol{r}_Q\right\|_2} dS_Q \quad (3)$$

For computational efficiency, each mascon is treated as a point-mass at its center, simplifying to

$$V_P = \frac{Gm_Q}{l_{P,Q}} = \frac{G\sigma_Q S_Q}{l_{P,Q}} \quad (4)$$

where $l_{P,Q} = \|\boldsymbol{r}_P - \boldsymbol{r}_Q\|_2$. When the Earth's surface is discretized into p mascon blocks, the potential V at any point in space due to surface density variations $\delta\sigma_j$ in each mascon j is

$$V = G \sum_{j=1}^{p} \frac{\sigma_j S_j}{l_j} \quad (5)$$

This formulation allows the mascon solution to solve for surface density variations within blocks, providing a practical means of modeling gravitational changes on an ellipsoidal Earth surface.

3 Gravity Field Estimation Process

3.1 Simulation Environment

We employ a numerical closed-loop simulator, the Reduced Scale Simulator (RSS), based on Rummel's (1979) "acceleration method" and further refined by Hauk (2020), see also Sect. 3.3. This simulator was extended to solve optionally for mascons in addition to spherical harmonic coefficients. Figure 1 provides an overview of this process. Simulated observations are derived from input gravity signals based on

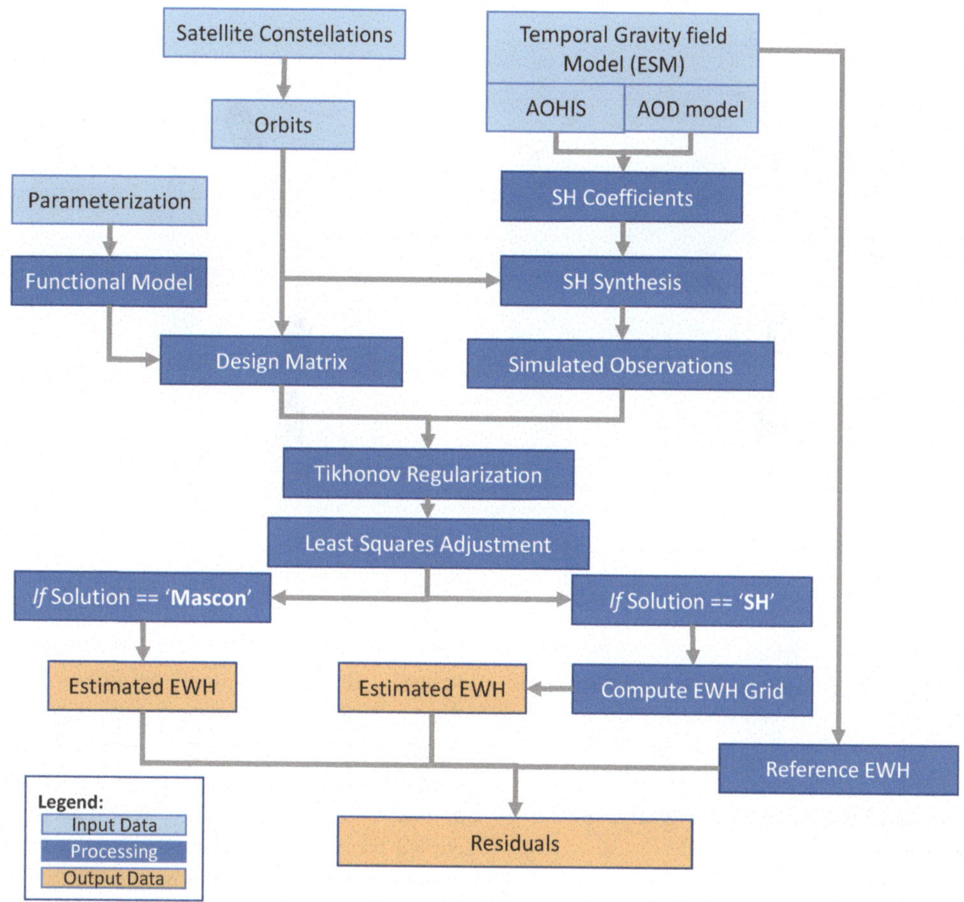

Fig. 1 Gravity field simulation flow chart of the Reduced Scale Simulator

a "true" Earth model, incorporating errors from background models to account for short-period signals. We simulate mass variations using the updated ESA Earth System Model (Dobslaw et al. 2015) and apply a realistic atmosphere-ocean de-aliasing (AOD) model (Dobslaw et al. 2016). Instrumental errors from the inter-satellite ranging system, accelerometers, and GNSS receivers are not explicitly simulated in this study. This simplification allows the analysis to focus on evaluating the impact of temporal aliasing errors, which are typically dominant in GRACE-type gravity field recovery.

The main steps in the simulation procedure are:

1. Generate simulated satellite-to-satellite acceleration observations and satellite position observations
2. Construct the linear equation system between the unknown parameters and observation vectors, which is based on gravity acceleration differences
3. Estimate the gravity field parameters (SH coefficients and mascons, respectively) and evaluate the performance at $1° \times 1°$ resolution of EWH grid.

3.2 Mission Scenarios

For this study, a single-pair GRACE-like and a double-pair Bender-type mission are evaluated for their gravity retrieval performance. Figure 2 shows the two mission architectures.

The GRACE-like mission uses a pair of in-line satellites in a near-polar low Earth orbit, measuring microwave-based gravity-induced inter-satellite distance changes. This configuration provides high sensitivity to along-track accelerations but limited sensitivity in the across-track direction. The satellites are assumed to orbit at an altitude of 450 km with an inclination of 89°.

The Bender-type configuration includes an additional inclined pair at 70° inclination. The inclined pair, orbiting at a lower altitude of about 400 km, adds cross-track measurements. This geometry reduces error anisotropy and temporal aliasing, as highlighted by Wiese et al. (2012). Both configurations assume a maximum inter-satellite distance of about 200 km, a 30-day repeat cycle,

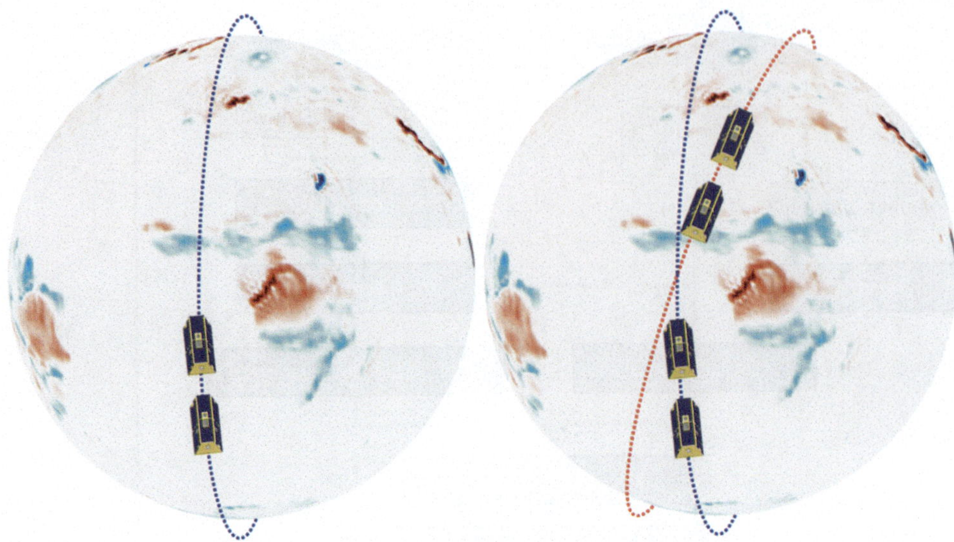

Fig. 2 Mission architectures

and a 60-s sampling interval, which remains sufficient to avoid spatial undersampling in along-track direction while reducing computational load (Schlaak et al. 2023).

3.3 Observation Equation and Parameter Estimation

In the RSS, gravity observations are derived from differences in gravitational acceleration between satellite pairs, measured along their line of sight (LoS). This approach relies on the relationship between gravitational acceleration and the gradient of the Earth's potential $a = \nabla V$. The observation model, as described by Hauk (2020), is

$$\boldsymbol{b} = \langle \Delta a_{grav}, \Delta r_0 \rangle = \nabla V_2 e_{2,1} - \nabla V_1 e_{1,2} \tag{6}$$

where $\boldsymbol{b}$ is the gravity field observation, Δa_{grav} is the gravitational acceleration difference, and Δr_0 is the unit vector pointing from one satellite to the other. With the functional model defined by Eq. (6), the estimation process for gravity field parameters utilizes a least-squares adjustment method. For spherical harmonics, the estimated parameter vector $\hat{x}$ includes coefficients $\left[\overline{C}_{nm}, \overline{S}_{nm}\right]^T$, and for mascon approach, it includes surface densities $[\sigma_j]^T$.

Due to insufficient global sampling in time, temporal aliasing effects particularly impact the higher degrees of the SH spectrum. After the inversion procedure this results in north-south striping artifacts in the EWH maps. Tikhonov regularization technique, combined with the L-curve method, is developed to reduce these errors. By incorporating a penalty term into the least-squares problem, Tikhonov regularization stabilizes the normal equation system, reducing sensitivity to noise or small perturbations, and improving robustness. The L-curve method determines the appropriate regularization parameter α, providing a clear, intuitive balance between data reliability and solution stability. This approach is also applicable to real data processing, and its detailed formulation is given by Save et al. (2012). In addition to stabilization, Tikhonov regularization also implicitly acts as a filter, suppressing high-frequency noise. This filtering effect arises from modifications to the singular values of the system matrix, reducing the influence of small singular values typically associated with high-frequency components. Unlike a-posteriori filtering, Tikhonov regularization integrates the filtering effect within the inversion process, ensuring system stability during computation.

To optimize computational efficiency and comparability, while minimizing biases from a-prior knowledge, we implement generalized Tikhonov regularization with an identity matrix for regularization in both SH and mascon solutions. The L-curve method, as described by Save et al. (2012), effectively selects regularization parameters in Tikhonov-regularized least squares problems, but it tends to produce overly smooth results. This approach significantly reduces striping in the solutions but at the cost of greater signal attenuation than desired. To counteract the effect of over-smoothing, this study introduces an additional scaling factor to undervalue the regularization parameter α obtained from L-curve. The scaling factor is determined by minimizing the differences between the 'true' world and the monthly mean global EWH map. By scaling α, signal preservation is improved while maintaining reasonable error reduction. The relationship between the estimated parameter and the observations is expressed through design matrix A, leading to the estimated parameter formulation

$$\hat{x} = \left(A^T A + \text{scaling factor} \cdot \alpha \cdot I\right)^{-1} A^T b \tag{7}$$

4 Simulation Solution Analysis

The numerical closed-loop simulation results are analyzed for each scenario of the satellite mission constellations and for each form of gravity field basis functions used in the gravity field retrieval procedure. The scenarios include (1) a single-pair GRACE-like mission and (2) a Bender-type constellation with one polar pair and a second pair in a 70° inclined orbit. The gravity field basis functions include (1) SH up to d/o 60 and (2) 3° equal-area point-mass mascon approach, with the rationale for both the SH truncation and mascon resolution discussed in Sect. 4.1.2.

For all solutions, the input AOHIS model's maximum degree of SH expansion is 60, and the high-frequency atmospheric and oceanic signals are modeled and removed beforehand. However, even in the simulation studies, de-aliasing products can only reduce the aliasing problem to a certain extent but cannot combat the original source, which is the under-sampling of high-frequency signals (Hauk 2020). The unconstrained solutions (see also Sect. 4.2) exhibit substantial striping in global EWH maps caused by background model errors, distorting the representations of mass variations in the gravity field and are thus not presented in spatial domain. Regularization further reduces these errors in the Earth's gravity field estimates significantly.

4.1 Spatial Comparisons

4.1.1 Spherical Harmonics

By quantifying differences with the global root mean square (RMS) between the reference and the solution, optimal solutions are obtained by applying a scale factor of 0.7 in the course of both SH single-pair and double-pair solutions.

The results are summarized in Figs. 3. Figure 3 illustrates the global EWH distribution in millimeters derived from SH solutions, comparing the reference dataset with single-pair and double-pair satellite configurations up to d/o 60. The top panel presents the reference global EWH map for January 1995, serving as the baseline for analysis and highlights hydrological features and mass distribution patterns. In general, large-scale variabilities are well captured by both satellite configurations. However, the single-pair solution exhibits visible north-south striping artifacts, especially (i) over the South Pacific sector south of South America (approximately 40 – 60°S, 150 – 60°W), (ii) across the South Indian Ocean and south of Australia (approximately 45 – 60°S, 60 – 150°E) and (iii) across the Siberia – North Pacific region (approximately 45 – 65°N, 80 – 170°E). This remaining aliasing effect mainly originates from the scaling of the regularization parameter, which mitigates signal attenuation but also introduces more striping artifacts. In contrast, the double-pair solution demonstrates a clear reduction of these striping artifacts, especially within the same regions, leading to a more physically consistent representation of geophysical signals. The oceanic regions also appear notably cleaner, with reduced stripping patterns and improved agreement with the expected large-scale background variability. This improvement can be attributed to the modified azimuth of the inter-satellite link provided by the inclined satellite pair. This adjustment effectively rotates the error anisotropy and leads to a better cross-over geometry of ascending and descending satellite tracks, enhancing the stability of monthly gravity field solutions.

Residual maps (Fig. 3, middle and bottom) highlight notable geospatial signals over regions such as Antarctica, the Himalayas, Greenland, and the Amazon. These residuals mainly arise from signal damping introduced by the applied regularization, which suppresses high-frequency components to stabilize the inversion. When considering an additional pair, the increased number of observations and the improved observation geometry enhance the recovery performance, reducing the global RMS differences by approximately 9 mm. Although the double-pair solution retrieves higher signal amplitudes than the single-pair configuration, it still underestimates the reference values. This means that even if the regularization parameter is undervalued, the estimations still suffer from signal attenuation.

4.1.2 Mascons

In contrast to SH, the use of mass concentration blocks as the native basis function allows for the primary advantage that each mascon has a specific known geophysical location. We can take advantage of this convenient property that mascon basis functions are local, rather than global, to apply constraints on the location and shape of surface mass variations. This provides a more convenient architecture in which the scaling factor for the regularization parameter can be derived to balance striping and signal attenuation.

The mascon solutions in this study are based on a global grid of equal-area cells, each covering an area roughly equivalent to a 3° × 3° rectangle at the equator. The equal-area property is implicitly reflected in the surface area term S_j in Eq. (5), ensuring nearly uniform spatial weighting of mascons and improving the numerical stability of the least-squares adjustment. This grid resolution was chosen to match the spherical harmonic solutions truncated at d/o 60, corresponding to a minimum wavelength of approximately 333 km. This truncation aligns with the effective spatial resolution of GRACE-type missions, as higher degrees are

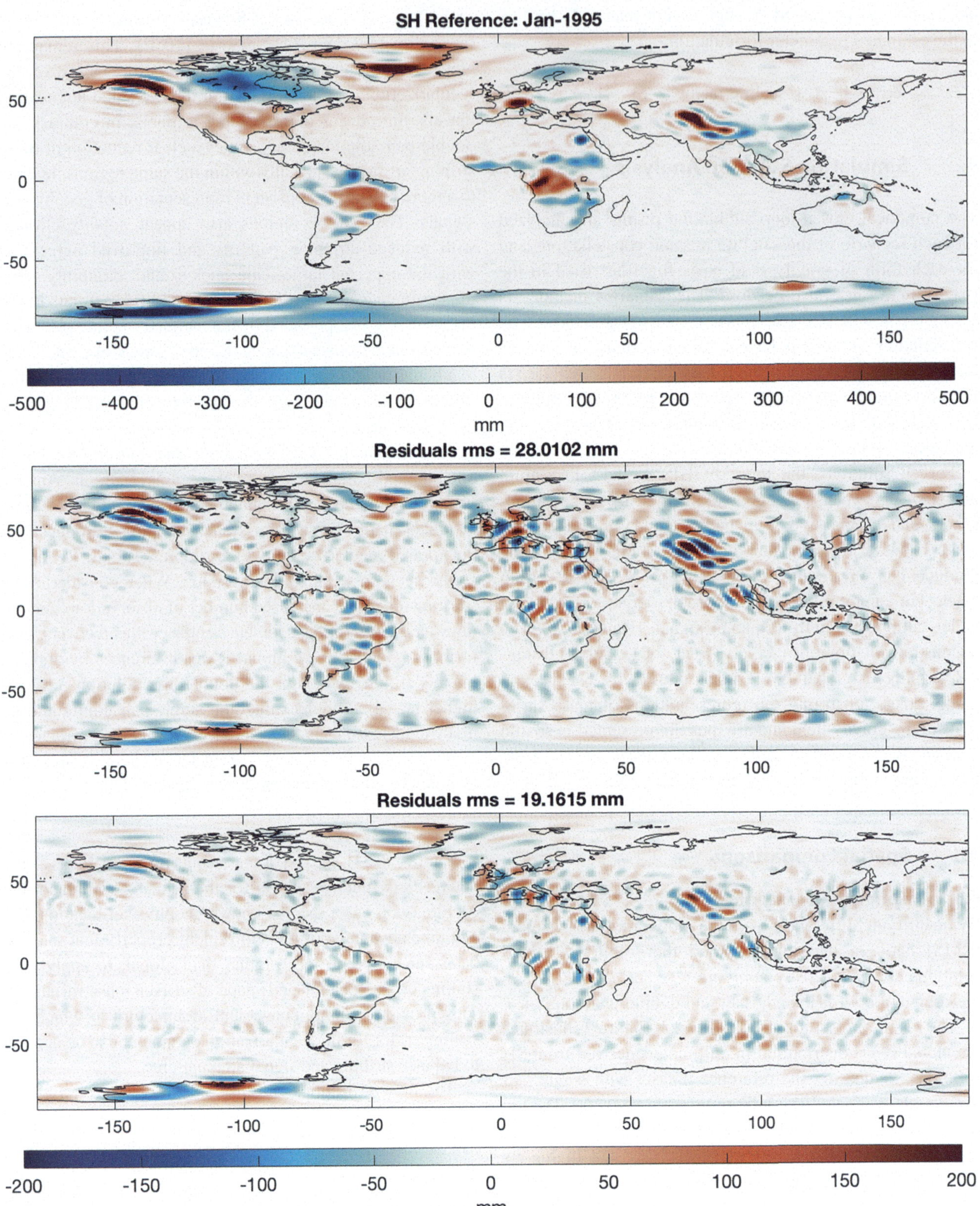

Fig. 3 Global equivalent water height reference (top) and residuals (reference – solution) from single-pair (middle) and double-pair (bottom) SH solutions

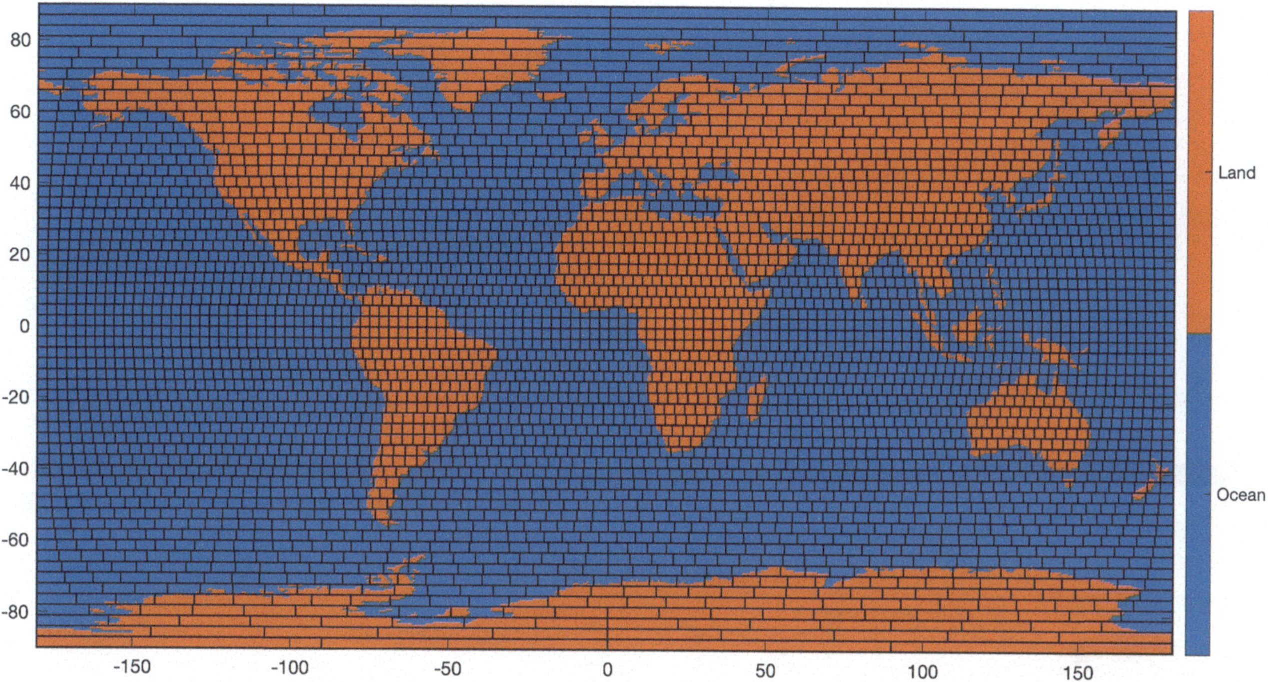

Fig. 4 Mascon locations: 3° equal-area grid

increasingly dominated by noise. In fact, at the equator, the $3° \times 3°$ cells correspond to about 333 km, ensuring comparable resolution between the two methods, while toward higher latitudes, the longitudinal extent of each cells increases. The global grid consists of 4,614 equal-area cells, whose displacements are visualized in Fig. 4.

In order to allow a fair comparison of the mascon solutions without introducing errors due to the differences in spatial resolution, the reference global map of EWH (Fig. 3 top) is converted into a 3°-averaged format (Fig. 5 top) according to the chosen mascon displacement. Although this averaging process may not perfectly represent reality, it captures the mean signal variations within each mascon block, allowing for a consistent comparison. The reference map reveals that land signals exhibit higher amplitudes and a more uneven spatial distribution, while ocean signals are characterized by lower amplitudes and greater uniformity. The global RMS of residuals is first used to define an initial scale factor. Subsequently, given the distinct regional characteristics of individual mascons, this scale factor is adjusted for each mascon classification to better accommodate local variations. For regions where signal smoothing is prioritized, such as the oceans, a relatively larger scale factor is used to reduce striping artifacts. Conversely, for areas with high-amplitude signals, a smaller scale factor is applied to better preserve geophysical features.

Figure 5 presents the EWH residuals for January 1995, illustrating the difference between the retrieved and reference for the single-pair (middle) and double-pair (bottom) mascon solutions. The residual patterns reveal that aliasing errors are substantially reduced in the mascon solutions compared to the regularized SH solutions. This reduction mainly results from the use of adjustable scale factors that mitigate the dominance of striping. In addition, the inherent spatial averaging of equal-area mascons becomes more effective toward higher latitudes, as the grid resolution decreases with increasing latitude, further mitigating striping effects.

In the single-pair configuration, the RMS error amounts to 20.77 mm, indicating relatively large discrepancies caused by the limited ability to resolve small-scale features. In contrast, the double-pair configuration achieves visible improvements, particularly in regions such as the Gulf of Alaska, southern Greenland, and the Himalayas, where the recovered signals show better consistency with the reference field. The RMS error decreases to 16.98 mm, demonstrating enhanced agreement with the reference dataset. This 4 mm improvement, although smaller than that of SH solutions, still demonstrates the advantages of the double-pair configuration in mitigating aliasing and enhancing the accuracy of mass change estimates.

4.2 Spectral Comparisons

In this section, the comparison is conducted in the spectral domain. Mascon parameters are decomposed into their spherical harmonic coefficient representations, which makes the comparison in the spectral domain possible. The decom-

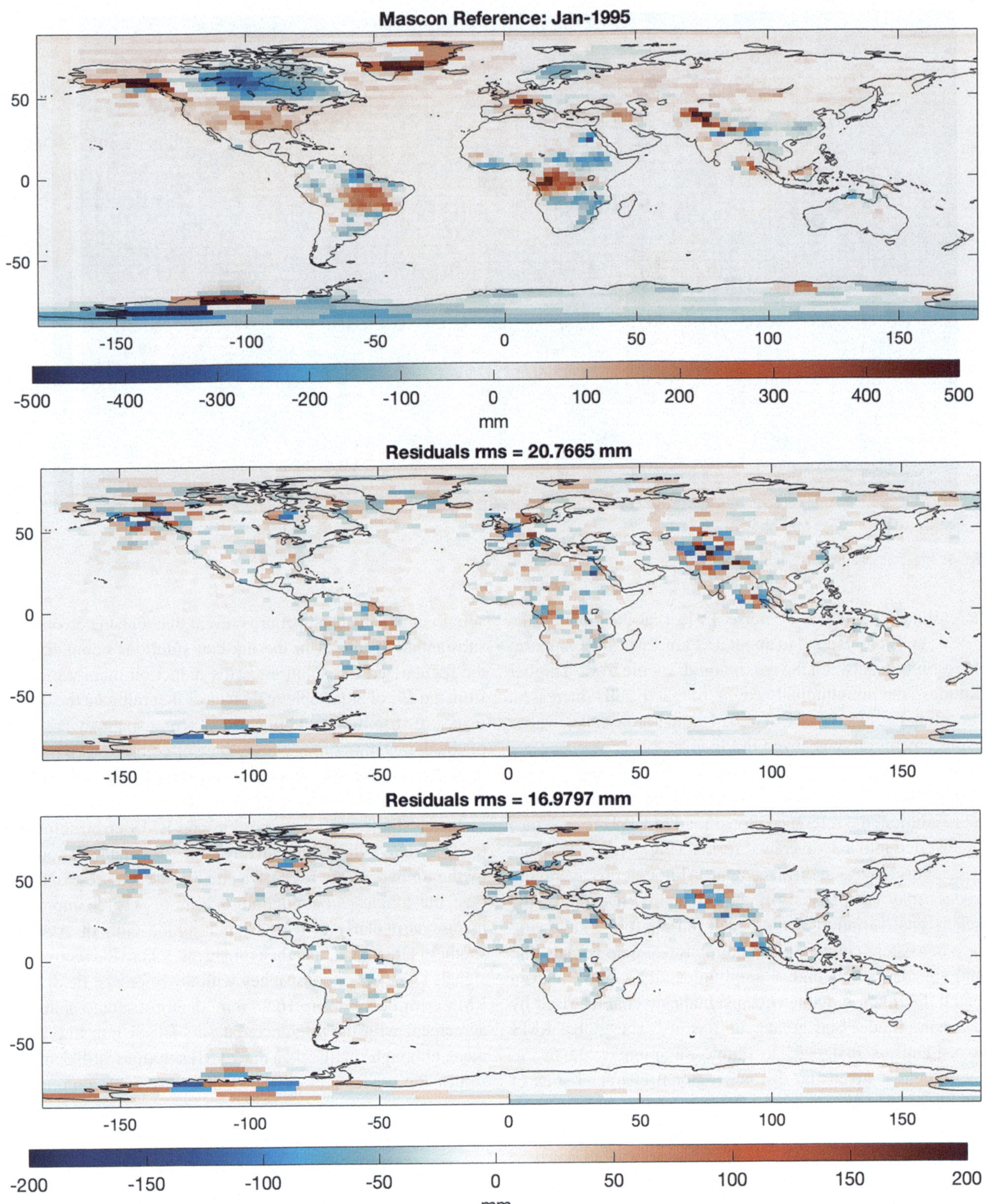

Fig. 5 Global equivalent water height reference (top) and residuals (reference – solution) from single-pair (middle) and double-pair (bottom) Mascon solutions

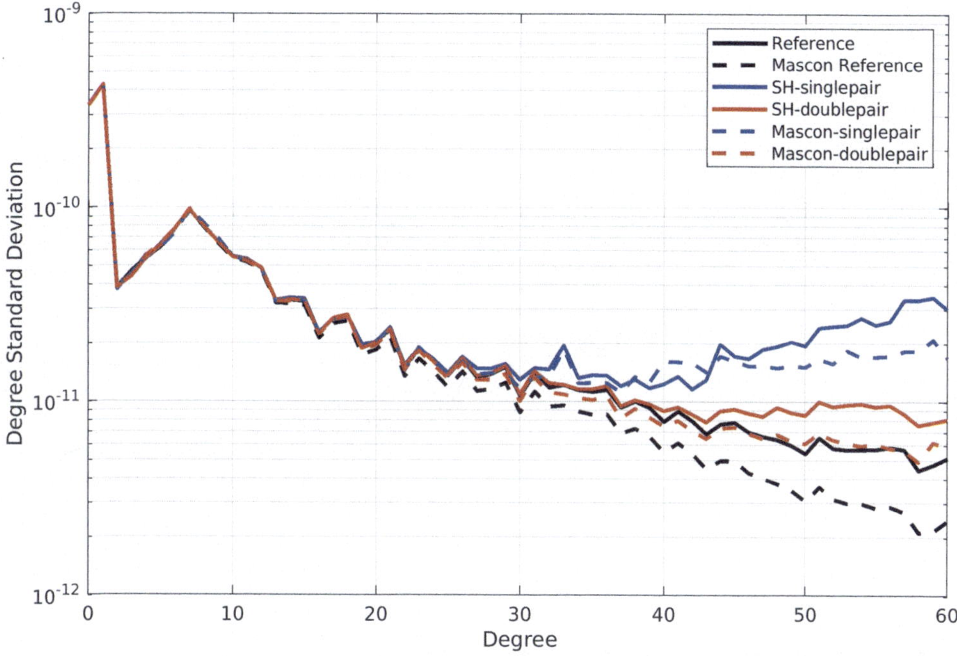

Fig. 6 Degree standard deviation of the reference compared with unconstrained spherical harmonic and mascon solutions

position is performed through a global spherical harmonic analysis applied to discrete datasets, following the least-squares formulation described by Sneeuw (1994). Assuming an equal-angular discretization in the longitudinal direction, the mascon-derived surface mass anomalies are expanded into spherical harmonics and truncated at a maximum degree and order of 60. Figure 6 provides strong evidence that the unconstrained global solutions (mascons and SH) are very similar. The power spectra of both solutions are almost identical from degree 0 to degree 15 and very similar from degree 16 to degree 25. At high degrees, the power of both single-pair scenario solutions greatly exceeds that of the reference. The excess power mainly arises from temporal aliasing, as no other error sources are included in this simulation. The second inclined pair of satellites improves the observation geometry, resulting in more isotropic error behavior, and thus reduces the temporal aliasing effect in the double-pair solutions. However, if we focus on the dashed red line, we can find that, due to the effect of averaging, the unconstrained double-pair mascon solutions show relatively less power.

Figure 7 presents the comparison of the regularized solutions. Both mascon solutions and SH solutions reveal comparable power levels, with both demonstrating lower power compared to the reference model, particularly as the degree increases. It is important to highlight that all solutions exhibit strong agreement with the reference model, albeit with slightly higher power in the spherical harmonic solutions compared to the mascon solutions at higher degrees. This discrepancy is attributed, in large part, to the damping effect induced by the 3° averaging methods. As a result, the mascon solutions exhibit a closer alignment with the 3°-averaged reference.

As described by Watkins et al. (2015), the mascon reference (3° averaging) represents the optimal solution achievable with mascon as basis function, given the true mass variations over areas by 1° × 1° resolution. However, when comparing the reference curves in Fig. 8, it is evident that this averaging introduces a damping effect at high degrees. This demonstrates that the damping effect arises from averaging the 1° × 1° signal into 3° × 3° mascon displacements. As shown in Fig. 8, employing a denser grid for mascon solutions (2° × 2°) reduces the disparity between the reference and averaged reference signal at high degrees.

As illustrated in Figs. 7 and 9, the introduction of a second pair of satellites has the potential to enhance performance. A noteworthy distinction emerges for degrees exceeding 30, where double-pair solutions consistently demonstrate superior power compared to single-pair solutions. The residual plot shows that the constrained mascon solutions exhibit smaller difference from the reference than the constrained SH solutions when considering all degrees. However, it should be noted that the mascon solutions are compared with their optimal achievable solution – the 3° averaged reference. In fact, due to the averaging effect, the 3° equal-area mascon solutions always exhibit less power than the SH solutions up to d/o 60.

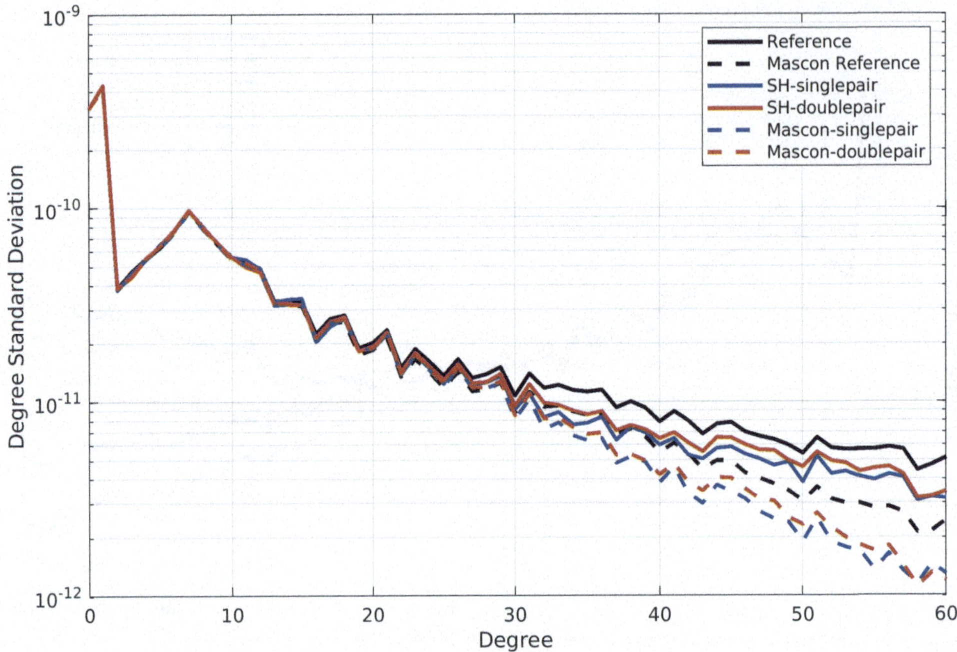

Fig. 7 Degree standard deviation of the reference compared with regularized spherical harmonic and mascon solutions

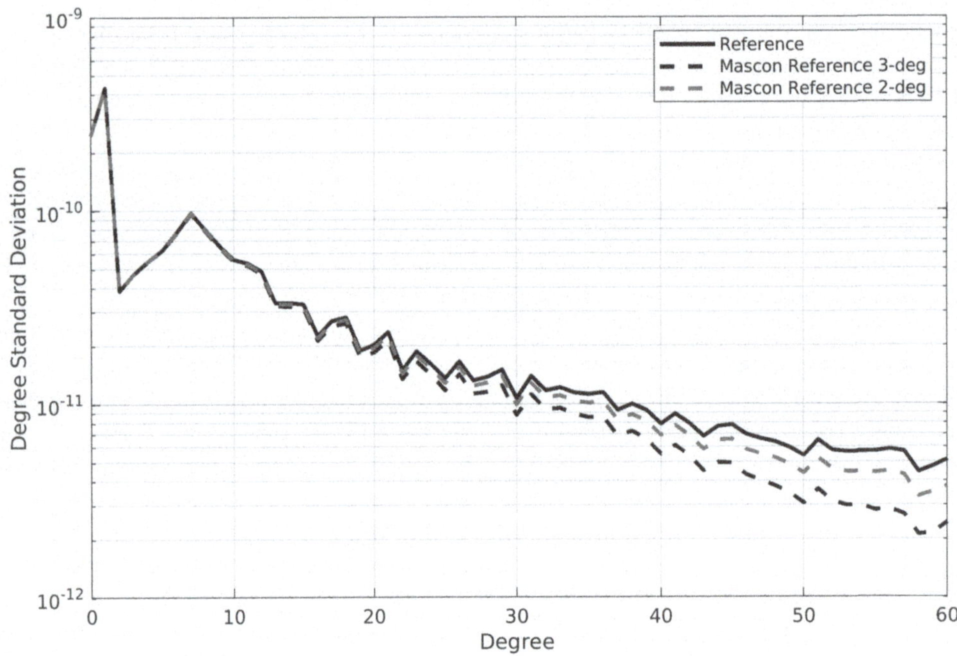

Fig. 8 Degree standard deviation of the reference compared with 3° averaged and 2° averaged references

5 Conclusions

We have inter-compared a gravity recovery method using point-mass mascon blocks as a basis function with the standard SH parameterization. The mascon approach, based on an equal-area rectangular grid with an approximate resolution of 3° at the equator, applies regularization tailored to the geophysical information of each individual mascon. This partitioning minimizes errors, eliminates the need for post-processing filters to mitigate striping artifacts, and balances signal recovery magnitude with error reduction. The global RMS of residuals for mascon solutions is 20.77 mm for single-pair and 16.98 mm for double-pair scenarios, with the retrieved signal closely approximating the method's optimal solution despite attenuation from averaging and regulariza-

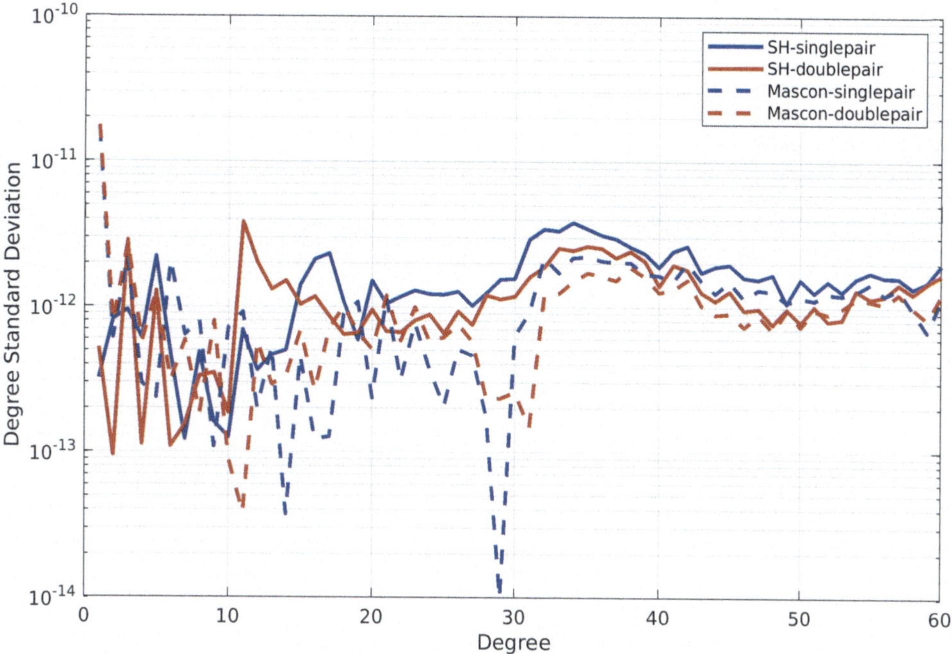

Fig. 9 Residual of degree standard deviation of regularized spherical harmonic and mascon solutions (SH: reference – solution; Mascon: Mascon reference – solution)

tion. Constrained spherical harmonic solutions also reduce aliasing errors, with a global RMS of residuals of 28.01 mm for single-pair and 19.16 mm for double-pair configurations. The scaling factor compensates for over-smoothing but introduces residual aliasing errors, constituting one of main sources of error in the residual map. While the regularized SH solutions experience signal attenuation from regularization process, they retain greater power at high degrees than mascon solutions.

The double-pair mission configuration demonstrates significant advantages due to its superior observation geometry. This configuration substantially reduces aliasing effects and enhances signal amplitude in both SH and mascon solutions, as evidenced by improved recovery of gravity signals. The additional inclined satellite pair enhances spatial sampling and reduces error anisotropy, improving the performance of monthly gravity field estimates.

The results underscore the trade-offs between the mascon and SH approaches for gravity recovery. The mascon solutions demonstrate superior performance in terms of global RMS of residuals, due to the averaging effect and convenient application of geophysical-based regularization. However, from a spectral perspective, this averaging suppresses high degree signal amplitude. To achieve signal power comparable to SH solutions, a slightly higher mascon resolution would be necessary, accompanied by appropriate adjustments to the regularization to control the ill-posed nature of the inversion. Additionally, in real data processing, the a-priori geophysical models or SH solutions can serve as a useful reference for determining appropriate scaling factors for the mascon solution, enabling improved amplitude recovery without relying on gain factors to restore signal amplitudes in post-processing.

Future work will build on the strengths of the geophysical-based a-priori conditioning applied in the mascon approach to design high-resolution, leakage-reducing mascon solutions. Furthermore, these advancements will be applied to real data processing to evaluate their effectiveness in temporal gravity field recovery from GRACE.

References

Dobslaw H, Bergmann-Wolf I, Dill R et al (2015) The updated ESA Earth System Model for future gravity mission simulation studies. J Geod 89:505–513. https://doi.org/10.1007/s00190-014-0787-8

Dobslaw H, Bergmann-Wolf I, Forootan E (2016) Modeling of present-day atmosphere and ocean non-tidal de-aliasing errors for future gravity mission simulations. J Geod 90:423–436. https://doi.org/10.1007/s00190-015-0884-3

Hauk M (2020) Simulation studies for gravity field retrieval in the context of a next generation satellite gravity mission, Dissertation, Technische Universität München, München

Heiskanen WA, Moritz H (1967) Physical geodesy. W.H. Freeman and Company, San Francisco, CA

Heller-Kaikov B, Pail R, Daras I (2023) Mission design and processing aspects for the mass change and geoscience international constellation (MAGIC). Geophys J Int 235:718–735. https://doi.org/10.1093/gji/ggad266

Landerer FW, Flechtner FM, Save H, Webb FH, Bandikova T, Bertiger WI et al (2020) Extending the global mass change data record: GRACE follow-on instrument and science data perfor-

mance. Geophys Res Lett 47:e2020GL088306. https://doi.org/10.1029/2020GL088306
Luthcke SB, Sabaka TJ, Loomis BD, Arendt AA, McCarthy JJ, Camp J (2013) Antarctica, Greenland, and Gulf of Alaska land-ice evolution from an iterated GRACE global mascon solution. J Glaciol 59:613–631. https://doi.org/10.3189/2013JoG12J147
Massotti L, Siemes C, March G, Haagmans R, Silvestrin P (2021) Next generation gravity mission elements of the mass change and geoscience international constellation: from orbit selection to instrument and mission design. Remote Sens 13:3935. https://doi.org/10.3390/rs13193935
Rowlands DD, Luthcke SB, Klosko SM, Lemoine FGR, Chinn DS, McCarthy JJ, Cox CM, Anderson OB (2005) Resolving mass flux at high spatial and temporal resolution using GRACE intersatellite measurements. Geophys Res Lett 32:L04310. https://doi.org/10.1029/2004GL021908
Rummel R (1979) Determination of short-wavelength components of the gravity field from satellite-to-satellite tracking or satellite gradiometry: an attempt to an identification of problem areas. Manuscr Geod 4:107–148. https://doi.org/10.1007/BF03654938
Sabaka TJ, Rowlands DD, Luthcke SB, Boy J-P (2010) Improving global mass flux solutions from gravity recovery and climate experiment (GRACE) through forward modeling and continuous time correlation. J Geophys Res 115:B11403. https://doi.org/10.1029/2010JB007533
Save H, Bettadpur S, Tapley BD (2012) Reducing errors in the GRACE gravity solutions using regularization. J Geod 86:695–711. https://doi.org/10.1007/s00190-012-0548-5
Save H, Bettadpur S, Tapley BD (2016) High resolution CSR GRACE RL05 mascons. J Geophys Res Solid Earth 121:7547–7569. https://doi.org/10.1002/2016JB013007
Schlaak M, Pail R, Jensen L, Eicker A (2023) Closed loop simulations on recoverability of climate trends in next generation gravity missions. Geophys J Int 232(2):1083–1098. https://doi.org/10.1093/gji/ggac373
Sneeuw N (1994) Global spherical harmonic analysis by least-squares and numerical quadrature methods in historical perspective. Geophys J Int 118(3):707–716. https://doi.org/10.1111/j.1365-246X.1994.tb03995.x
Swenson S, Wahr J (2006) Post-processing removal of correlated errors in GRACE data. Geophys Res Lett 33:L08402. https://doi.org/10.1029/2005GL025285
Tapley BD, Bettadpur S, Watkins M, Reigber C (2004) The gravity recovery and climate experiment experiment: mission overview and early results. Geophys Res Lett 31(9). https://doi.org/10.1029/2004gl019920
Tregoning P, McGirr R, Pfeffer J, Purcell A, McQueen H, Allgeyer S, McClusky SC (2022) ANU GRACE data analysis: characteristics and benefits of using irregularly shaped mascons. J Geophys Res Solid Earth 127:e2021JB022412. https://doi.org/10.1029/2021JB022412
Wahr J, Molenaar M, Bryan F (1998) Time variability of the Earth's gravity field: hydrological and oceanic effects and their possible detection using GRACE. J Geophys Res 103(B12):30205–30229. https://doi.org/10.1029/98JB02844
Watkins MM, Wiese DN, Yuan D-N, Boening C, Landerer FW (2015) Improved methods for observing Earth's time variable mass distribution with GRACE using spherical cap mascons. J Geophys Res Solid Earth 120:2648–2671. https://doi.org/10.1002/2014JB011547
Wiese DN, Nerem RS, Lemoine FG (2012) Design considerations for a dedicated gravity recovery satellite mission consisting of two pairs of satellites. J Geod 86:81–98. https://doi.org/10.1007/s00190-011-0493-8

Altimetry Data Analysis for Coastal Erosion: A Time Series Approach

R. N. Adam and G. S. Vergos

Abstract

Addressing coastal erosion is a long-standing issue that has been the focus of extensive research in recent years. Other fields of geosciences demand the determination of the Earth's gravity field and geoid with very high accuracy, on the order of ±1 cm for wavelengths of approximately 10 km along the coastal front, in order to integrate with other Earth observation data, whether satellite-based or terrestrial. Satellite altimetry has offered unprecedented opportunities to monitor sea level variations and study the marine gravity field over the past decades. This study presents the results of a multi-year analysis of satellite altimetry data, focusing on the Aegean Sea, Greece, employing both classical Low-Resolution Mode (LRM) and SAR/SARin data. Two main analysis methods are employed as tools to identify the primary modes of sea level variation and trends, i.e., Empirical Orthogonal Function/Principal Component Analysis (EOF/PCA) and Wavelet Transform. Based on these approaches, annual and semi-annual patterns are identified in the altimetry time series, while the trend of sea level changes is determined. Two scenarios, based on 50-year and 100-year simulations, are presented, highlighting the areas in the North Aegean Sea that are most susceptible to flooding and, consequently, more vulnerable to erosion due to sea level forcing.

Keywords

Coastal erosion · EOF analysis · Satellite altimetry · Sea level anomalies · Time-series analysis · Wavelet analysis

1 Introduction

Coastal areas are dynamic and complex systems (Šimac et al. 2023), which are characterized by high population density and significant economic activities. Therefore, it is reasonable to acknowledge that the environmental degradation of coastal areas is one of the most critical issues modern societies face. Sea level forcing and coastal erosion can be regarded as a natural evolutionary process, which is rapidly accelerating due to human activities that intensify climate change and consequently sea level rise (Jevrejeva et al. 2009). This fact leads to serious consequences for both natural ecosystems and human activities along coastal zones (Nicholls and Cazenave 2010). In the Mediterranean region, and particularly in the Aegean Sea, sea level rise is already noticeable, making coastal areas particularly vulnerable to flooding and environmental degradation (Bonaduce et al. 2016). According to similar studies conducted, global sea level is rising at a rate of approximately 3.3 mm/year during the period 1993–2018 (Chen et al. 2017; Nerem et al. 2018). This increasing trend is the result of polar ice melting and the thermal expansion of oceans as a direct consequence of the overall increase in temperature (Micheal et al. 2019).

R. N. Adam (✉) · G. S. Vergos
Laboratory of Gravity Field Research and Applications (GravLab), Department of Geodesy and Surveying, Aristotle University of Thessaloniki, Thessaloniki, Greece
e-mail: adamrafa@topo.auth.gr

J. T. Freymueller, L. Sànchez (eds.), *International Symposium on Gravity, Geoid and Height Systems 2024 (GGHS2024)*, International Association of Geodesy Symposia 158, https://doi.org/10.1007/1345_2025_301

To address the catastrophic effects of coastal erosion that have resulted as a direct consequence of sea level rise, geoscientists require high precision in determining the geoid and the Earth's gravitational field, especially in coastal areas. The geoid can be the ideal reference surface with respect to which we can measure sea level variations, as it can provide the connecting link to physical heights of the Earth's topography (Rummel and Colombo 1985; Sanchez et al. 2024; Vergos et al. 2024). This becomes of special interest as the geoid serves as the reference surface for all other geosciences and its accurate determination is of high importance especially in cases that height combination at the sea-land boundary is needed. Accurate mapping of the sea level, combined with terrestrial data, requires a precision of approximately ± 1 cm for wavelengths of about 10 km along the coast (Sandwell et al. 2014).

This study aims to analyse sea level time series from altimetry data, employing Sea Level Anomalies (SLA), for the period 1993–2024. The focus is on the detection and interpretation of sea level fluctuations in the area of the Aegean Sea, Greece. The analysis of altimetry data was performed using two main approaches to detect dominant spatiotemporal variability patterns. These refer to Wavelet and Empirical Orthogonal Function (EOF) analysis. The overall study of the results aims to draw conclusions regarding the impacts of climate change on coastal areas and determine sea level trends that can be used in order to predict vulnerable coastal areas. As a case study, the shoreline of Asprovalta (Northern Aegean Sea, Greece) was selected for a more detailed assessment of local sea level variability and potential coastal vulnerability.

2 Methodology

2.1 Time Series of Altimeter Data

Satellite altimetry data enable the systematic recording of sea level on a global as well as regional scales, providing critical information on long-term trends (Cipollini et al. 2017). Altimetry data from missions such as Jason-3 and CryoSat-2 offer highly accurate observations of the time-varying sea surface, effectively contributing to the study of sea level fluctuations and the study of local phenomena they are related to the climatic conditions and topography of the area (Pavlis et al. 2012). Furthermore, the contribution of techniques, such as the Low-Resolution Mode (LRM) and that of the more recent acquisitions by SAR/SARin, is vital. The improved ability to observe sea level near the coast is mainly provided by SAR/SARin altimetry. In this acquisition mode, the application of Delay-Doppler processing allows much better resolution along the trajectory ($\sim$300 m), while improving the recovery of physical parameters from the altimeter waveform. However, as we approach the coastal zone, both SAR/SARin and conventional LRM operations generally require special retracking procedures to obtain reliable sea level estimates. These techniques, especially at small geographical scales, facilitate the prediction of the impacts of sea level rise (Benveniste et al. 2020).

Satellite altimetry is used to determine sea surface heights relative to a defined geodetic reference system using the ellipsoid or geoid as a reference, based on the measurement of the distance between the satellite and the instantaneous sea surface (Chelton et al. 2001; Deng 2020). In the computation of the satellite-to-instantaneous sea surface distance (R), the dominant errors originate from changes in the propagation velocity of the radar pulse caused by atmospheric and ionospheric conditions. Therefore, to obtain usable results, the distance R is corrected for various errors, which are classified into four general categories: orbital errors, altimeter errors, atmospheric effects, and model errors. After applying the necessary corrections, the fundamental equation for calculating the corrected distance R takes the form:

$$\begin{aligned} R &= R_{obs} - \textstyle\sum_i \Delta R_i \\ &= R_{obs} - \left(\Delta R_{tdry} + \Delta R_{twet} + \Delta R_{iono} + \Delta R_{dyn} + \ldots\right), \end{aligned} \tag{1}$$

where the ΔR_i , i = 1 . . . , represent the corrections. Specifically, ΔR_{tdry} and ΔR_{twet} correspond to the effects of the dry and wet components of the troposphere, ΔR_{iono} represents the ionospheric effect, and ΔR_{dyn} accounts for dynamic factors such as ocean currents, tides, Earth's polar motion, solid Earth tides, and sea state dynamics (Chelton et al. 2001; Deng 2020).

Equally important is the definition of the Mean Sea Surface (MSS). The MSS is obtained by calculating the average value of the sea surface heights for at least 1 year and refers to a specified set period (Dehant 1990). Sea level can also be expressed through SLAs, which are calculated in relation to the MSS as follows:

$$SLA = SSH - MSS, \tag{2}$$

In this study, the Sea Surface Height (SSH) represents the height of the instantaneous sea surface above the reference ellipsoid along the ellipsoidal normal, as derived from the satellite altimeter range measurements and precise orbit data.

For the construction of a 1993–2024 altimetry data time series for the area under study, all available altimetric observations have been collected in the form of SLAs considering the necessary aforementioned corrections. In the present work, satellite altimetry data from the missions ERS-1, ERS-2, ENVISAT, GEOSAT Follow-On, TOPEX/Poseidon, Jason-1, Jason-2, Jason-3, CryoSat-2, SARAL, Sentinel-3A/3B, and SWOT, have been employed, to construct a multi-satellite time-series of SLA variations in the area under study. As various altimetry databases are available, we have relied on the geophysical data records from AVISO+ (Ssalto/Duacs and distributed by Aviso+, with support from Cnes; https://www.aviso.altimetry.fr; SSALTO/DUACS 2025), RADS (Radar Altimeter Database System; Scharroo et al. 2012), and OpenADB (Open Altimeter Database; Schwatke et al. 2024). All altimetry data used refer to a common geodetic reference frame as well as a common MSS. For the former, they have been referred to the ITRF2014 (Altamimi et al. 2016) and the CLS2021 single combined-hybrid MSS (CLS 2021). The altimetric orbits and SLA measurements are expressed with respect to the ITRF2014 frame, using the WGS84 ellipsoid as the geometric reference surface (NIMA 2000). The standard geophysical and atmospheric corrections included in each mission's processing chain were applied according to the corresponding GDR-F altimetry data standards, providing consistency in the handling of tropospheric, ionospheric, tidal and sea-state-related effects across missions. It should be noted that for the geophysical corrections among multi-mission altimetry data to be consistent, those provided in the AVISO User Handbooks have been applied (AVISO 2025). The along-track data collected, were sorted by month for the entire time series. Despite these data having already undergone preprocessing to remove gross errors, an additional 3σ test was applied to each monthly data record to identify and remove any potential blunders in the dataset. In months where data from a single satellite mission was recorded, the data were used as is, while in months where there was coverage from more than one mission, the data were combined to calculate the average SLA values for the month, ensuring consistency in processing.

From the analysis of the time series, an SLA variation of approximately 3 ± 0.1 mm/year was observed for the period 1993–2024 after applying a simple linear regression. Figure 1 shows the estimated average monthly values, giving a clear picture of the increasing trend followed by the SLA values. According to the time series statistics (see Table 1), the maximum value of SLA reaches 13.146 cm, while the minimum value is −21.098 cm. The average value of the time series is −2.501 cm, even though since 2013 a positive average value is recorded. The standard deviation of the time series is 2.62 cm. These results give a clear picture of variation around the mean value, which reflects the early periods with negative anomalies, affecting the overall time series mean.

With the completion of the preparation and processing of the altimetric data, the next phase of the study focuses on the analysis of trends and the main modes of sea level variation. For this purpose, the use of two methods was chosen: the EOF Analysis and Wavelet Analysis. The combination of

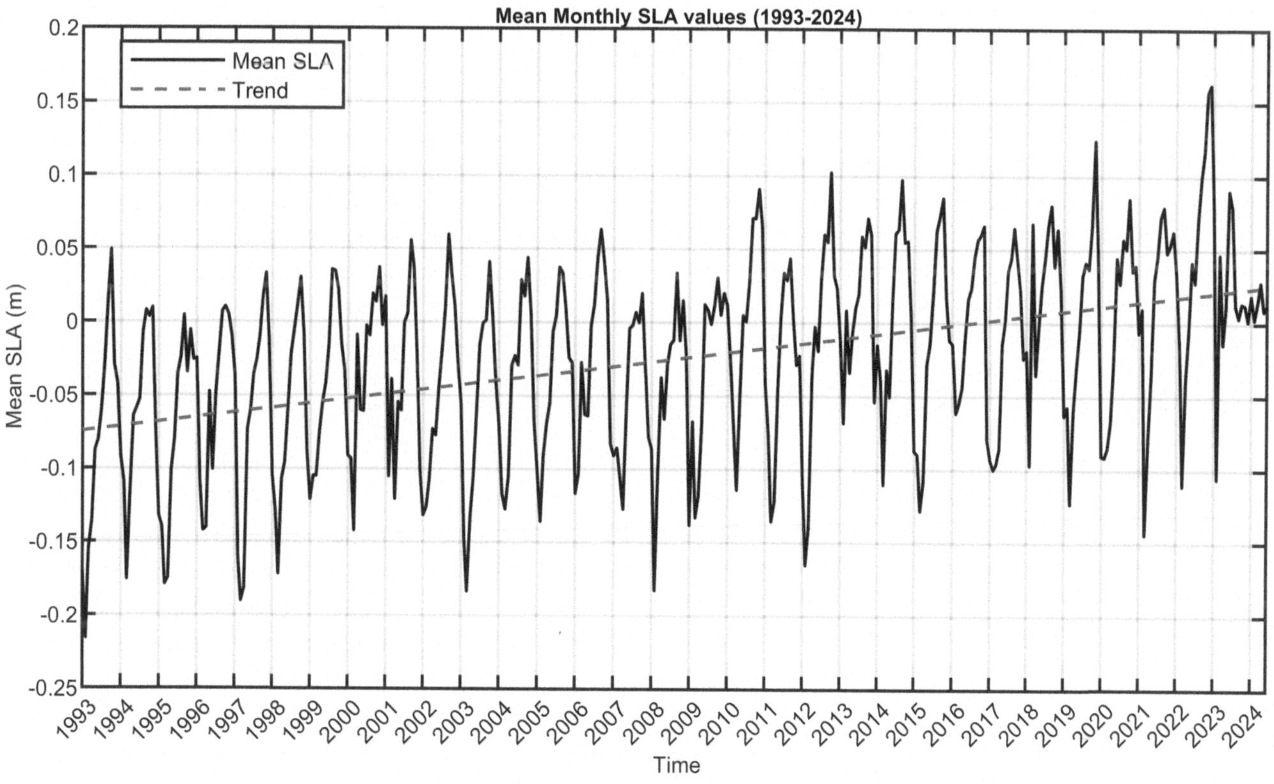

Fig. 1 Mean monthly SLAs and trend for the study area. Units: [m]

Table 1 Statistics of the time series 1993–2024. Units: [cm]

Max	Min	Mean	Std
13.146	−21.098	−2.501	2.620

these two approaches provides a complete picture of the long-term and short-term trends in the study area.

2.2 Empirical Orthogonal Function Analysis – EOF Analysis

EOF analysis is a powerful tool for data compression and dimension reduction used broadly in meteorology and oceanography (North et al. 1982). In this study, the constructed altimetric time series has been analyzed using EOF analysis, enabling the decomposition of the time series into a series of orthogonal functions, known as modes, which capture the dominant variability (Hannachi 2004; Zhong et al. 2019). To enhance the results using the EOF analysis, the monthly SLA data have been averaged over quarterly annual time periods, so as to correspond also to the Wavelet analysis Levels. This method helps to reduce the uncertainty that can result from short-term fluctuations and allows for a clearer and more stable pattern of sea level changes throughout the period considered.

Initially, the first EOF is calculated (see Fig. 2), which represents the main mode of sea level variation and explains the largest percentage of the total variance (66.08% of total variance). In the present study, the First EOF corresponds to long-term sea level rise trends mostly in areas of the Northern Aegean and south of Athens, similar to observations in other parts of the Mediterranean (Tsimplis and Shaw 2010). The main variability is seen close to the Sporades islands in Central Aegean, and in the area between Greek mainland and the island of Crete to the South. The highest positive values of the EOF (up to 4.17 mm) are mainly over Central, Eastern and Southeastern Aegean, while the lowest values (0.14 mm) are recorded in the northern Ionian Sea and parts of the Central-North Aegean. The average value of the first EOF for the entire study area is 2.26 mm, with a standard deviation of 0.45 mm (see Table 2) which represents the variance of sea level variability for the area, but with significant spatial variations. The spatial distribution of the first EOF is closely linked to the influence of ocean currents and atmospheric systems on sea level variability, as well as phenomena related to climate change, such as sea temperature rise and polar ice melt (Church et al. 2013).

To remove the long-term variability from the dataset, it is necessary to quantify the underlying trend in the SLAs. Figure 3 depicts the spatial distribution of the estimated SLA trend for the period 1993–2024, expressed in mm/year. The data clearly demonstrate a significant long-term trend across different locations within the study region, indicating a continuing rise in sea level, although its magnitude and spatial pattern vary substantially. More specifically, the highest positive trends (up to ∼12 mm/year) are observed in parts of the northeastern Aegean and southeastern regions, while localized areas in the northern Aegean and Ionian Sea showing weaker or even negative trends, indicating a small decline in sea level. Nevertheless, the whole study area seems to show an upward trend of ∼3–4 mm/year (see Table 3). These patterns highlight the influence of both regional oceanographic processes and broader climatic parameters on SLA variability. Given the existence of this non-negligible long-term trend, it is necessary to remove it before further analysis to highlight dominant patterns of variability that are not related to gradual sea-level rise.

The next step in the EOF analysis refers to detrending the data, as this allows to focus on dynamic changes and seasonal features associated with regional or short-term processes. The analysis follows three main steps. First, the calculated trend is removed from the SLA time series at each geographic location to retain only the most rapid or seasonal variations. This step is critical to highlight short-term and possibly intra-annual factors contributing to volatility. Then, EOF analysis is re-applied to the detrended data, with the aim of identifying the main components of the variation, free from the influence of the long-term trend. This procedure involves the calculation of the eigenvectors and eigenvalues of the covariance matrix, allowing the categorisation of spatio-temporal variation into independent functional patterns.

The outcome of this process is the dominant Principal Components (e.g. PC1, PC2, PC3), which represent the basic patterns of sea level change without the influence of the trend. The PCs in combination with the spatial distributions of the first EOF make it possible to draw reliable conclusions about sea level variability and the causes leading to these changes (Hannachi et al. 2007). The PC1 (see Fig. 4) is characterised by strong annual periodicity, reflecting a clear seasonal cycle. It also interprets the largest part of the SLA signal in the area since it contains the most useful information. In contrast, PC2 and PC3 reveal longer than annual variability, and a much smaller variances in the PC modes. It should be noted that the presented PC1 is normalized, hence its variation from −1 to 1, in contrast to the main modes, where no normalization has been performed to retain units. All three SLA modes were calculated as well, with the first one, as in the first PC, contributing to 85.09% of the total variation with positive anomalies across most of the study area. The highest values occur in the southern and central Aegean Sea, decreasing towards the northeast. This mode probably reflects the dominant seasonal signal. The first eigenvalue corresponding to the first Mode is representative of the dominant temporal patterns which

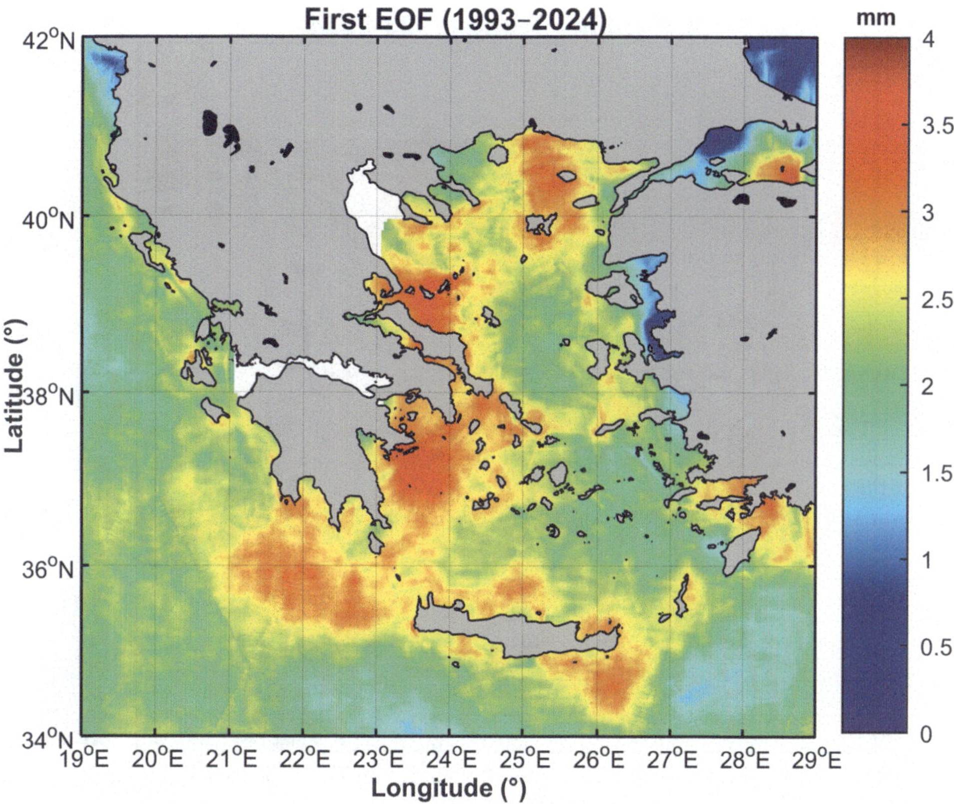

Fig. 2 First EOF map. Units: [mm]

Table 2 Statistics of the first EOF. Units: [mm]

Max	Min	Mean	Std
4.174	0.139	2.261	0.447

is confirmed by the percentage it occupies of the total variability of the phenomenon. For this reason, the focus on PC1 and Mode 1 allows simplifying the analysis without losing useful information. The other Modes explain smaller percentages of variation and may include more noise or be associated with more local and secondary processes.

2.3 1-D Discrete Wavelet Transform

Wavelet Transform (WT) analysis is another useful tool for analyzing phenomena that vary both in space and time, which makes wavelets particularly useful for the study of SLA time series which have significant space and time variations. Based on this, it becomes possible to detect short-term and long-term fluctuations in the data, interpreting seasonal phenomena, climate trends and extreme weather events (Torrence and Compo 1998).

In this study, Wavelet analysis was applied to analyse the time series of altimetry data covering the period 1993–2024. The data to be analysed were averaged over quarterly time periods, firstly in order to have a clearer and more stable pattern of sea level changes and secondly to have a common time scale of analysis with the preceding EOF analysis. In addition, the analysis was performed using six Levels of decomposition, allowing the investigation of events occurring from a few months to several years. In this context, Level 1 captures variability at 3–6 month scales, Level 2 at 6–12 months, Level 3 at 12–24 months, Level 4 at 2–4 years, Level 5 at 4–6 years, and Level 6 at 6–12 years. A class 10 Daubechies orthogonal wavelet has been chosen for the analysis, due to its ability to accurately represent both high and low frequency components, while maintaining good localizability over time (Daubechies 1988). The analysis aims to extract significant seasonal variations in the time series, which, in combination with the long-term trends, can be related to the steady rise in sea level (Marcos and Tsimplis 2007).

Wavelet analysis indicated strong variability in the entire SLA time series. To quantitatively estimate the energy distribution across time scales, the Root Mean Square (RMS) was calculated for each Level separately (see Table 4). At the lower Levels, higher RMS value were found, being at the 0.036 m for Level 1, 0.034 m for Level 2, 0.012 m for Level 3, 0.014 m for Level 4 0.014 m for Level 5 and 0.017 m for Level 6. The approximation coefficients, i.e., the signal remaining after the decomposition to Level 6, has an RMS of

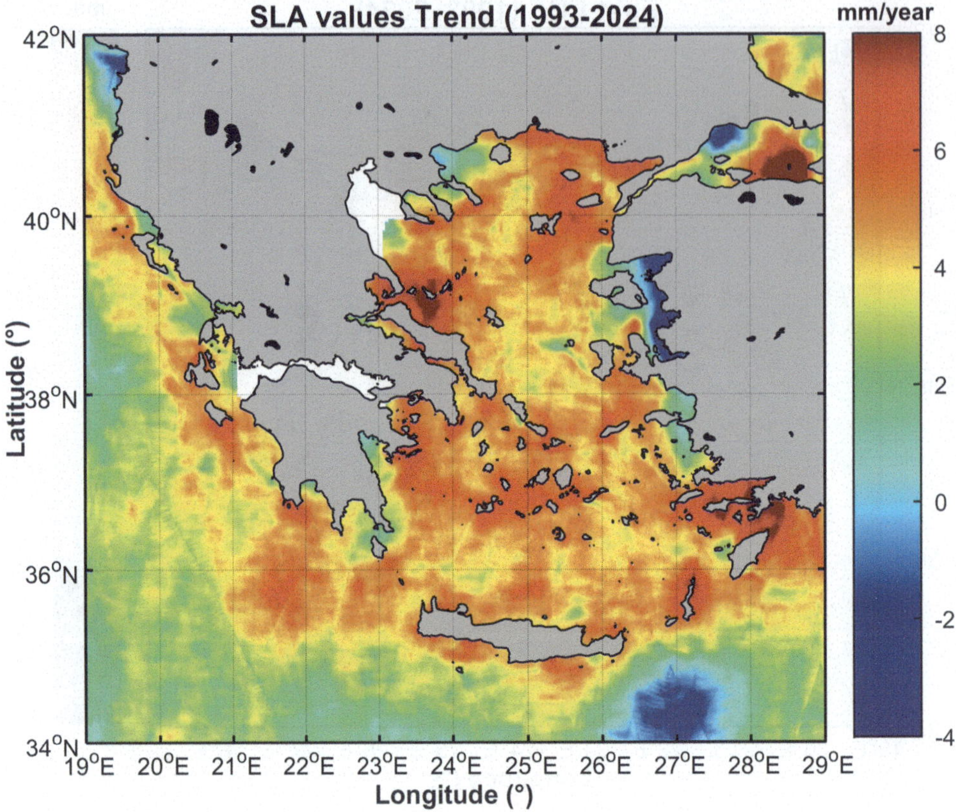

Fig. 3 SLA trend in the Aegean Sea. Units: [mm/year]

Table 3 Statistics of the SLA trend. Units: [mm/year]

Max	Min	Mean	Std
12.733	−4.732	3.696	1.636

0.028 m. These results demonstrate that short-term seasonal variability dominates the lower Levels, whereas longer-term trends become evident in the higher levels and in the final approximation component.

Particular attention should be paid to Levels 1 and 2 as they capture the dominant short-term temporal variability of SLAs (see Fig. 5). In the bottom panel of Fig. 5, the Level 2 component (shown as a black solid line), which corresponds to variability at 6–12 month scales, is compared to the PC1, shown as a blue dashed line. The left y-axis represents mean SLA in meters (not normalized), while the right y-axis corresponds to the PC1 amplitude, which is normalized between −1 and 1. The high correlation recorded between Level 2 and PC1 indicates that both decomposition methods effectively isolate the annual cycle of sea level variation in the Aegean Sea with a similar pattern of variability. Additionally, the observed changes are consistent with corresponding results derived from studies that highlight the importance of temperature and atmospheric factors on surface water levels (Marcos and Tsimplis 2007). At Levels 3, 4, 5 and 6 patterns are identified that are related to longer-term phenomena such as changes in sea currents and the effects of atmospheric systems. These patterns presented could be argued to demonstrate the influence of phenomena that cause fluctuations in pressure and winds, especially in the Eastern Mediterranean (Kaskaoutis et al. 2023).

2.4 Forecasting Sea Level Changes Over 50 and 100 Years

As it has already become clear, sea level rise is ongoing in the area under study hence the ability to predict future sea level change poses an important tool for assessing the future state of the seas and for sustainable coastal management. In this context, the present study attempts to estimate the future evolution of SLA in the Aegean region, based on the observed trends over the last decades.

A methodology based on observed trends in the Aegean region was applied to predict future trends in SLAs, using March 2008, being the reference epoch of the used dataset, as the reference month. This approach forms the basis of the SLA projections, where the spatially distributed trend estimated from the EOF analysis is scaled by the corresponding number of months between the reference epoch (March

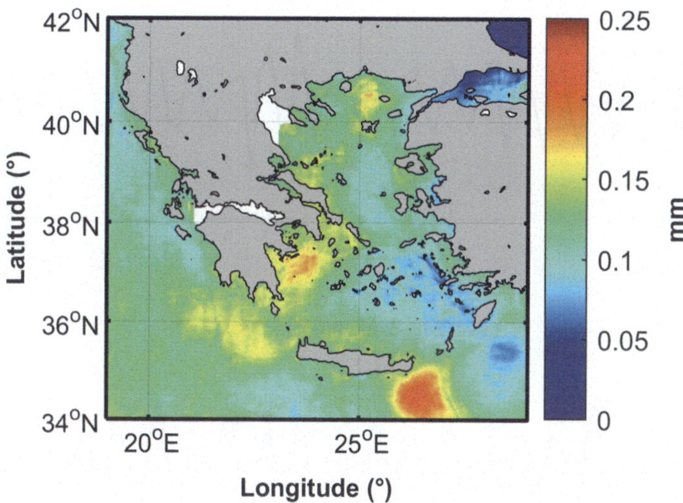

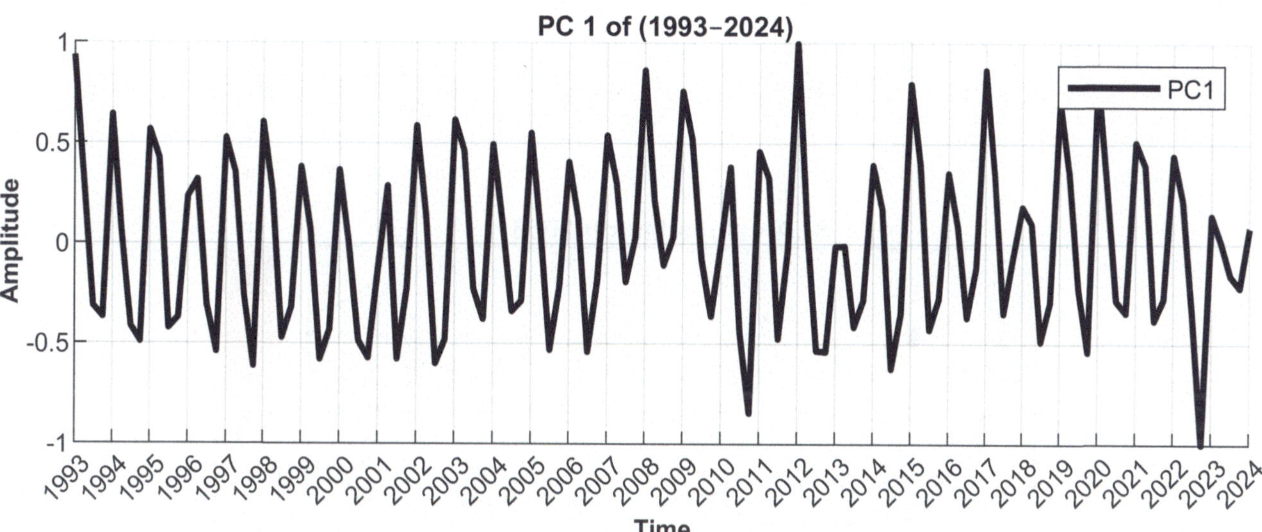

Fig. 4 The first EOF (top) and PC (bottom) from the detrended 1993–2024 SLA (EOF/Modes Units [mm] and PC is normalized)

Table 4 RMS values for each Level. Units: [m]

Level 1	Level 2	Level 3	Level 4	Level 5	Level 6	Approx. coeff.
0.036	0.034	0.012	0.013	0.014	0.017	0.028

2008) and the forecast year. Estimates of the evolution of SLA values were made for the next 50 and 100 years from the reference month, more specifically for 2058 (see Fig. 6) and 2108 (see Fig. 7). The analysis suggests that by 2058, sea level in the Aegean region is expected to rise in line with global rising trends recording an average value of 0.399 m, while the situation in 2108 is expected to be even more severe with average value of SLA rise to be 0.905 m (see Table 5).

The projected SLAs in 2058 and 2108 have been used to predict flooded coastal areas, under the pressure of sea level rise, in the area under study. This was achieved by combining the projected SLAs with information about land topography, based on the Copernicus DEM-GLO-30 Digital Terrain Model (DTM) (Anon 2022; Airbus 2022). In order to carry our such a height combination among SLAs and orthometric heights from a DEM, they need to be in consistent horizontal and vertical reference systems and frames, while they should also follow the same considerations for the permanent deformation of the Earth's crust under the influence of tides. Traditionally, satellite altimetry data refer to the mean-tide (MT) system, while the datasets used in this study employ the WGS84/ITRF2014 as a reference frame. SSHs are a purely geometric quantity, therefore no assumption about a reference equipotential surface is needed. In same cases, SLAs can be constructed relative to a geoid model rather than and MSS, but this is just a convention as when constructing the SSHs the influence of either the MSS

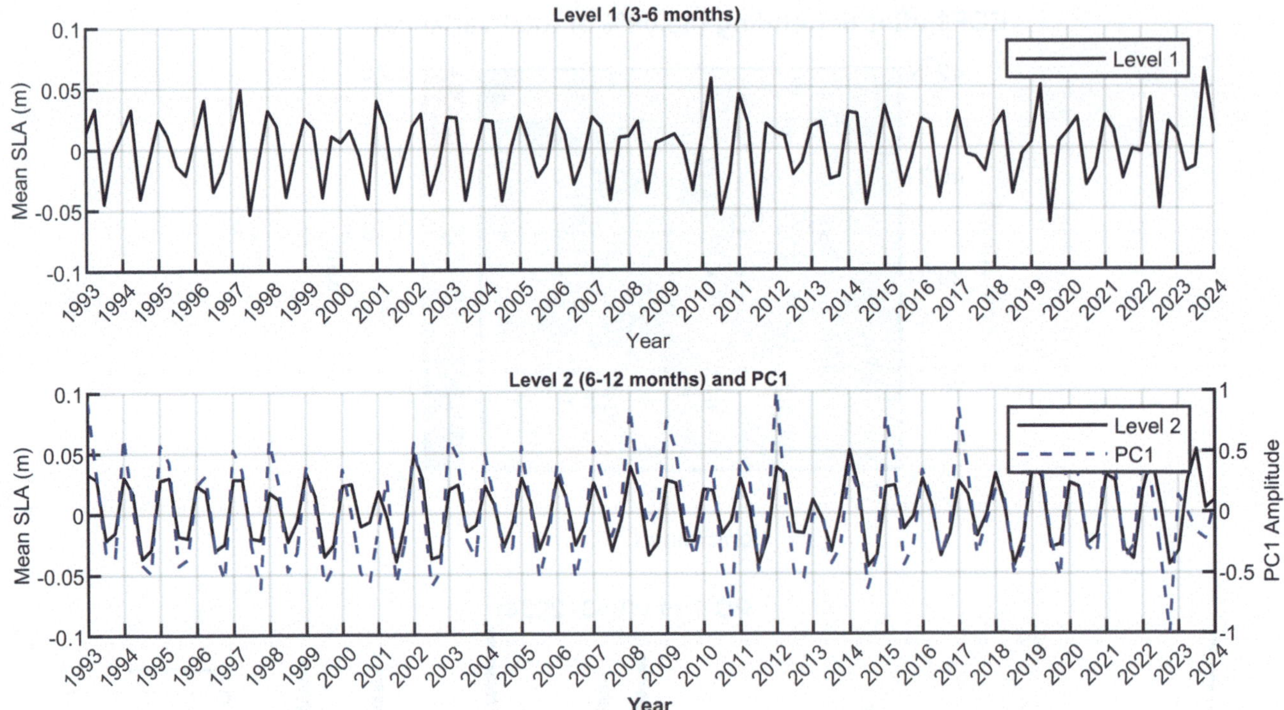

Fig. 5 1D–discrete wavelet transform – level 1 (top), level 2 with PC1 (bottom) (levels 1–2 units [m] and PC is normalized)

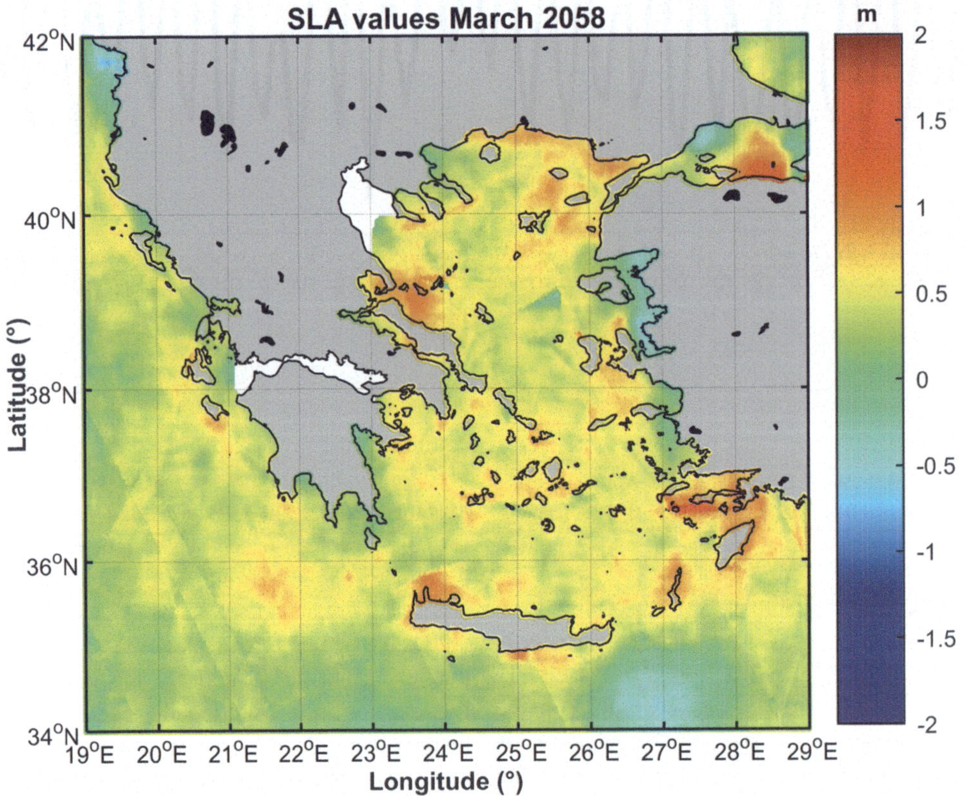

Fig. 6 SLA in the Aegean Sea for 2058. Units: [m]

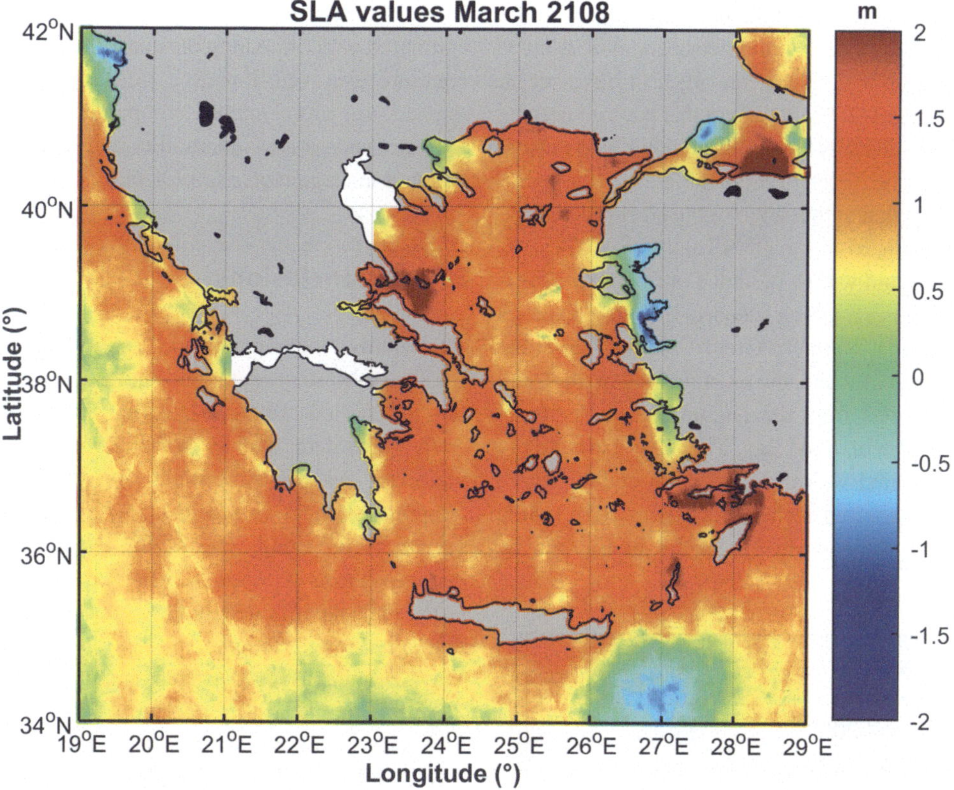

Fig. 7 SLA in the Aegean Sea for 2108. Units: [m]

Table 5 Statistics of SLAs for March 2008, March 2058, March 2108. Units: [m]

	Max	Min	Mean	Std
March 2008	0.998	−1.275	−0.107	0.080
March 2058	2.094	−0.833	0.399	0.232
March 2108	3.191	−1.268	0.905	0.449

or the geoid is restored. On the other hand, the orthometric heights from the Copernicus DEM, which provide a realization of physical heights, refer to the WGS84/ITRF2014 reference frame as well (Airbus 2022), while EGM2008 (Pavlis et al. 2012) is used as the reference equipotential surface with respect to which these orthometric heights are computed from. As no specific tide system considerations are provided in the Copernicus data handbook (Airbus 2022) it is assumed that since the online EGM2008 calculator service at the National Geospatial Agency (NGA) has been used, the EGM2008 spherical harmonic coefficients employed refer to the tide-free (TF) system, to be also consistent with GNSS observations. Hence, the such-derived orthometric heights have been transformed from the TF free to the MT system, following Mäkinen (2021) and Ekman (1989), in order to be consistent with the altimetric observations. For the SSHs to be comparable to orthometric heights on land, they need to be referenced to a geoid model so that the dynamic ocean topography is defined, being the marine equivalent of orthometric heights at land, as:

$$DOT = SSH - N, \tag{3}$$

For the flooding analysis to be consistent with local GNSS/Levelling benchmarks, which are commonly used in land surveying studies to determine orthometric heights for, e.g., hydrological and hydraulic analysis, we employed a national gravimetric geoid model (Tziavos et al. 2010), as it is more representative of the local and regional gravity field variations and peculiarities of the area under study, compared to a global geopotential model like EGM2008. This is a selection performed intentionally in the present study, instead of just using EGM2008 that was used in the Copernicus DEM, to support local studies tied to some GNSS/Levelling and tide-gauge benchmarks in the area (Grigoriadis et al. 2014). Hence in the Eq. (3), a geoid model is needed to convert the SSHs to a meaningful quantity that can be used to directly compare the sea level variations with physical heights, or rather their realization in terms of orthometric heights. The gravimetric geoid model used (Tziavos et al. 2010), uses WGS84/ITRF2014 as a reference frame, while it refers to the tide-free system, hence a conversion of the derived geoid heights to the mean-tide system has been performed following Mäkinen (2021) and Ekman (1989). Moreover, as far as the land topography is concerned, the determined orthometric heights have been referred from the EGM2008 geoid to the national gravimetric geoid of Tziavos et al. (2010), by using a simple mean offset (Tziavos et al. 2012; Vergos et al. 2014, 2015, 2018). In that way, all heights use a consistent geometric reference frame and a consistent geoid model and can be directly comparable.

It should be noted that at this stage of the current research, no vertical land motion, from e.g. GNSS and/or SAR/InSAR, is taken into account and will be included in future investigations. Figure 8 presents a flooding scenario resulting from sea level rise in Asprovalta, Northern Greece. In this figure, the sea surface is depicted in white, the existing land in grey, and the coastline, as derived from the Digital Elevation Model (DEM), is delineated with a red line. The areas estimated to be flooded are presented with different color coding: green refers to the areas expected to be affected by 2058, while blue indicates the further expansion of the flood zone by 2108. The flooded area shows a significant increase in 2108 compared to 2058. Along the coastline, increases in the flood zone are observed, however, in this part of Asprovalta the impacts are more intense compared to other areas, which indicates increased vulnerability and highlights the need for targeted management and protection of the coastal area.

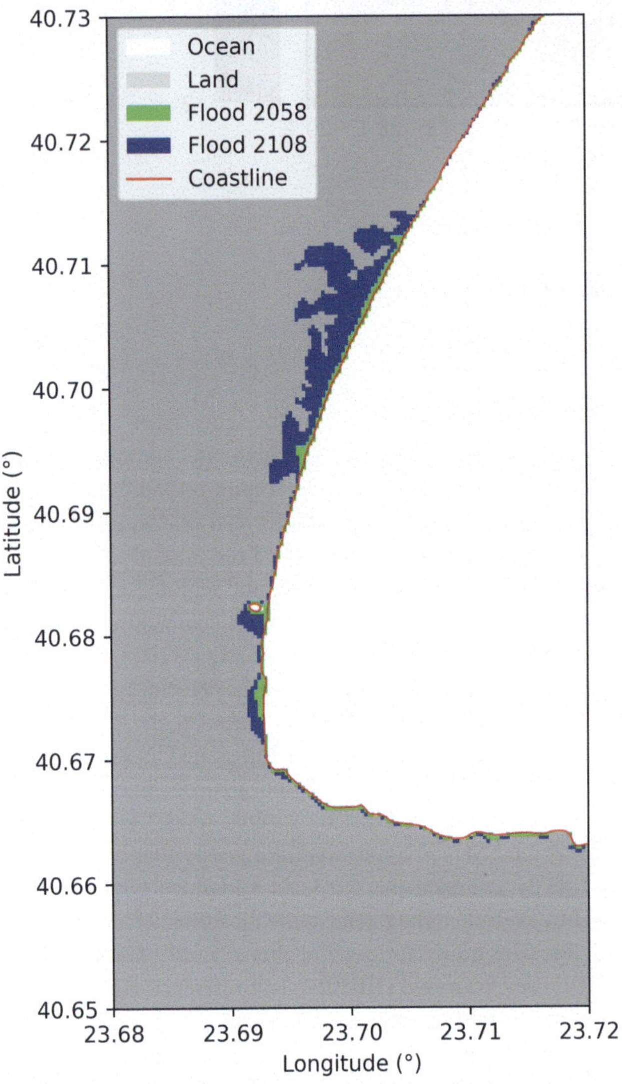

Fig. 8 Flooding scenario map for Asprovalta in 2058 and 2108

3 Conclusions

This study focused on the spatio-temporal analysis of satellite altimetry data and more specifically SLA data for the period 1993–2024. Two methods were used to analyze the data, namely Wavelet and EOF analysis. These methods revealed important information regarding the temporal changes in sea level, with WT showing seasonal variations on small time scales and the second Level of decomposition depicted well the annual sea level variation. EOF analysis revealed both spatial and temporal patterns that agreed with the corresponding temporal patterns of the Wavelet analysis. It identified spatial patterns in variability, identifying areas mostly in Central and Southeastern Aegean Sea with increased sea level variations and increased sea level rise trend. These findings underline the need for a multifactorial approach to understanding sea level dynamics. Both the projections for 2058 and 2108 highlight the need for continuous monitoring and the implementation of coastal management strategies to closely monitor sea level in the Aegean and on a global scale. From the projection of the sea level rise to the future, areas with increased vulnerability to sea level rise can be identified, to derive and formulate coastal management measures.

Overall, the study confirms that both Wavelet analysis methods and EOF analysis are equally effective tools for investigating sea level change trends. The results obtained can be used for the proper management of coastal areas, especially in areas with complex climatic and geophysical conditions such as the Aegean Sea.

References

Airbus Defence and Space GmbH (2022) Copernicus digital elevation model – product handbook. Version 5.0. https://doi.org/10.5270/ESA-c5d3d65

Altamimi Z, Rebischung P, Métivier L, Collilieux X (2016) ITRF2014: a new release of the international terrestrial reference frame modeling nonlinear station motions. J Geophys Res Solid Earth 121(8):6109–6131. https://doi.org/10.1002/2016JB013098

Anon (2022) COPERNICUS Digital Elevation Model (DEM) for Europe at 30 meter resolution derived from Copernicus Global 30 meter dataset. http://data.europa.eu/88u/dataset/f576cda8-d598-478c-b8fe-ad2634c927e8. Accessed 27 Apr 2025

AVISO+ (2025) AVISO+ user handbooks.https://www.aviso.altimetry.fr/en/data/product-information/aviso-user-handbooks.html. Accessed 6 Nov 2025

Benveniste J, Birol F, Calafat F, Cazenave A, Dieng H, Gouzenes Y, Legeais J, Leger F, Niño F, Passaro M, Schwatke C, Shaw A (2020) Coastal sea level anomalies and associated trends from Jason satellite altimetry over 2002–2018. Sci Data 7:357. https://doi.org/10.1038/s41597-020-00694-w

Bonaduce A, Nadia P, Oddo P, Spada G, Gilles L (2016) Sea-level variability in the Mediterranean Sea from altimetry and tide gauges. Clim Dyn 47:2851–2866. https://doi.org/10.1007/s00382-016-3001-2

Chelton DB, Ries JC, Haines BJ, Fu L-L, Callahan PS (2001) Chapter 1. Satellite altimetry. In: Fu L-L, Cazenave A (eds) Satellite altimetry and earth sciences, vol 69. Academic Press, pp 1–ii. https://doi.org/10.1016/S0074-6142(01)80146-7

Chen X, Zhang X, Church J, Watson C, King M, Monselesan D, Legresy B, Harig C (2017) The increasing rate of global mean sea-level rise during 1993–2014. Nat Clim Chang 7:492–497. https://doi.org/10.1038/NCLIMATE3325

Church J, Clark P, Cazenave A, Gregory J, Jevrejeva S, Levermann A, Merrifield M, Milne G, Nerem R, Nunn P, Payne A, Pfeffer W, Stammer D, Alakkat U (2013) Climate change 2013: the physical science basis. Contribution of Working Group I to the Fifth Assessment Report of the Intergovernmental Panel on Climate Change. Sea Level Change, pp 1138–1191

Cipollini P, Calafat FM, Jevrejeva S, Melet A, Prandi P (2017) Monitoring sea level in the coastal zone with satellite altimetry and tide gauges. Surv Geophys 38(1):33–57. https://doi.org/10.1007/s10712-016-9392-0

CLS (2021) 2021 hybrid mean sea surface (version 2021) dataset. CNES. https://doi.org/10.24400/527896/A01-2021.004

Daubechies I (1988) Orthonormal bases of compactly supported wavelets. Commun Pure Appl Math 41(7):909–996. https://doi.org/10.1002/cpa.3160410705

Dehant V (1990) On the nutations of a more realistic earth model. Geophys J Int 100(3):477–483. https://doi.org/10.1111/j.1365-246X.1990.tb00700.x

Deng X (2020) Satellite altimetry. In: Sideris MG (ed) Encyclopedia of geodesy. Springer International Publishing, pp 1–7. https://doi.org/10.1007/978-3-319-02370-0_58-2

Ekman M (1989) Impacts of geodynamic phenomena on systems for height and gravity. Bull Géod 63:281–296. https://doi.org/10.1007/BF02520477

Grigoriadis VN, Kotsakis C, Tziavos IN, Vergos GS (2014) Estimation of the reference geopotential value for the local vertical datum of continental Greece using EGM08 and GPS/leveling data. In: Marti U (ed) Gravity, geoid and height systems. International Association of Geodesy Symposia, vol 141. Springer, Cham. https://doi.org/10.1007/978-3-319-10837-7_32

Hannachi A (2004) A primer for the eof analysis of climate data. Department of Meteorology, University of Reading, UK. http://ncascms.nerc.ac.uk

Hannachi A, Jolliffe I, Stephenson D (2007) Empirical orthogonal functions and related techniques in atmospheric science: a review. Int J Climatol 27:1119–1152. https://doi.org/10.1002/joc.1499

Jevrejeva S, Grinsted A, Moore JC (2009) Anthropogenic forcing dominates sea level rise since 1850. Geophys Res Lett 36(20). https://doi.org/10.1029/2009GL040216

Kaskaoutis D, Liakakou E, Grivas G, Gerasopoulos E, Mihalopoulos N, Alastuey A, Dulac F, Dumka U, Pandolfi M, Pikridas M, Sciare J, Titos G (2023) Interannual variability and long-term trends of aerosols above the Mediterranean, pp 357–390. https://doi.org/10.1007/978-3-031-12741-0_11

Mäkinen J (2021) The permanent tide and the international height reference frame IHRF. J Geod 95(9):106. https://doi.org/10.1007/s00190-021-01541-5

Marcos M, Tsimplis M (2007) Forcing of coastal sea level rise patterns in the North Atlantic and the Mediterranean Sea. Geophys Res Lett 34:L01604. https://doi.org/10.1029/2007GL030641

Micheal O, Glavovic B, Hinkel J, van Roderik, Magnan A, Abd-Elgawad A, Rongshu C, Cifuentes M, Robert D, Ghosh T, Hay J, Ben M, Meyssignac B, Sebesvari Z, Smit AJ, Dangendorf S, Frederikse T (2019) Sea level rise and implications for low lying Islands, Coasts and Communities, pp 321–445. https://doi.org/10.1017/9781009157964.006

National Imagery and Mapping Agency (NIMA) (2000) Department of Defense World Geodetic System 1984: its definition and relationships with local geodetic systems. TR8350.2. NIMA, Bethesda

Nerem RS, Beckley BD, Fasullo JT, Hamlington BD, Masters D, Mitchum GT (2018) Climate-change–driven accelerated sea-level rise detected in the altimeter era. Proc Natl Acad Sci 115(9):2022–2025. https://doi.org/10.1073/pnas.1717312115

Nicholls R, Cazenave A (2010) Sea-level rise and its impact on coastal zones. Science (New York, N.Y.) 328:1517–1520. https://doi.org/10.1126/science.1185782

North G, Bell T, Cahalan R, Moeng F (1982) Sampling errors in the estimation of empirical orthogonal functions. Mon Weather Rev 110(7):699–706. https://doi.org/10.1175/1520-0493(1982)110<0699:SEITEO>2.0.CO;2

Pavlis NK, Holmes SA, Kenyon SC, Factor JK (2012) The development and evaluation of the earth gravitational model 2008 (EGM2008). J Geophys Res Solid Earth 117(B4). https://doi.org/10.1029/2011JB008916

Rummel R, Colombo OL (1985) Gravity field determination from satellite gradiometry. Bull Géod 59(3):233–246. https://doi.org/10.1007/BF02520329

Sanchez L, Barzaghi R, Vergos G (2024) Operational infrastructure to ensure the long-term sustainability of the international height reference system and frame (IHRS/IHRF). https://doi.org/10.1007/1345_2024_250

Sandwell D, Müller D, Smith W, Garcia E, Francis R (2014) New global marine gravity from CryoSat-2 and Jason-1 reveals buried tectonic structure. Science 346:65–67. https://doi.org/10.1126/science.1258213

Scharroo R, Leuliette E, Lillibridge J, Byrne D, Naeije M, Mitchum G (2012) RADS: consistent multi-mission products. In: Proceeding of the Symposium on 20 Years of Progress in Radar Altimetry. European Space Agency Special Publication, Venice, Italy. ISBN 978-92-9221-274-2

Schwatke C, Dettmering D, Passaro M, Hart-Davis MG, Scherer D, Müller FL, Bosch W, Seitz F (2024) OpenADB: DGFI-TUM's open altimeter database. Geosci Data J 11(4):573–588. https://doi.org/10.1002/gdj3.233

Šimac Z, Lončar N, Faivre S (2023) Overview of coastal vulnerability indices with reference to physical characteristics of the Croatian Coast of Istria. Hydrology 10:14. https://doi.org/10.3390/hydrology10010014

SSALTO/DUACS (2025) Multi-mission altimeter products (Global Ocean – Multimission altimeter satellite gridded sea surface heights and derived variables) [Sea Level Anomalies Along-track Level-2+ (L2P) for other missions]. AVISO+. Accessed 12/11/2025

Torrence C, Compo GP (1998) A practical guide to wavelet analysis. Bull Am Meteorol Soc 79(1):61–78. https://doi.org/10.1175/1520-0477(1998)079<0061:APGTWA>2.0.CO;2

Tsimplis M, Shaw A (2010) Seasonal sea level extremes in the Mediterranean Sea and at the Atlantic European coasts. Nat Hazards Earth Syst Sci 10. https://doi.org/10.5194/nhess-10-1457-2010

Tziavos I, Vergos GS, Grigoriadis V (2010) Investigation of topographic reductions and aliasing effects on gravity and the geoid over Greece based on various digital terrain models. Surv Geophys 31:23–67. https://doi.org/10.1007/s10712-009-9085-z

Tziavos IN, Vergos GS, Grigoriadis VN, Andritsanos VD (2012) Adjustment of collocated GPS, geoid and orthometric height observations in Greece. Geoid or orthometric height improvement? In:

Kenyon S, Pacino M, Marti U (eds) Geodesy for planet earth. International Association of Geodesy Symposia, vol 136. Springer, Berlin, Heidelberg. https://doi.org/10.1007/978-3-642-20338-1_58

Vergos GS, Andritsanos VD, Grigoriadis VN, Pagounis V, Tziavos IN (2015) Evaluation of GOCE/GRACE GGMs over Attica and Thessaloniki, Greece, and wo determination for height system unification. In: Jin S, Barzaghi R (eds) IGFS 2014. International Association of Geodesy Symposia, vol 144. Springer, Cham. https://doi.org/10.1007/1345_2015_53

Vergos GS, Erol B, Natsiopoulos DA, Grigoriadis VN, Işık MS, Tziavos IN (2018) Preliminary results of GOCE-based height system unification between Greece and Turkey over marine and land areas. Acta Geod Geophys 53:61–79. https://doi.org/10.1007/s40328-017-0204-x

Vergos GS, Grigoriadis VN, Tziavos IN, Kotsakis C (2014) Evaluation of GOCE/GRACE global geopotential models over Greece with collocated GPS/levelling observations and local gravity data. In: Marti U (ed) Gravity, geoid and height systems. International Association of Geodesy Symposia, vol 141. Springer, Cham. https://doi.org/10.1007/978-3-319-10837-7_11

Vergos GS, Tziavos IN, Mertikas S, Piretzidis D, Frantzis X, Donlon C (2024) Local gravity and geoid improvements around the Gavdos satellite altimetry Cal/Val site. Remote Sens 16(17). https://doi.org/10.3390/rs16173243

Zhong J, Lei J, Yue X, Wang W, Burns AG, Luan X, Dou X (2019) Empirical orthogonal function analysis and modeling of the topside ionospheric and Plasmaspheric TECs. J Geophys Res Space Physics 124(5):3681–3698. https://doi.org/10.1029/2019JA026691

Analysis of Extreme Flood Events Using GRACE-FO and SAR Data: A Case Study of Short-Term Flood in Thessaly, Greece

A. I. Triantafyllou, E. M. G. Mamagiannou, G. S. Vergos, and E. A. Tzanou

Abstract

Extreme weather events have increasingly resulted in significant disasters, impacting human life and economic development. Recently, short-term floods have caused substantial damage worldwide. Such was the case of the catastrophic flood that struck the region of Thessaly, Greece, in September 2023. In this study, we investigate the application of Earth Observation satellite data – optical, SAR, and gravimetric – for mapping flood events, focusing on the catastrophic flood that occurred in Thessaly. The rationale for incorporating both EO imagery and gravimetric observations lies in their complementary strengths: while optical and SAR sensors provide detailed, high-resolution snapshots of surface hydrological changes, gravimetry captures changes in the total water mass, enabling an assessment of broader hydrological conditions that may not be visible in conventional imagery. Our analysis was divided into three main components. Firstly, we utilized post-event satellite images from the Sentinel-2 platform, filtered by area of interest (AOI) and date range, with a constraint of less than 20% cloud cover. The Normalized Difference Water Index (NDWI) was applied to identify water bodies in optical imagery, serving as a reference for subsequent flood mapping methodologies. Secondly, we employed Sentinel-1 Synthetic Aperture Radar (SAR) data from pre- and post-event acquisitions. SAR images underwent edge masking and were merged to generate two mosaics. Flooded areas were detected by identifying regions with low backscatter values, specifically where the vertical vertical polarisation (VV) band exhibited values below a threshold of -17 in post-event mosaics relative to pre-event counterparts. Visual inspection confirmed the accuracy of flood detection against Sentinel-2 imagery. Lastly, we integrated GRACE and GRACE-FO Level-2 spherical harmonic coefficients provided by the Jet Propulsion Laboratory (JPL) to derive Liquid Equivalent Water Thickness (EWT), representing water thickness per grid cell. Despite the spatial resolution limitation of monthly GRACE products, EWT effectively captured spatial characteristics during short-term flood events. Simulating water level rise involved calculating flow direction, accumulation, and depressions using SRTM DEM data. The water spread simulation incorporated EWT values, modeling water movement and accumulation across terrain. This approach considered filled depressions and flow dynamics, providing a realistic representation of flood scenarios by accounting for terrain elevation and water volume. Modeled flooded areas closely resembled the spatial extent of

A. I. Triantafyllou (✉) · E. M. G. Mamagiannou · G. S. Vergos
Laboratory of Gravity Field Research and Applications – GravLab, Department of Geodesy and Surveying, Aristotle University of Thessaloniki, Thessaloniki, Greece
e-mail: anastria@topo.auth.gr

E. A. Tzanou
Department of Surveying and Geoinformatics Engineering of the School of Engineering, International Hellenic University, Terma Magnesias, Serres, Greece

J. T. Freymueller, L. Sànchez (eds.), *International Symposium on Gravity, Geoid and Height Systems 2024 (GGHS2024)*, International Association of Geodesy Symposia 158, https://doi.org/10.1007/1345_2026_307

the actual event, although volumetric accuracy was constrained by the spatial resolution of GRACE mission data.

Keywords

Flooding · GRACE-FO · Greece · SAR · Sentinel-1/2

1 Introduction

Mapping and analysing the extent of natural hazards have always posed significant challenges due to limited observational capabilities and the dynamic nature of such events. Traditionally, such assessments also required extensive field surveys, which were both costly and time-consuming, often taking weeks or even months to produce comprehensive results (de Moel et al. 2015; Albano et al. 2020). However, recent disasters like Storm Daniel (2023) and Storm Ianos (2020) in Greece, the devastating floods in Emilia-Romagna in Italy (2023), and those in Valencia in Spain (2024) have shown that the increasing prevalence of these hazards has significant social, economic, and environmental impact. As a result, it is becoming essential to map the extent of these disasters and assess their effects employing hetergogenouns data and methods (European Environment Agency 2010). Given its global coverage and excellent temporal precision, satellite remote sensing has become a crucial tool for real-time hazard mapping.

The most commonly used remote sensing methods for mapping hazards are satellite-based observations, particularly when supplemented with publicly accessible data from the Copernicus Sentinel missions. For instance, Sentinel-2, which has a 5-day return period and a 10-m spatial resolution (ESA 2012b) can offer valuable data by distinguishing between land and water utilizing the Shortwave Infrared (SWIR) and Near-Infrared (NIR) bands (Ji et al. 2009). Cloud cover during severe weather events, however, usually decreases its effectiveness (Ji et al. 2009). The Synthetic Aperture Radar (SAR) capabilities of Sentinel-1, on the other hand, overcome this limitation by enabling day-and-night imaging at a comparable spatial resolution in all weather circumstances (ESA 2012a). Sentinel-1 can consistently identify flooded areas by detecting variations in backscatter intensity, which is caused by signal absorption, which makes open water surfaces seem darker (Ji et al. 2009; Monti et al. 2024).

In this study, we utilized Sentinel-1 SAR Level – 1 GRD and Sentinel-2 MSI Level-2A optical data, between September 6 and September 11, 2023, to accurately measure the extent of the floods caused by Storm Daniel in Thessaly, Greece. Moreover, we have employed information from the Gravity Recovery and Climate Experiment Follow-On (GRACE-FO) mission, our method went beyond conventional satellite-based flood detection and added a new element. We incorporated GRACE-FO because, despite its coarse spatial and temporal resolution, it uniquely captures changes in total water storage, enabling us to estimate the volumetric component of the flood (Rodell and Li 2023; Tapley et al. 2019), an information that cannot be derived from optical or SAR data and is essential for simulating realistic water accumulation and spread. Specifically, we have employed Level-2 GRACE and GRACE-FO data in the form of monthly spherical harmonics coefficients from JPL (NASA Jet Propulsion Laboratory (JPL) 2018) to compute the Equivalent Water Thickness (EWT) following the recommendations after Loomis et al. (2020). The results have been derived in the form of equivalent water thickness per grid cell over the computation area.

2 Hydrological Context of the Study Area

The Region of Thessaly, located in the center of Greece, spans approximately 14,000 km^2 and is one of the country's most significant agricultural and hydrological regions. The landscape is predominantly characterized by extensive plains, which make up about 36% of its total area, but it is also defined by major mountain ranges: Mount Olympus to the north, the Pindus range to the west, and Mount Pelion to the east (HVA International 2024). These natural barriers significantly influence the region's climate and hydrology, trapping warm air masses and contributing to high humidity and frequent rainfall, particularly in winter (Nastos et al. 2015). It also receives a mean annual precipitation of about 44.6 mm, though rainfall varies significantly depending on altitude and local geography (HVA International 2024).

Thessaly's hydrology is primarily shaped by the Pineios River Basin (GR16) and the Almyros-Pelion River Basin (GR17) (see Fig. 1), as defined by the Management Plan of the River Basins of Thessaly (Special Secretariat for Water 2014). These two basins are characterized by a mix of natural floodplains, artificial drainage infrastructure, and groundwater-dependent ecosystems, all of which play a crucial role in the region's water management and flood risk

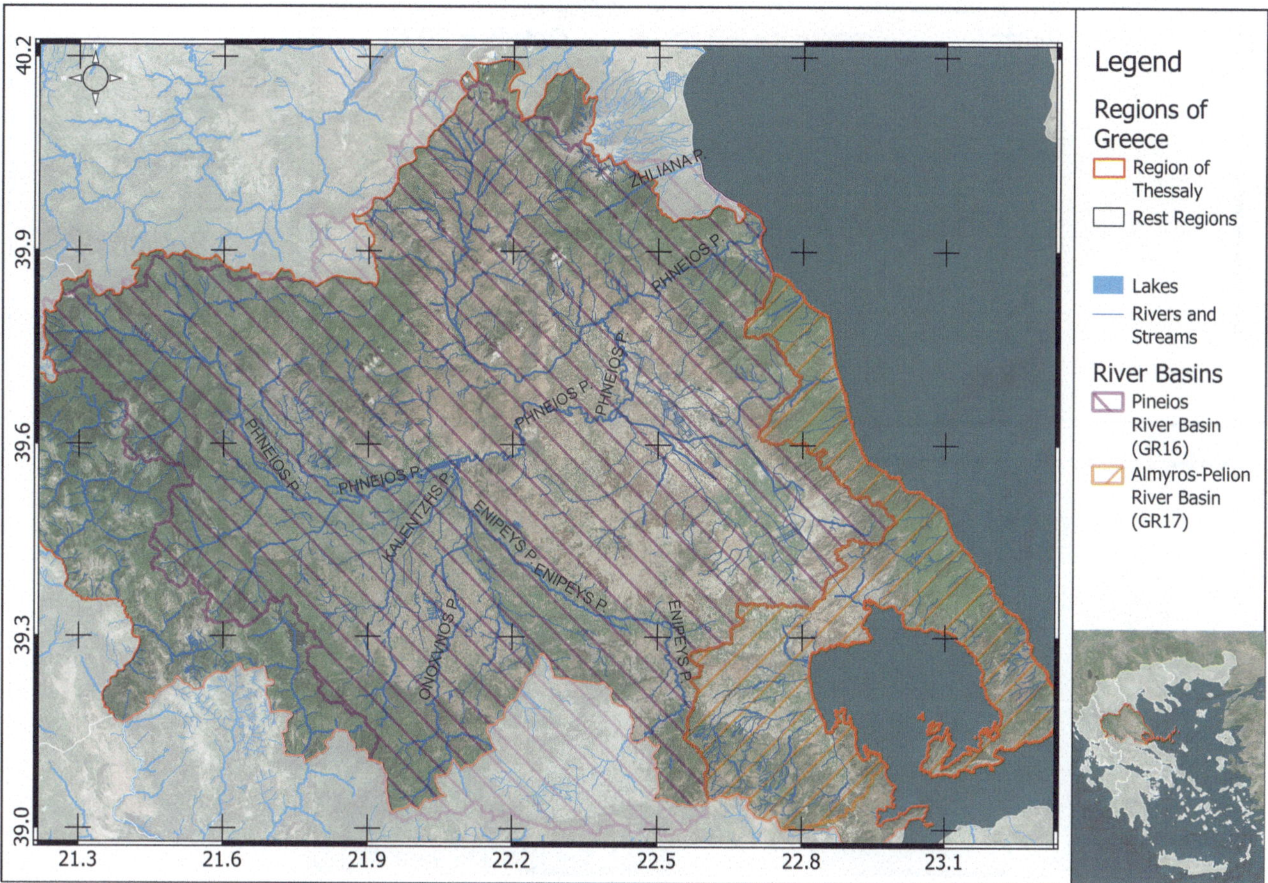

Fig. 1 The Region of Thessaly in Greece, showing the two major river basins: Pineios (yellow hatch) and Almyros-Pelion (orange hatch)

mitigation (HVA International 2024). The Pineios River Basin, covering approximately 10,700 km^2, is the largest and most hydrologically significant river system in Thessaly.

According to the Flood Risk Management Plan, Thessaly's river systems are becoming increasingly vulnerable to extreme weather events due to a combination of the increase in heavy rainfall, outdated drainage infrastructure, and human-induced land modifications (HVA International 2024). The vulnerability of Thessaly's flood management system was starkly exposed during Storm Daniel, which struck central Greece in early September 2023. The storm brought unprecedented rainfall to the Thessaly region, resulting in catastrophic flooding. Between September 4–7, cumulative precipitation reached an average of 360 mm over the Pineios River basin, overwhelming the region's drainage capacity and submerging vast areas (Dimitriou et al. 2024).

3 Data and Methods

To map the flood event in Thessaly, we utilized heterogeneous satellite datasets, such as optical, SAR and gravimetric ones, each contributing uniquely to the methodological framework. The first step of the applied methodology employed GRACE and GRACE-FO data in capturing the hydrological impact of the event. This was achieved by utilizing monthly global geopotential models in the form of spherical harmonic coefficients for the period April 2002 to May 2024, from the International Centre for Global Earth Models (ICGEM) to derive EWT (Ince et al. 2019).

EWT variations represent changes in the total water storage, including surface water, groundwater, and soil moisture (Han et al. 2005; Seoane et al. 2013; Scanlon et al. 2016), for the area under study.

From the monthly EWT estimates (see Fig. 2), mean values of the EWT variations have been derived (see Fig. 3), so as to represent the average EWT per month for the entire area. For Daniel's flood detection in Thessaly (which occurred in September 2023), EWT values were analyzed before, during, and after the event. In Fig. 2, the EWT variations have been interpolated onto a 20 arcmin resolution so as to try to map the variations at a finer scale. Prior to the event (see Fig. 3), Thessaly exhibited EWT values ranging from −8 to +10 cm, and the mean monthly EWT for August was −4.512 cm, reflecting typical seasonal conditions. During the flood, EWT values increased significantly to a monthly

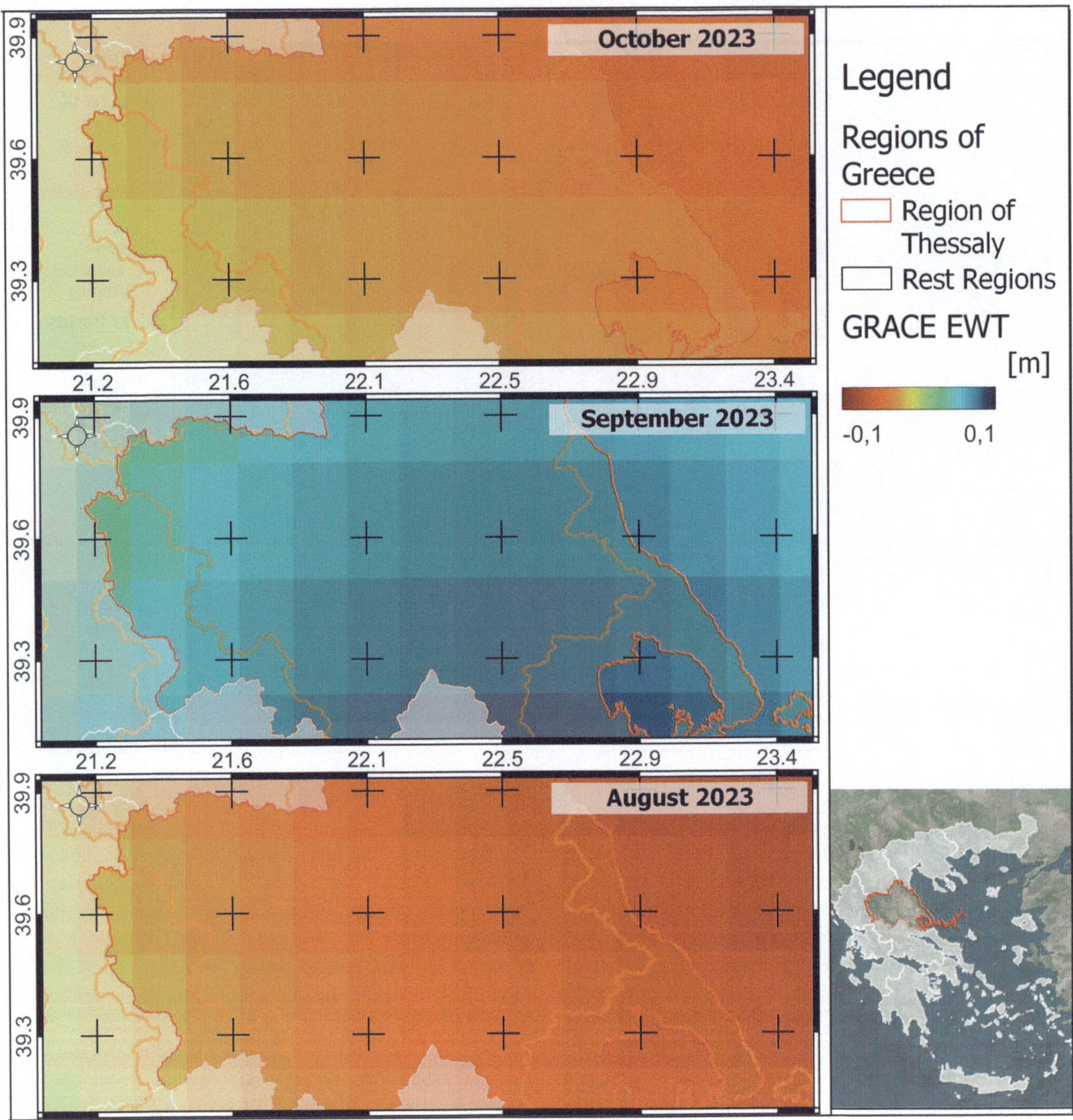

Fig. 2 Equivalent Water Thickness (EWT) in the Thessaly region, d August 2023 (**a**), September 2023 (**b**) and October 2023 (**c**)

mean of +6.518 cm. Within the EWT of September 2023 (see Fig. 2 and Table 1) only positive EWT values were found with a peak value of +10.017 cm (see Table 1), demonstrating a substantial accumulation of water across the region. After the event, the EWT values returned to the expected ranges, as floodwaters receded and the excess water drained. The monthly mean for October 2023 declined to −2.786 cm, indicating partial recession of floodwaters while still remaining elevated relative to pre-flood levels. Table 1 presents for the above mentioned maps, the statistics showing the significant differences in EWT between the month of the flood event and the previous and following months. Especially for September 2023, the range of EWT variations is positive, with minimum values of 0.9 cm and a maximum of 10 cm, with a strongly postive mean value compared to August and October 2023. These sharp month-to-month differences highlight GRACE-FO's capability to detect and quantify regional-scale flood impacts, affirming its value in hydrological monitoring despite inherent resolution constraints.

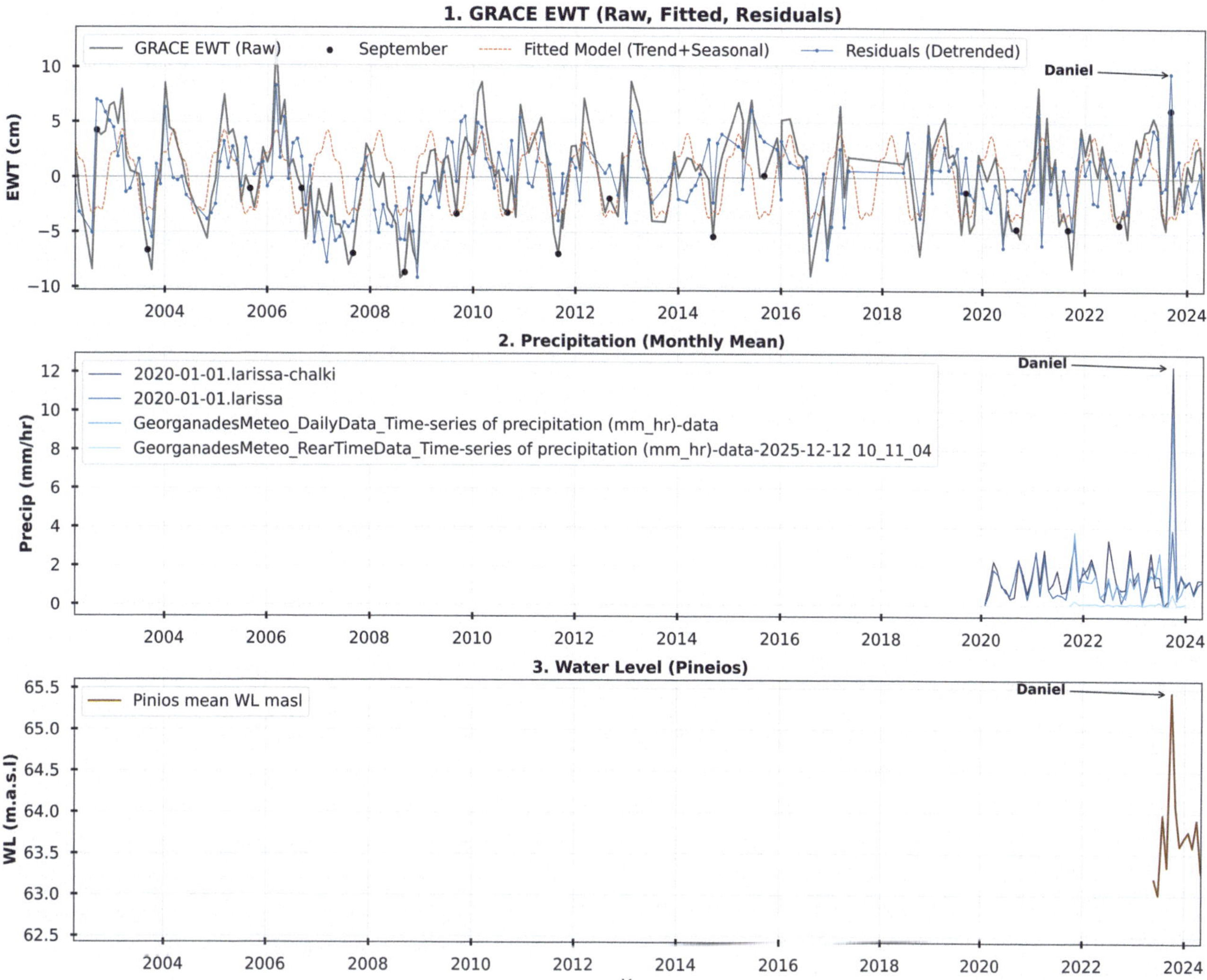

Fig. 3 EWT over the region of Thessaly derived from GRACE and GRACE-FO data, showing the raw signal (gray), the fitted seasonal–trend model (red dashed), and the detrended residuals (blue) (top). Also, monthly mean precipitation for the stations of Larissa, Chalki and Georganades in the area are depicted (middle) from METEO Greece and HCMR. The bottom panel shows mean river depth for the Pinios river as derived from the HCMR gauge sensor data in Larisa

Table 1 Statistics of EWT for the study area (units: cm)

	Min	Max	Mean	Std	RMS
August '23	−7.903	0.844	−4.512	0.054	5.011
September '23	0.938	10.017	6.518	0.019	6.822
October '23	−6.519	1.495	−2.786	0.006	3.322

In order to better quantify the contribution of GRACE and GRACE-FO to monitoring such extreme flood events like Daniel, given also the small spatial scale of the area under study, in-situ sensory data for the region have been employed. First, the EWT signal has been decomposed through a Fourier into annual, semi-annual and components. The seasonal–trend component was estimated using a least-squares fit that includes a linear trend, annual and semiannual and seasonal harmonic terms, so that after their subtraction from the original EWT signal residuals are estimated. The scope is to isolate short-term anomalies associated with extreme hydrological events, so that the such-derived residuals, depicted with a blue line in Fig. 3, should primarily depict episodic cases. Indeed, as it can be seen from the top panel of Fig. 3, the residual signal captured by the GRACE/GRACE-FO time-series gets its maximum value (∼10 cm) exactly for the month of September 2023, hence satellite gravimetry data manage to capture the flooding event due to Daniel, even for small scale catchment areas. This is further supported from in-situ precipitation and water level data from meteorological stations in the area and Pinios river. These data have been provided by the National Observatory of Athens (IERSD/NOA) and its meteo.gr service and the Hellenic Centre for Marine Research (HCMR) Institute of Marine Biological Resources and Inland Waters (IMBRIW) through its automatic inland water monitoring network. The precipitation data are depicted in Fig. 3 (middle panel), where the light blue lines represent the timeseries from stations in Larisa and Larisa-Chalki, while the dark blue lines represent the real time and the daily averaged percipitation data from the Geogranades station. Moreover, the bottom panel in Fig. 3 depicts Pinios river water level observations provided by the HCMR from a dedicate gauge station in the city of Larissa. It is clear that in September 2023 rainfall data exhibit their overall maximum value of more than 12 mm/h precipitation, while the water level of Pinios reaches it maximum of ~64.5 m. These extremes, depicted by the in-situ data are fully in-line with the information captured by the GRACE and GRACE-FO timeseries. This extreme clearly exceeds the range of typical September variations observed in previous years as depicted by the black dots in the EWT timeseries (Fig. 3 top plot).

Figure 4 shows how we further integrated optical and radar datasets from the Copernicus Sentinel program using Python and Google Earth Engine (Gorelick et al. 2017), for processing and analysis, so as to provide a comprehensive and spatially explicit depiction of the flood extent. The methodology begins with the acquisition of Google Earth Engine Sentinel-1 SAR data, using the Python's API which was used to delineate flooded areas through a backscatter-based classification approach. First, the Area of Interest (AOI) is defined, and Sentinel-1 imagery from sixth of September to 12th of September, is fetched and filtered (spatially and temporally) to isolate both pre-event and post-event datasets. These are then preprocessed to remove noise, correct for terrain and radiometric distortions, and standardize the data for analysis. Following preprocessing, the workflow applies flood detection techniques by focusing on low backscatter values in the VV band, as water-covered surfaces exhibit low backscatter values due to their smooth surface which causes minimal radar reflection (Martinis et al. 2018).

Preprocessing was carried out, which included radiometric calibration, terrain correction, and the application of a speckle filter (Filipponi 2019), to ensure data quality and minimize inherent SAR noise. After preprocessing, pre-event and post-event SAR images were compared pixel-by-pixel to detect floods. A thresholding technique was used to identify flooded areas; pixels in the post-event image with backscatter values less than −17 dB were categorized as inundated (Gulácsi and Kovács 2020). The reflecting characteristics of water-covered surfaces show noticeably lower backscatter values than terrestrial surfaces and correlate to this threshold (see Fig. 5).

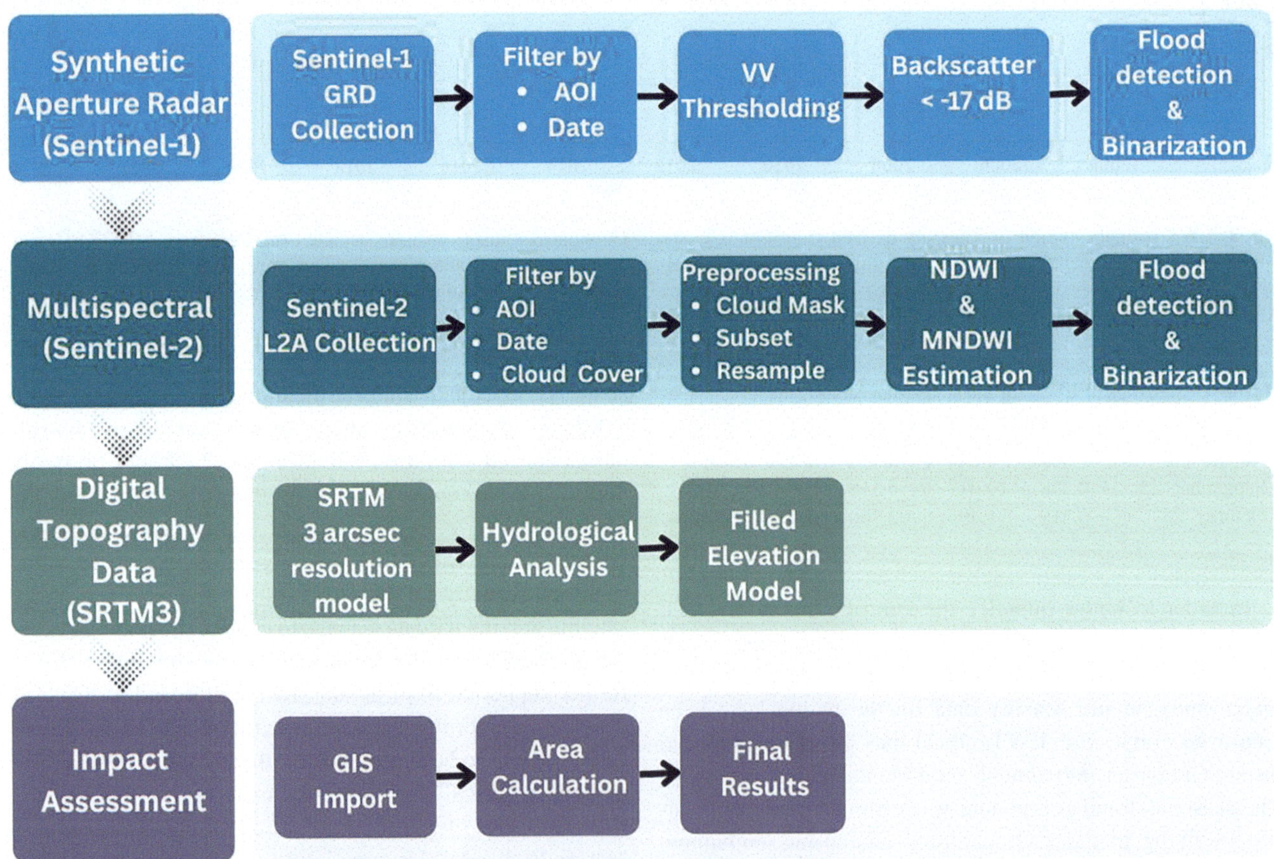

Fig. 4 Methodology workflow overview

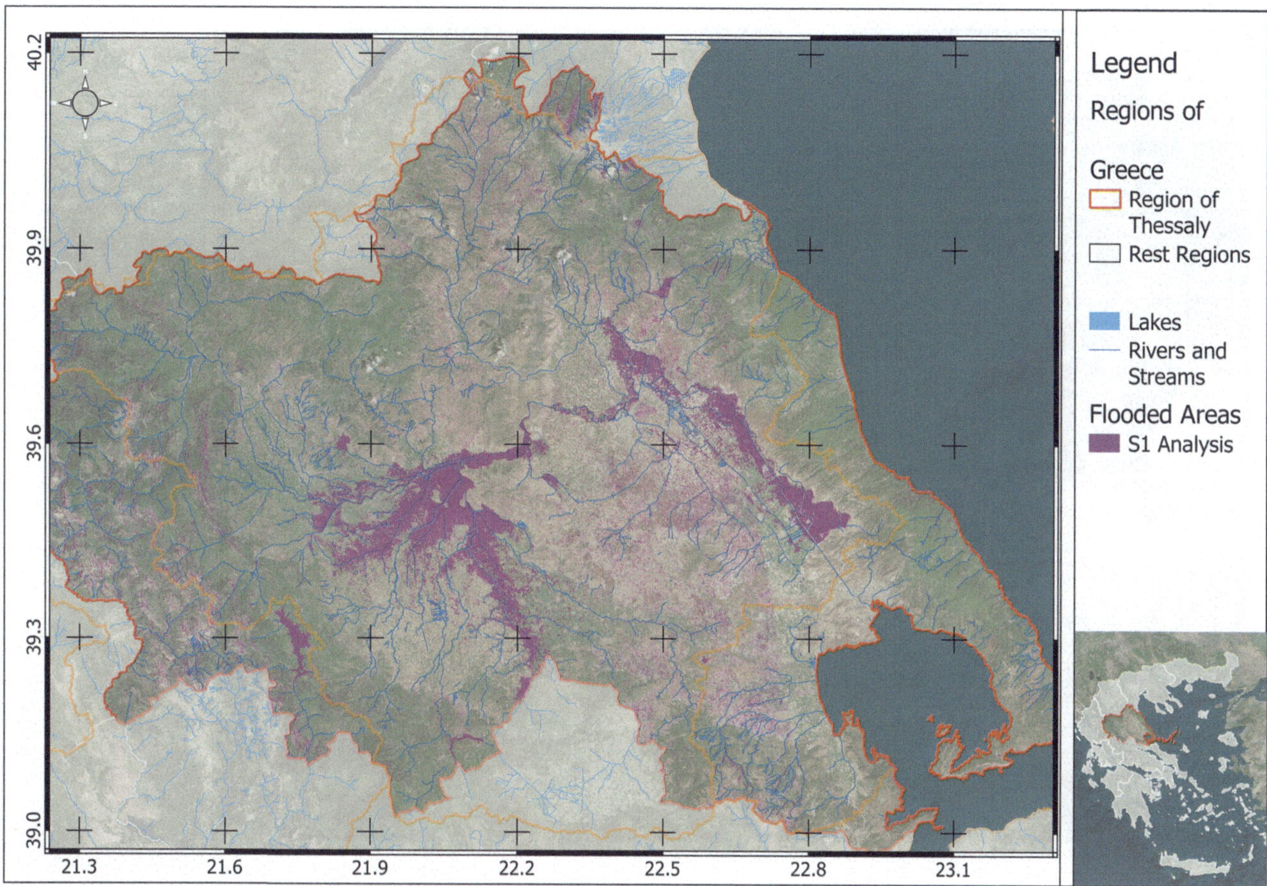

Fig. 5 Flood extent as mapped from the backscattered values of the Sentinel-1 analysis

Based on the results acquired from the SAR data processing, the total flooded area accumulates to 814.13 km^2 (see purple areas in Fig. 5) and cover a large portion of the Thessaly hydrological basin, characterized by low-land areas in the west, close to Karditsa Regional unit (146.15 km^2) and Lake Karla and Larisa town in the east (399.88 km^2).

This classification produced the final flood extent map shown in Fig. 5, and was verified further by using processed optical data from the Sentinel-2 satellites, acquired from Google's Earth Engine API. To guarantee that the data covered the immediate aftermath of the disaster, these optical datasets only contained acquisitions made between September 6 and September 11, 2023. Images with more than 20% cloud cover were not included in order to reduce atmospheric interference (Clerc et al. 2018; Tiede et al. 2021) By utilizing the unique spectral reflectance characteristics of the green (Band 3) and near-infrared (Band 8) bands, the remote sensing analysis was combined with the Normalized Difference Water Index (NDWI) to differentiate water bodies from other types of land cover (Gao 1996). The filtered scenes were converted into a median composite NDWI image, which improved the visibility of consistent surface water throughout the post-event period while lessening the impact of outliers and transitory air noise.

In order to get the boundaries of the flooded areas the Modified Normalized Difference Water Index (MNDWI) was also computed since it is more sensitive to the soil moisture retrieval (Gulácsi and Kovács 2020; Zhang and Chen 2015). The index calculation is identical to NDWI, but instead of the near-infrared band (Band 8) it uses the Short-wave infrared (Band 11) (see Fig. 6). From the MNDWI analysis of the optical Sentinel 2 data, before and after the event, a total area of 727.50 km^2square km is identified as flooded.

Figure 7 depicts an overlay of the identified flooded areas in Thessaly based on the SAR data and optical images. As it can be seen, the correlation between the two is high since they equally identify the main flooded areas. The calculated flooded area difference of 86.63 km^2, as calculated from the S1 and S2 flooding areas estimations, is mainly due to the two lakes that were identified as flooded areas during the S1 analysis with false positives over wet fields. Regarding the overall S1 backscattered pixels found within the flooded polygons of the multispectral analysis, just 24.4% have true agreement. This was expected due to the spatial resolution differences (the rasterization of the S2 data introduces mismatch at pixel level) and the S1 data noise.

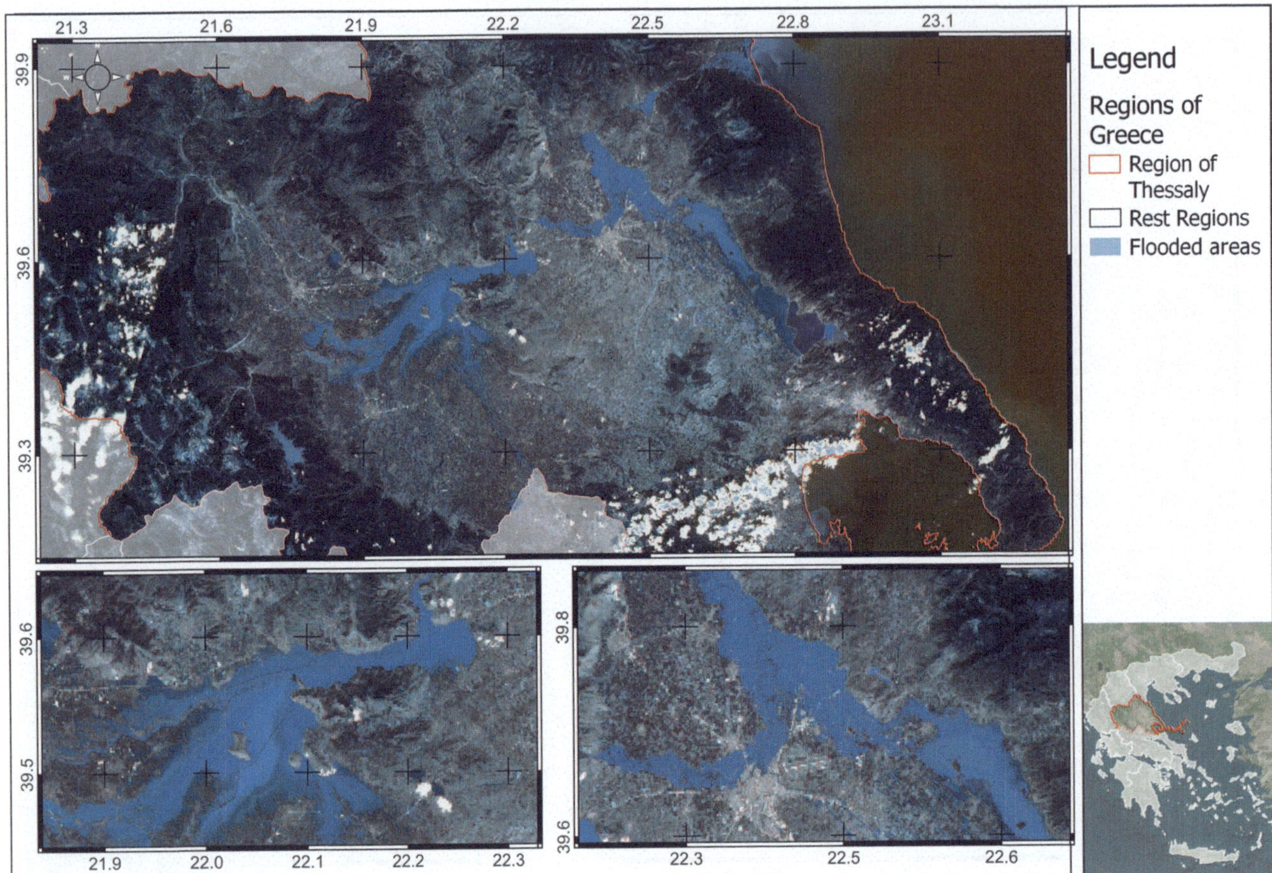

Fig. 6 Flood extent as mapped from the Sentinel-2 analysis

The correlation between the two datasets reaches 46.6% level. The correlation level was considered good taking into account the different acquisition dates. Finally, the optical data also confirmed the derived flooded areas from the SAR and the MNDWI analysis. Only some parts of the area in the west are not identified from the SAR data processing.

Furthermore, using digital topography data from the SRTM 3 arcsec resolution model (Farr et al. 2007), we also performed a hydrological analysis to evaluate the dynamics of floods in Thessaly. The analysis was performed in GRASS GIS (GRASS Development Team 2024) using the TerraFlow module (Arge et al. 2003), and confirmed that flow direction in a potential flood is concentrated in low-lying areas such as the region Karditsa and Lake Karla, being in the southern and eastern part of the region respectively, which are also identified by GRACE-FO data (please see the middle panel of Fig. 2).

Based on the SRTM data, a filled elevation model was generated representing the depth of depressions that would retain water during extreme precipitation events, ranging from 0.1 m to 30 m. As illustrated in Fig. 8, shallow depressions (1–5 m) correspond to localized pooling areas, typically found in agricultural fields and low-lying urban zones. Intermediate depressions (5–15 m) indicate natural floodplains that can sustain more prolonged flooding. Deeper depressions (15–30 m) highlight major flood-prone basins, such as areas surrounding Karditsa in the southwest and Lake Karla in the east, where large-scale water accumulation occurs. The results align partially with observed flood extents from Storm Daniel, as calculated from the remote sensing analysis.

4 Results

The EWT measurements and SAR data were combined to estimate the amount of floodwater stored in the area after Storm Daniel. The mean EWT data, which averaged 0.06518 m, represented the average depth of the water accumulated over the region during the month of the flood event and the Sentinel-1 mapping of the flood extent was estimated to be roughly 814.13 km^2, which is consistent with other comparable studies (He et al. 2024) (Qiu et al. 2023). This combined dataset yielded a floodwater volume of roughly 53.06 million m^3, which is within the range determined by Dimitriou et al. (2024) using the data from

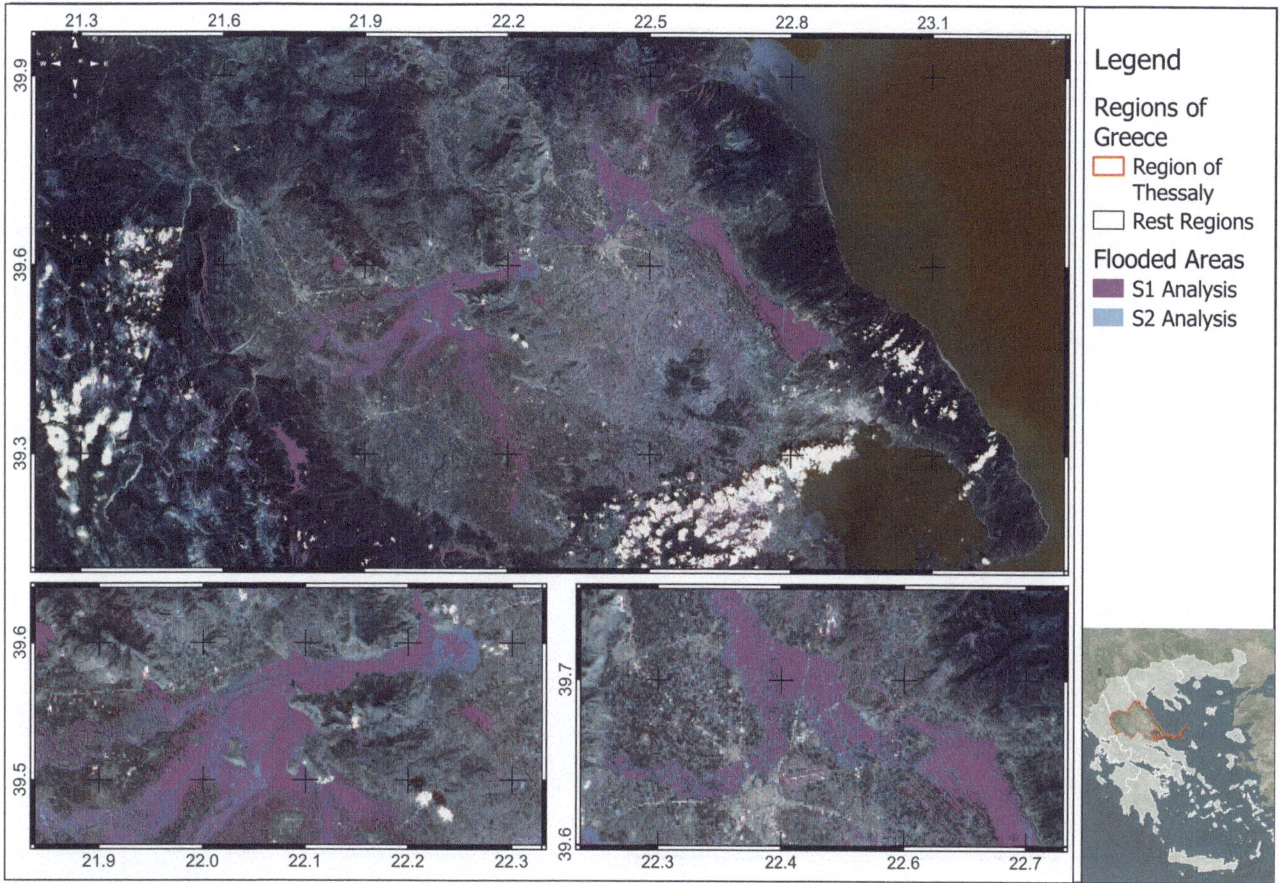

Fig. 7 Comparison of calculated Flood extent as mapped from the Sentinel-1 and Sentinel-2 analysis

two stations along the main cause of Peneus river that have water-level and discharge measurements (412 and 1,538 hm^3 at the two stations accordingly) Considering the heavy event, this estimation may be underestimated due to terrain's relief retentions and overflows across the drainage system and the soil moisture which was not taken into account in the methodology.

5 Conclusions

In this work, a combination scheme of gravimetric, optical and SAR data for the analysis of extreme flood events has been presented. The technique does have certain fundamental drawbacks as the area under study should include a combination of multispectral and SAR coverages in order to minimize the noise from Sentinel-1 backscattered values and enhance the results of the Sentinel-2 imagery analysis. Regarding the flooding volume calculation, GRACE-FO's coarse spatial resolution of roughly 300 km limits its capacity to precisely depict localized flood conditions and identify fine-scale changes in water thickness. Furthermore, the monthly temporal resolution of GRACE-FO data makes it difficult to completely capture the quick dynamics of short-term flood events like the ones that occurred in Thessaly. Additionally, while the computation is made simpler by assuming that the water is distributed uniformly throughout the flooded area, this assumption does not accurately reflect actual changes in water depth.

Looking ahead, the upcoming Mass-Change and Geosciences International Constellation (MAGIC) mission is anticipated to offer significantly enhanced spatial and temporal resolution. This advancement is expected to improve the accuracy of flood volume estimates and enable more precise real-time monitoring, thereby enhancing flood management and response strategies for extreme weather events.

Acknowledgments Precipitation data for the stations Larissa and Larissa-Chalki has been provided by the meteo.gr service, National Observatory of Athens (https://www.meteo.gr/meteomaps). Precipitation data for the station of Georganades, and Pinios river water level data have been provided by the Hellenic Centre for Marine Research, Institute of Marine Biological Resources and Inland Waters (IMBRIW) water-related data from the automatic inland water monitoring network of IMBRIW–HCMR (https://hydro-stations.hcmr.gr/). We greatfully acknowledge their support in providing in-situ observations. (sources: (a) Weather forecasts: (b) Hellenic Centre for Marine Research, Institute of Marine Biological Resources and Inland

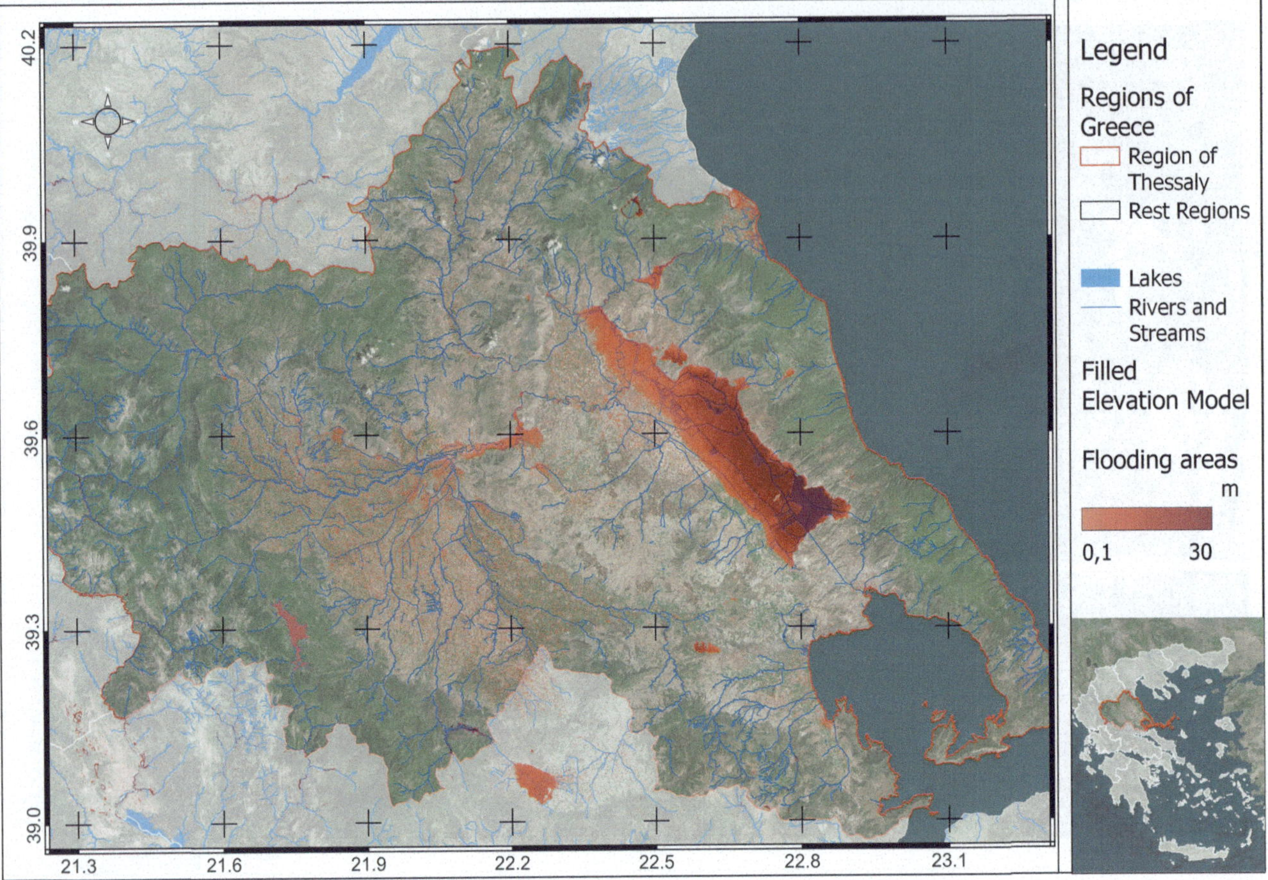

Fig. 8 Filled (flooded) elevation model

Waters (IMBRIW). 2025. Water-related data from the automatic inland water monitoring network of IMBRIW–HCMR. Available at: Plot 3: River water level observations (source: Hellenic Centre for Marine Research - Department of Inland Waters, 2025, Water level data from the automatic water monitoring network of IMBRIW-HCMR, https://hydro-stations.hcmr.gr/)

References

Albano R, Samela C, Craciun I, Manfreda S, Adamowski J, Sole A, Sivertun Å, Ozunu A (2020) Large scale flood risk mapping in data scarce environments: an application for Romania. Water (Switzerland) 12(6). https://doi.org/10.3390/w12061834

Arge L, Chase JS, Halpin P, Toma L, Vitter JS, Urban D, Wickremesinghe R (2003) Efficient flow computation on massive grid terrain datasets. GeoInformatica 7(4):283–313. https://doi.org/10.1023/A:1025526421410

Clerc S, Bévy C, Jackson J, Papadopoulou T, Garcia-Soto A (2018) S2check: a tool to for quality control of sentinel-2 products. Eur J Remote Sensing 51(1):777–784. https://doi.org/10.1080/22797254.2018.1485469

de Moel H, Jongman B, Kreibich H, Merz B, Penning-Rowsell E, Ward PJ (2015) Flood risk assessments at different spatial scales. Mitig Adapt Strateg Glob Chang 20(6):865–890. https://doi.org/10.1007/s11027-015-9654-z

Dimitriou E, Efstratiadis A, Zotou I, Papadopoulos A, Iliopoulou T, Sakki GK, Mazi K, Rozos E, Koukouvinos A, Koussis AD, Mamassis N, Koutsoyiannis D (2024) Post-analysis of Daniel extreme flood event in Thessaly, Central Greece: practical lessons and the value of state-of-the-art water-monitoring networks. Water 16(7):980. https://doi.org/10.3390/w16070980

ESA (2012a) Sentinel-1: ESA's radar observatory mission for GMES operational services. In K. Fletcher (Ed.), ESA Special Publication (Vol. 1, Issue 1322). ESA Communications ESTEC, AG Noordwijk, AG Noordwijk, The Netherlands. www.esa.int. https://sentinel.esa.int/documents/247904/349449/S1_SP-1322_1.pdf

ESA (2012b) SENTINEL – 2:ESA's Optical High-Resolution Mission for GMES Operational Services. Https://sentinel.esa.int/web/sentinel/missions/sentinel-1/satellite-description/orbit (ed). ESA Communications ESTEC, AG Noordwijk, The Netherlands. www.esa.int

European Environment Agency (2010) Mapping the impacts of natural hazards and technological accidents in Europe – European Environment Agency (Issue 13). https://doi.org/10.2800/62638

Farr TG, Rosen PA, Caro E, Crippen R, Duren R, Hensley S, Kobrick M, Paller M, Rodriguez E, Roth L, Seal D, Shaffer S, Shimada J, Umland J, Werner M, Oskin M, Burbank D, Alsdorf D (2007) The Shuttle Radar Topography Mission Farr. NASA 45(2) http://www2.jpl.nasa.gov/srtm/SRTM_paper.pdf

Filipponi F (2019) Conference Paper.Pdf. The 3rd International Electronic Conference on Remote Sensing (ECRS 2019), 3(June), 2–6

Gao BC (1996) NDWI - a normalized difference water index for remote sensing of vegetation liquid water from space. Remote Sens Environ 58(3):257–266. https://doi.org/10.1016/S0034-4257(96)00067-3

Gorelick N, Hancher M, Dixon M, Ilyushchenko S, Thau D, Moore R (2017) Google earth engine: planetary-scale geospatial analysis for everyone. Remote Sens Environ 202:18–27. https://doi.org/10.1016/j.rse.2017.06.031

GRASS Development Team (2024) Geographic Resources Analysis Support System (GRASS) Software, Version 8.4. Open Source Geospatial Foundation. (8.4). Open Source Geospatial Foundation. https://grass.osgeo.org

Gulácsi A, Kovács F (2020) Sentinel-1-imagery-based high-resolution water cover detection on wetlands, aided by google earth engine. Remote Sens 12(10):1–20. https://doi.org/10.3390/rs12101614

Han SC, Shum CK, Jekeli C, Alsdorf D (2005) Improved estimation of terrestrial water storage changes from GRACE. Geophys Res Lett 32(7):1–5. https://doi.org/10.1029/2005GL022382

He K, Yang Q, Shen X, Dimitriou E, Mentzafou A, Papadaki C, Stoumboudi M, Anagnostou EN (2024) Brief communication: storm Daniel flood impact in Greece in 2023: mapping crop and livestock exposure from synthetic-aperture radar (SAR). Nat Hazards Earth Syst Sci 24(7):2375–2382. https://doi.org/10.5194/nhess-24-2375-2024

HVA International (2024) Master plan water management in thessaly in the wake of storm Daniel How, vol I. (Issue February)

Ince ES, Barthelmes F, Reißland S, Elger K, Förste C, Flechtner F, Schuh H (2019) ICGEM – 15 years of successful collection and distribution of global gravitational models, associated services, and future plans. Earth Syst Sci Data 11(2):647–674. https://doi.org/10.5194/essd-11-647-2019

Ji L, Zhang L, Wylie B (2009) Analysis of dynamic thresholds for the normalized difference water index. Photogramm Eng Remote Sens 75(11):1307–1317

Loomis BD, Rachlin KE, Wiese DN, Landerer FW, Luthcke SB (2020) Replacing GRACE/GRACE-FO C30 with satellite laser ranging: impacts on antarctic ice sheet mass change. Geophys Res Lett 47(3). https://doi.org/10.1029/2019GL085488

Martinis S, Plank S, Ćwik K (2018) The use of Sentinel-1 time-series data to improve flood monitoring in arid areas. Remote Sens 10(4):583. https://doi.org/10.3390/rs10040583

Monti R, Rossi L, Reguzzoni M (2024) The Nova Kakhovka dam collapse flooding as seen from Sentinel-1 SAR satellite images. Adv Geodesy Geoinform 50. https://doi.org/10.24425/agg.2023.146162

NASA Jet Propulsion Laboratory (JPL) (2018) Grace static field geopotential coefficients JPL release 6.0. NASA Physical Oceanography Distributed Active Archive Center. https://doi.org/10.5067/GRGSM-20J06

Nastos PT, Kapsomenakis J, Dalezios NR, Kotsopoulos S (2015) Drought variability over Thessaly plain, Greece. Present Future Changes 17:7349

Qiu J, Zhao W, Brocca L, Tarolli P (2023) Storm Daniel revealed the fragility of the Mediterranean region. Innovat Geosci 1(3):100036. https://doi.org/10.59717/j.xinn-geo.2023.100036

Rodell M, Li B (2023) Changing intensity of hydroclimatic extreme events revealed by GRACE and GRACE-FO. Nature Water 1(3):241–248. https://doi.org/10.1038/s44221-023-00040-5

Scanlon BR, Zhang Z, Save H, Wiese DN, Landerer FW, Long D, Longuevergne L, Chen J (2016) Global evaluation of new GRACE mascon products for hydrologic applications. Water Resour Res 52(12):9412–9429. https://doi.org/10.1002/2016WR019494

Seoane L, Ramillien G, Frappart F, Leblanc M (2013) Regional GRACE-based estimates of water mass variations over Australia: validation and interpretation. Hydrol Earth Syst Sci 17(12):4925–4939. https://doi.org/10.5194/hess-17-4925-2013

Special Secretariat for Water (2014) Management plan of the River Basins of Thessalia River Basin District. Summary. September 2014, 103. http://wfdver.ypeka.gr/wp-content/uploads/2017/04/files/GR08/GR08_P26b_Perilipsi_En.pdf

Tapley BD, Watkins MM, Flechtner F, Reigber C, Bettadpur S, Rodell M, Sasgen I, Famiglietti JS, Landerer FW, Chambers DP, Reager JT, Gardner AS, Save H, Ivins ER, Swenson SC, Boening C, Dahle C, Wiese DN, Dobslaw H et al (2019) Contributions of GRACE to understanding climate change. Nat Clim Chang 9(5):358–369. https://doi.org/10.1038/s41558-019-0456-2

Tiede D, Sudmanns M, Augustin H, Baraldi A (2021) Investigating ESA Sentinel-2 products' systematic cloud cover overestimation in very high altitude areas. Remote Sens Environ 252:112163. https://doi.org/10.1016/j.rse.2020.112163

Zhang HW, Chen HL (2015) The application of modified normalized difference water index by leaf area index in the retrieval of regional drought monitoring. Int Arch Photogram Remote Sens Spatial Inform Sci ISPRS Archiv 40(7W3):141–147. https://doi.org/10.5194/isprsarchives-XL-7-W3-141-2015

Strapdown Airborne Gravimetry for Geophysical Applications: Latest Results with New Lomonosov MSU Software and AGP-Grav Gravimeter

Vadim Vyazmin and Andrey Golovan

Abstract

In 2020–2023, Lomonosov MSU has been developing postprocessing algorithms and software for airborne gravimeters based on navigation-grade strapdown inertial measurement units (IMUs). The focus was made on the potential of strapdown gravimeters to be used under dynamic flight conditions. This is of high demand in the present-day aerogeophysical surveys conducted using the so-called draped flights (that is, following the terrain). The developed postprocessing algorithms were implemented in the software suite of programs, which were widely tested using state-of-the-art strapdown airborne graivmeters. In the paper, gravimetry results from a new strapdown system (AGP-Grav) are presented. The cross-over analysis of the results shows that the new strapdown gravimeter can achieve a measurement precision of 1.4 mGal (root-mean-square error) under the draped flight conditions given a smooth drape flight surface. We also compare the gravity disturbances derived from AGP-Grav with gravity data obtained from another IMU (iMAR), which operated side-by-side with AGP-Grav in one of the flights. The results from both sensors are similar, with a root-mean-square error of the differences of 0.9 mGal. The paper also describes the developed postprocessing software programs, which cover all postprocessing stages from the quality control of raw IMU and GNSS data to calculation of the gravimetry solutions on the aircraft flight path.

Keywords

Airborne gravimetry · Gravity disturbance · Postprocessing software · Strapdown IMU

1 Introduction

In recent years, airborne gravimeters based on strapdown inertial measurement units (IMUs) are increasingly involved in aerogeophysical surveys (Brovkin et al. 2021; Simav 2020). The lower technical limitations of strapdown gravimeters for operation under dynamic flight conditions (compared to classical meters based on gyro-stabilized platforms (Studinger et al. 2008)) allow them to be installed in small multi-instrumented aircrafts (Peshekhonov et al. 2022; Johann et al. 2020; Ayres-Sampaio et al. 2015). This results in a higher spatial resolution of the collected airborne gravity data achieved due to the ability of a small aircraft to fly at slow speed and low altitude. Moreover, this allows airborne gravimetry to be conducted simultaneously with other aerogeophysical methods such as magnetometry and/or electromagnetics without restricting the aircraft's maneuvering during flights. The latter two methods are known to be sensitive to the distance to the terrain and therefore requiring so-called draped flights (that is, following the terrain at the lowest possible altitude) (de Barros Camara and Guimarães 2016). Depending on the complexity of the terrain in the survey area, this may lead to harsh dynamics conditions (high accelerations, rapid changes of the pitch and roll angles of the aircraft), which is generally avoided

V. Vyazmin (✉) · A. Golovan
Faculty of Mechanics and Mathematics, Lomonosov Moscow State University, Moscow, Russian Federation
e-mail: vadim.vyazmin@math.msu.ru

J. T. Freymueller, L. Sànchez (eds.), *International Symposium on Gravity, Geoid and Height Systems 2024 (GGHS2024)*,
International Association of Geodesy Symposia 158, https://doi.org/10.1007/1345_2025_299

when using classical stabilized-platform gravimeters (in this case, a compromise smooth drape flight surface is usually preplanned) (Olson 2010; Foroughi et al. 2024). The advantage of strapdown gravimeters is in the possibility to use them under such conditions (i.e., without restricting the aircraft's maneuvering required for other aerogeophysical methods) without significant deterioration of collected airborne gravity data (Vyazmin and Golovan 2023).

In the paper, we present the results from an ongoing aerogeophysical campaign based on a multi-instrumented fixed-wing aircraft using a new strapdown airborne gravimeter (AGP-Grav) manufactured by Aerogeophysica JSC (Russia) and commercially distributed by AGP Consulting (UAE). The results obtained with the new instrument show an accuracy of 1.4 mGal (root-mean-square error), which confirms the satisfactory performance of the new instrument under moderate flight dynamics conditions (flights based on a smooth drape surface). The accuracy was assessed using cross-over analysis.

We also present the results from a repeated line flight in which AGP-Grav operated side-by-side with another IMU (iCORUS by iMAR). We show that the gravimetry results from both sensors are similar, with a root-mean-square error of the differences between two datasets of gravity disturbances of 0.9 mGal.

The processing of the raw gravimeter data (which comprises specific force and angular rate measured by the gravimeter's IMU as well as raw GNSS observations) was performed using the methodology and postprocessing software package (the INS-GNSS and IMU-GRAV programs) developed at Lomonosov Moscow State University (MSU) for strapdown gravimeters and tested in dozens of airborne gravimetry campaigns carried out using iMAR strapdown systems and prototypes of the AGP-Grav system (Golovan and Vyazmin 2023). The commercial version of the developed software was released in 2023 and is now distributed under a proprietary license for each individual IMU. In the paper, we also briefly describe the postprocessing methodology and main subroutines of the developed software.

2 Survey Overview and Instrumentation

The airborne gravimetry campaign considered in the paper started in early summer of 2024 and is scheduled to be completed by the end of 2025. The campaign is based on a multi-instrumented Cessna 206 aircraft. In total, about 150 flights are planned, which cover a survey area of 300 by 300 km. The campaign is carried out for exploration purposes (by Aerogeophysica JSC) in the Republic of Sakha, Russia.

A new strapdown airborne gravimeter (AGP-Grav, Fig. 1) is installed in the Cessna 206 aircraft and has been in continuous use since the begininng of the campaign. This is the first commercial use of this instrument developed by GT LLC (Russia) in cooperation with Lomonosov MSU (data exchange protocols, calibration, and raw data postprocessing) and Aerogeophysica JSC (organization of flight tests and manufacturing).

The AGP-Grav instrument is based on a navigation-grade IMU (consisting of quartz-flexible accelerometers and fiber-optic gyroscopes) and a high-precision thermal stabilization system reliably ensuring that temperature variations inside the IMU do not exceed 0.1 K. Geodetic-grade global navigation satellite system (GNSS) receivers on board the aircraft and on the ground complement the gravimetry system.

3 Postprocessing Methodology and Software

Estimation of gravity disturbance δg is based on the fundamental equation of airborne scalar gravimetry:

$$\ddot{h} = \dot{V}_{up} = g_{etv} - \gamma - \delta g + \mathbf{n}^T \mathbf{f}_b, \qquad (1)$$

where h is the height above the ellipsoid, V_{up} is the vertical velocity of the proof mass of the accelerometers, g_{etv} is the Eötvös correction term, γ is the magnitude of the normal gravity vector (including the height correction term), $\mathbf{f}_b$ is the specific force with the components in the IMU body-frame and measured by three accelerometers, $\mathbf{n}$ is the unit vector of the sensitive axis of the IMU vertical accelerometer, expressed in the navigation (geodetic) frame.

To solve the fundamental equation and derive gravity disturbance from raw gravimeter data, we follow the postprocessing strategy presented in (Golovan and Vyazmin 2023):

1. Calculation of the GNSS position and velocity solutions in the carrier-phase differential mode from raw GNSS data (dual-frequency pseudorange, Doppler pseudorange rate, and carrier-phase measurements) recorded by the onboard and ground-based receivers (Golovan and Vavilova 2007).
2. Estimation of the linear drift of the vertical accelerometer bias using preflight and postflight IMU measurements.
3. IMU/GNSS integration based on the equations for the dynamics of strapdown inertial navigation system errors (position, velocity, and attitude errors as well as the inertial sensor biases, GNSS antenna offsets relative to the IMU, and time-synchronization errors in the IMU and GNSS data) (Vavilova et al. 2021; Jekeli 2001). The integration solutions (estimates of the inertial navigation system errors at each time instant) are provided by Kalman filtering and smoothing.

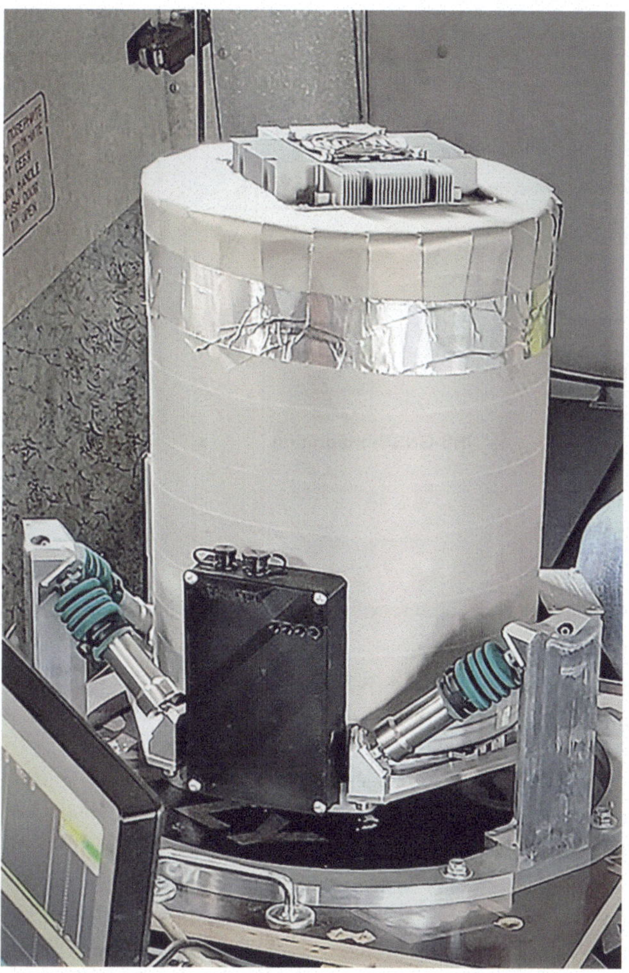

Fig. 1 Strapdown airborne gravimeter AGP-Grav on board the aircraft. The gravimeter includes a two-circuit thermal stabilization system and is mounted on shock absorbers (visible in the image)

4. Estimation of gravity disturbance on the flight path.

The gravity estimation process is based firstly on computing the vertical velocity using the accelerometer data, integrated IMU/GNSS solutions, and Eq. (1) as follows:

$$\dot{V}'_{up} = g_{etv} - \gamma + \mathbf{n}^T \mathbf{f}'_b, \tag{2}$$

where V'_{up} is the computed vertical velocity and $\mathbf{f}'_b$ is the specific force (expressed in the IMU body frame) measured by the accelerometers. The Eötvös correction term g_{etv}, the normal gravity γ and the unit vector $\mathbf{n}$ are assumed to be computed using the integrated IMU/GNSS solutions.

Then the equation for the dynamics of the computed vertical velocity error $\Delta V_{up} = V_{up} - V'_{up}$ is introduced as follows:

$$\Delta \dot{V}_{up} = -\delta g + k_E f'_N - k_N f'_E + \tau \mathbf{n}^T \dot{\mathbf{f}}'_b + \mathbf{n}^T \mathbf{q}_f, \tag{3}$$

where $\mathbf{q}_f$ is the accelerometer noise vector, f'_E, f'_N are the projections of the specific force in the east and north directions (computed from the accelerometer data using the rotation matrix obtained at the IMU/GNSS integration stage), k_E, k_N are the residual attitude errors in the east and north directions, and τ is the time-synchronization error in the accelerometer data relative to the GNSS data.

To derive gravity disturbance from Eq. (3), we use the Kalman filter framework by introducing well-known stochastic models for the gravity disturbance δg (the double-integral of a white noise process) and IMU errors k_E, k_N and τ (random walk processes). The GNSS-derived vertical velocity data are used as aiding observations. Then the estimates of δg, k_E, k_N and τ are obtained using the Kalman filter, which is known as the optimal (in the mean square error sense) stochastic estimation algorithm (Kailath et al. 2000; Krasnov et al. 2022).

Note that it is also possible to design a Kalman filter using GNSS-derived height or both GNSS-derived vertical velocity and height as aiding observations (Bolotin and Popelensky 2007; Krasnov et al. 2022).

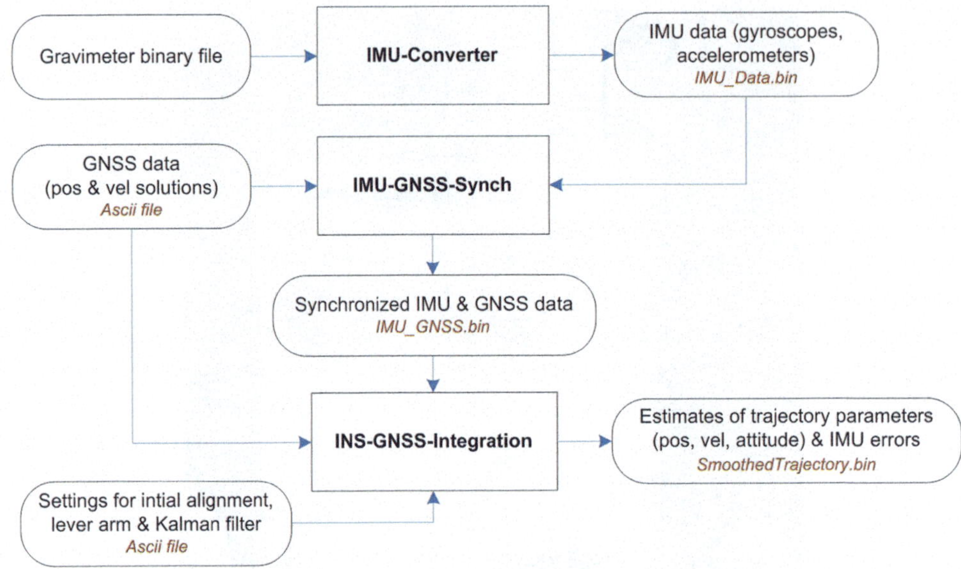

Fig. 2 Flow chart of the INS-GNSS software suite developed for calculating integrated IMU/GNSS solutions

The postprocessing software package consists of the INS-GNSS and IMU-GRAV programs and was developed to process raw gravimeter data (recorded by IMU sensors and GNSS receivers) resulting in gravity disturbance estimates on the aircraft flight trajectory. Additional information, including the initial position of the aircraft, gravity tie value at the aerodrome, and GNSS antenna offsets (lever arm), is required by the programs.

The user can choose different configurations, including calibration of the IMU accelerometers, the methods of computing the GNSS solutions and integrated IMU/GNSS solutions, and design of the gravimetric Kalman filter, to obtain final gravity values on the flight trajectory. The programs are designed to be easy to use for users without experience in inertial and integrated navigation. For instance, since 2023, the programs have been used by the staff of Aerogeophysica JSC in their daily work to process raw data from the iMAR and AGP-Grav sensors owned by this company.

The INS-GNSS programs were developed in C, consist of several executables and configuration files, and do not require installation. The programs include such subroutines as raw IMU data conversion (consisting of raw data quality control (QC) and inertial sensor calibration), IMU and GNSS data synchronization, calculation of GNSS solutions and integrated IMU/GNSS solutions (Fig. 2). A graphical user interface for the INS-GNSS programs was recently designed using C++.

The IMU-GRAV program was developed in the MATLAB environment as a stand-alone application and consists of only one executable file and a configuration file (MATLAB installation is not required).

4 Results

We first present the results from a flight test with the AGP-Grav gravimeter, which was installed side-by-side with an iMAR's IMU (iCORUS) on board a Cessna aircraft. Four repeated lines at a constant altitude of 760 m above the ellipsoid were flown during the test. The aircraft velocity is 70 m/s. The flight test was conducted by Aerogeophysica in Krasnoyarsk Krai (Russia) on December 17, 2022.

Figure 3 shows the gravity disturbance estimates at the repeated lines derived from the AGP-Grav and iCORUS data. Raw data postprocessing was performed independently for each instrument using the same software described in Sect. 3. The gravimetric filter cutoff frequency (half-transmission point) is 0.01 Hz. The corresponding temporal resolution (full width half maximum) of the obtained gravity disturbance estimates is 100 s, which results in a half-wavelength spatial resolution of 3.5 km (Jensen 2022).

Figure 4 shows the difference between the airborne gravity disturbances obtained from two gravimeters. The statistics for the differences (mean values, standard deviation (STD), and root-mean-square (RMS) values, respectively) are given in Table 1.

Next, we present the results from 18 flights of the ongoing airborne gravimetry campaign described in Sect. 2 using the AGP-Grav strapdown gravimeter. The survey lines were flown based on a gentle (smooth) drape flight surface. The survey lines were flown in the east-west and west-east directions with a line spacing of 0.5 km. The crossing lines (north-south/south-north directions) are spaced at a distance of 5 km. The lengths of survey and crossing lines range from

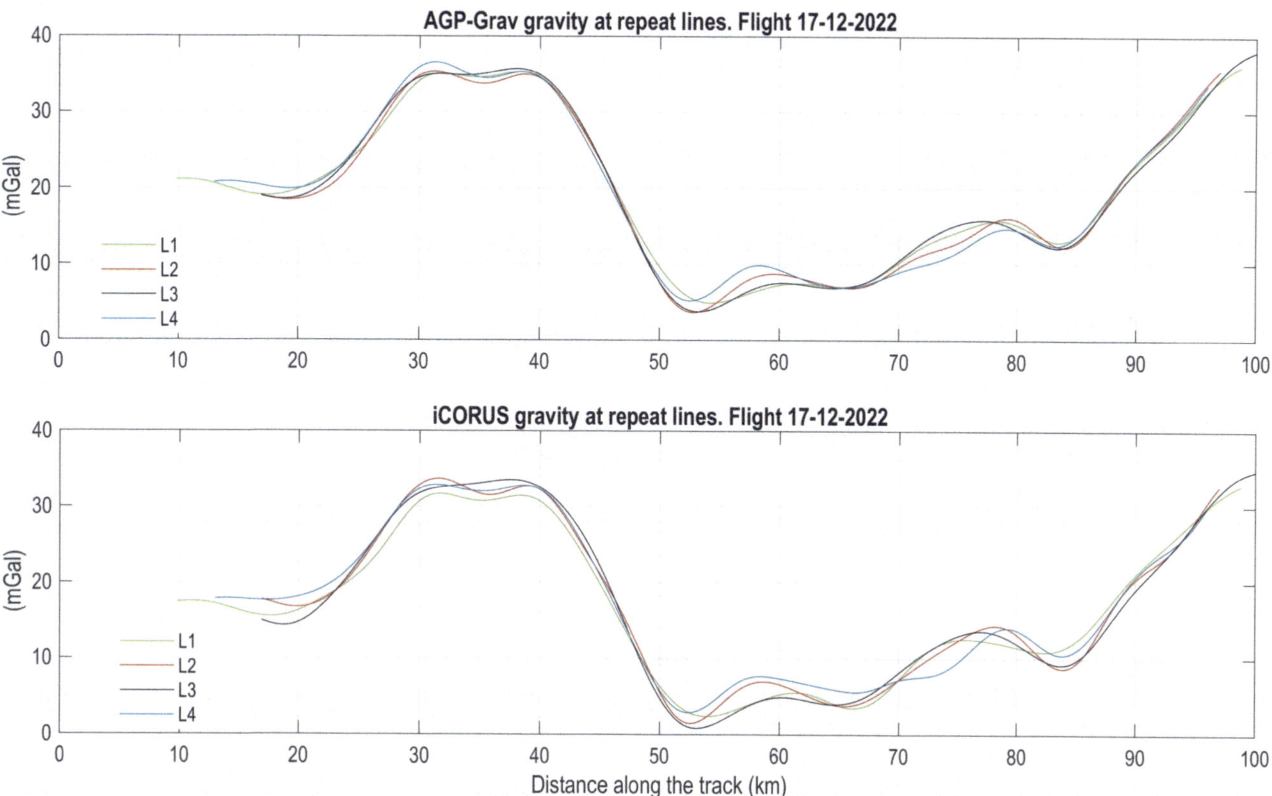

Fig. 3 The gravity disturbances at repeated lines derived from the AGP-Grav data (above) and iCORUS data (below), mGal. The gravimetric filter cutoff frequency (half-transmission point) is 0.01 Hz

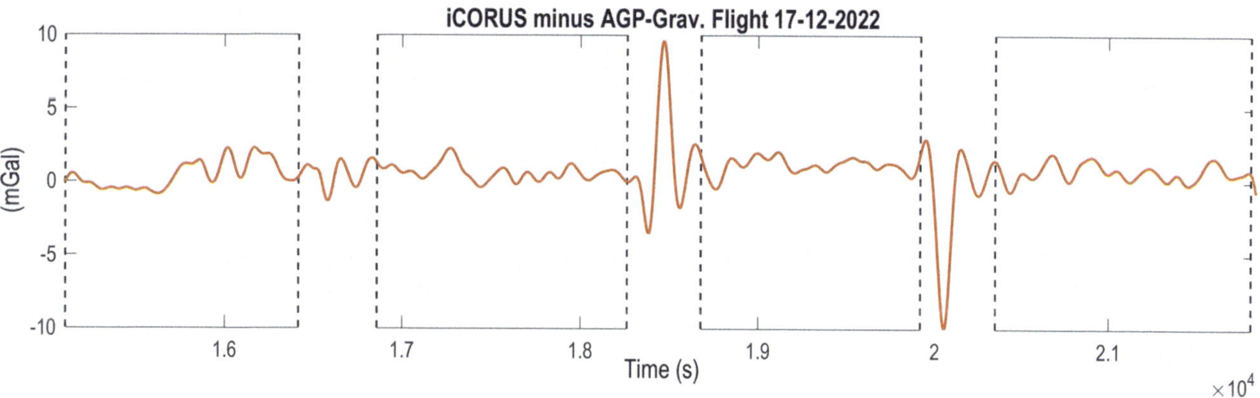

Fig. 4 The differences between the gravity disturbances derived from the AGP-Grav and iCORUS data, mGal. The vertical lines indicate the time intervals of the repeated flight lines

200 to 300 km. The flight height above the terrain is 250–450 m (the average height is 350 m). The average flight velocity at the lines is 55 m/s.

The raw IMU data QC is based on the accuracy assessment for the IMU position and velocity solutions in the autonomous (stand-alone) IMU operating mode. The GNSS solutions (position and velocity) serve as a reference solution. Figure 5 shows the IMU position and velocity solution errors (i.e., the difference between the IMU and the reference solutions) over the flight time of one flight. The statistics of the stand-alone IMU performance from several flights are given in Table 2.

The statistics of the vertical accelerometer bias drift over 18 flights and the mean values of temperature variations (measured by the vertical accelerometer temperature sensor) are shown in Fig. 6. The STD of temperature variations in each flight is the same (0.018 °C).

The temperature profiles from the accelerometer temperature sensors from one of the flights are shown in Fig. 7. The STD values are 0.021 °C and 0.020 °C for the horizontal

Table 1 Statistics of the differences between the airborne gravity disturbances derived from two sensors (AGP-Grav and iCORUS)

Line	Mean, mGal	STD, mGal	RMS, mGal
Line 1	0.42	0.96	1.05
Line 2	0.61	0.54	0.81
Line 3	1.17	0.51	1.28
Line 4	0.71	0.59	0.93
All lines	0.75	0.64	1.23

Table 2 Statistics of the stand-alone IMU position and velocity errors (maximum absolute values of differences between the IMU and GNSS solutions) for the AGP-Grav gravimeter

Flight date	Pos. error (east), km	Pos. error (north), km	Vel. error (east), m/s	Vel. error (north), m/s
02.07.2024	4.5	2.5	1.0	1.0
04.07.2024	2.2	1.7	1.5	2.5
05.07.2024	4.0	3.8	1.7	1.4
06.07.2024	3.0	4.1	1.7	1.2
09.07.2024	3.7	5.7	1.2	1.2
10.07.2024	5.4	4.5	1.7	1.9
11.07.2024	4.5	7.0	1.5	1.6

accelerometer sensors (X and Y) and 0.018 °C for the vertical accelerometer sensor (Z).

Figure 8 shows the accelerations (derived from the accelerometer data from one of the flights) projected onto the geodetic vertical with the Eötvös correction and normal gravity subtracted (smoothed by a two-pass Butterworth filter with a half-transmission point at 0.01 Hz). Figure 8 also shows the vertical velocity computed for the same flight using the accelerometer data, integrated IMU/GNSS solutions, and Eq. (2).

The estimates of gravity disturbance at two repeated flight lines (flown at the constant altitude above the control track located near the survey area) are shown in Fig. 9 along with the elevation profile. The statistics for the gravity disturbance estimates obtained with various values of the gravimetric filter cutoff frequency are presented in Table 3.

The gravity disturbance estimates at the survey and crossing lines are shown in Fig. 10. The gravimetric filter cutoff frequency (half-transmission point) is 0.01 Hz, which results in a half-wavelength resolution of the gravity disturbance estimates of 2.8 km.

In Fig. 10, the elevation and the gravity data derived from the global model XGM2019 (Zingerle et al. 2019) (in the spherical harmonic expansion to d/o 2190 computed using the ICGEM service) are also shown for comparison with the airborne gravity data.

The histogram derived from the gravity disturbance differences at cross-over points (intersections of the survey and crossing lines) is shown in Fig. 11.

5 Discussion

We first evaluate the AGP-Grav gravity disturbances obtained during the 2022 flight test (Figs. 3 and 4). The repeated line analysis shows the internal accuracy of 0.59 mGal (STD) and 0.81 mGal (RMS) for the AGP-Grav data. For the iCORUS data, the corresponding values are slightly larger and are 0.80 mGal (STD) and 0.94 mGal (RMS). Overall, the gravity datasets from both sensors are similar: the STD and RMS values derived from the differences between two datasets are 0.64 mGal and 1.23 mGal, respectively (Table 1). The corresponding value of the RMS error (defined as RMS/$\sqrt{2}$) is 0.87 mGal.

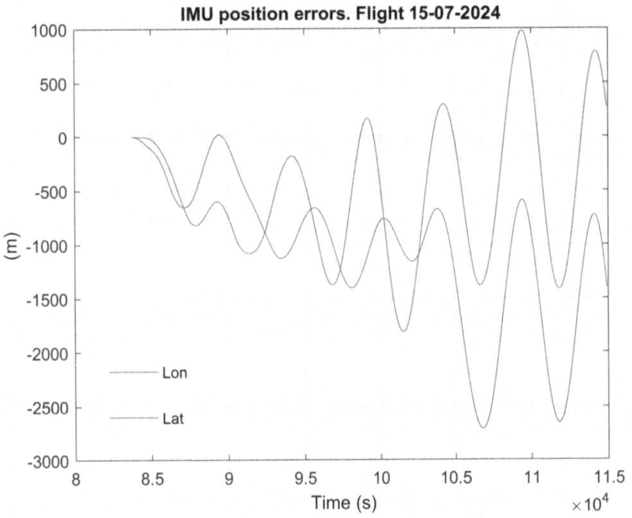

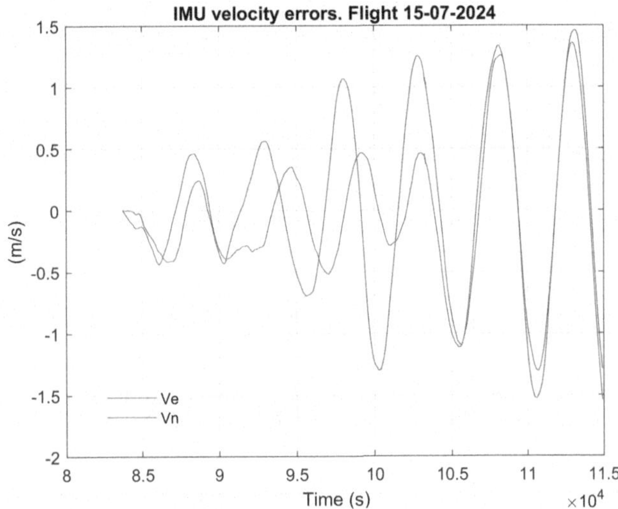

Fig. 5 The errors of the autonomous AGP-Grav's IMU solutions (relative to the GNSS solutions serving as reference data): IMU position errors, m (left) and IMU velocity errors, m/s (right)

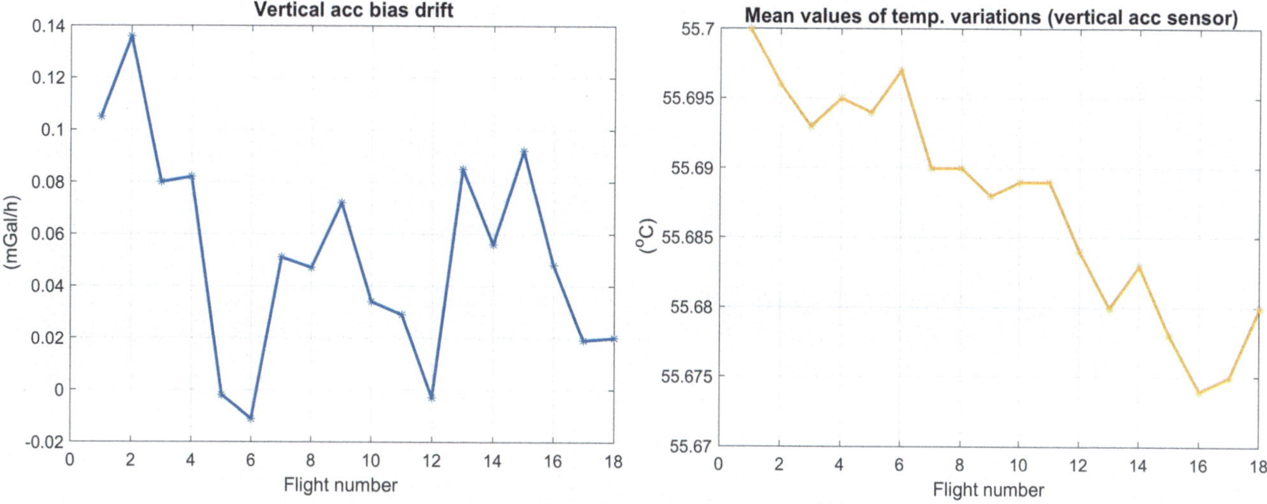

Fig. 6 The values of the AGP-Grav vertical accelerometer bias drift in each flight of the ongoing campaign, mGal/h (left). The mean values of temperature variations measured by the vertical accelerometer sensor, °C (right)

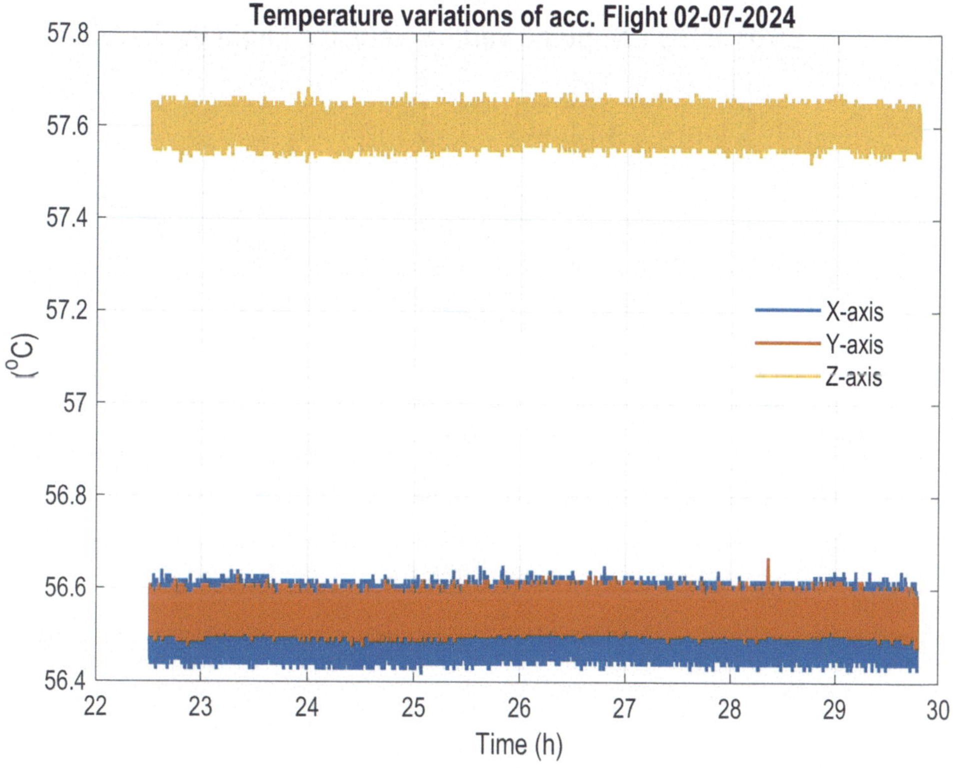

Fig. 7 Temperature profiles from the AGP-Grav accelerometer sensors for one of the flights, °C

Next, we evaluate the AGP-Grav gravity disturbances from the ongoing campaign. It can be seen from Fig. 5 that the IMU position and velocity errors reach 2.5 km and 1.5 m/s, respectively, after 8 h of flight. This is a good result confirming the navigation grade of the calibrated IMU. The oscillations in the position and velocity errors shown in Fig. 5 are the so-called Schuler oscillations, which are well-known in inertial navigation (Jekeli 2001). The stability of the stand-alone IMU performance is confirmed by the statistics over multiple flights (Table 2).

As Fig. 6 shows, the linear drift of the IMU vertical accelerometer bias does not exceed 0.09 mGal/h for most flights. The stability of the vertical accelerometer bias is reached by the reliable thermal stabilization of the AGP-Grav instrument (better than 0.1 K) (Figs. 6 and 7).

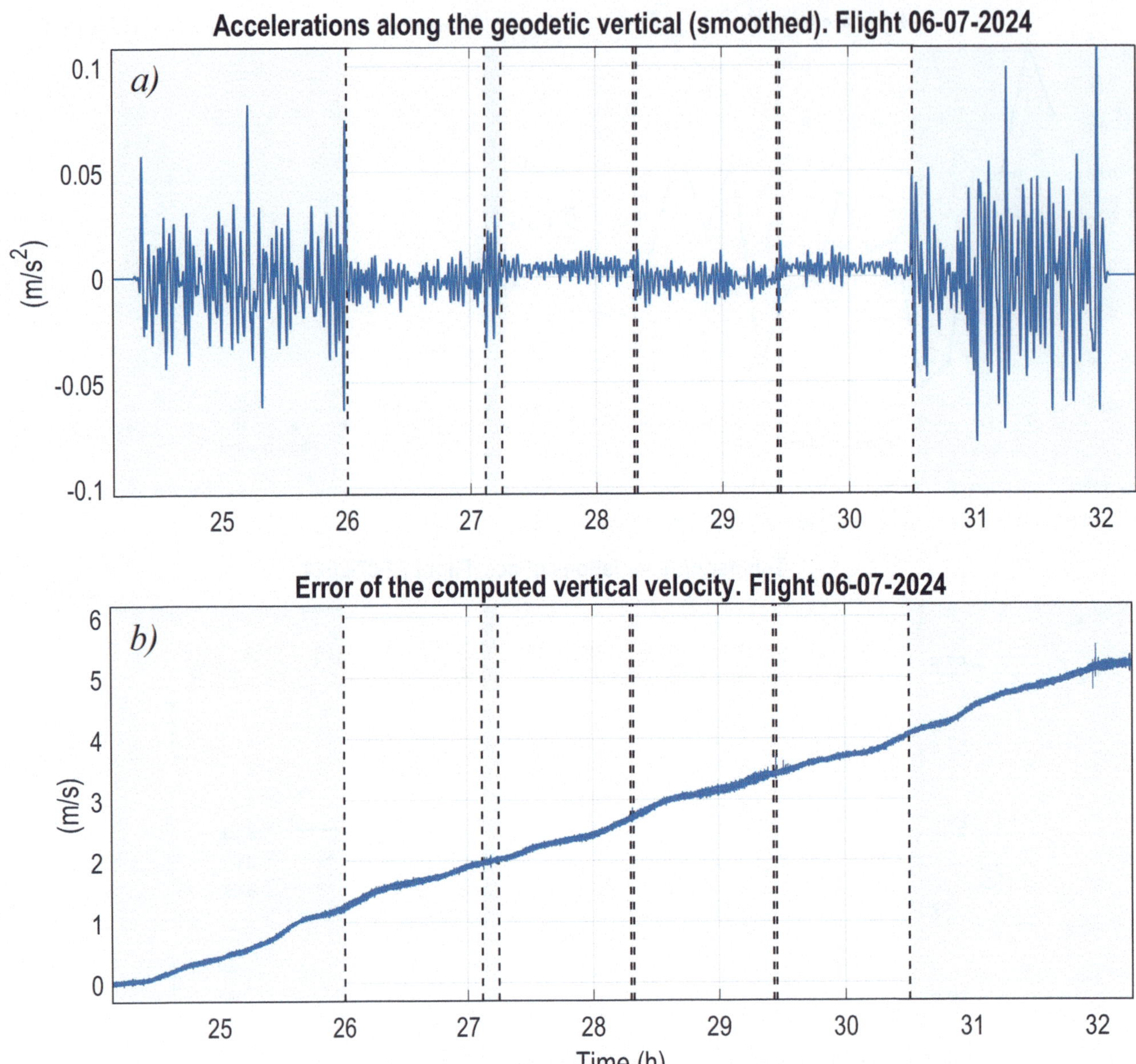

Fig. 8 (**a**) Accelerations along the geodetic vertical computed from the AGP-Grav accelerometer data (with the Eötvös correction and normal gravity subtracted), smoothed by a Butterworth filter with a half-transmission point at 0.01 Hz, m/s^2. (**b**) The vertical velocity derived from the AGP-Grav accelerometer data and IMU/GNSS solutions using Eq. (2) minus GNSS-derived vertical velocity, m/s

Figure 8 shows the range of vertical accelerations typical for the flights in the campaign under consideration. As shown in the figure, the vertical accelerations do not exceed 0.015 m/s^2 at the survey lines (after smoothing by a Butterworth filter with a half-transmission point at 0.01 Hz) and can reach 0.070 m/s^2 at the aircraft turns. This shows, on the one hand, the relatively benign flight conditions at the survey lines, and on the other hand, this confirms that when using a strapdown gravimeter, intensive maneuvering of the aircraft is permissible (in contrast to the case of classical stabilized-platform gravimeters).

The error of the vertical velocity derived from the accelerometer data and IMU/GNSS solutions using Eq. (2) (Fig. 8) confirms its expected monotonic behaviour (caused by the absence of δg in Eq. (2)). The vertical velocity error reaches 5 m/s after 8 h, which shows an average δg value of roughly 17 mGal for this particular flight.

Figure 9 shows good repeatability of the gravity disturbance estimates at two repeated lines. As can be concluded from the statistics shown in Table 3, the gravity estimation accuracy (in terms of STD) at a sub-mGal level is reached with a filter cutoff frequency (half-transmission

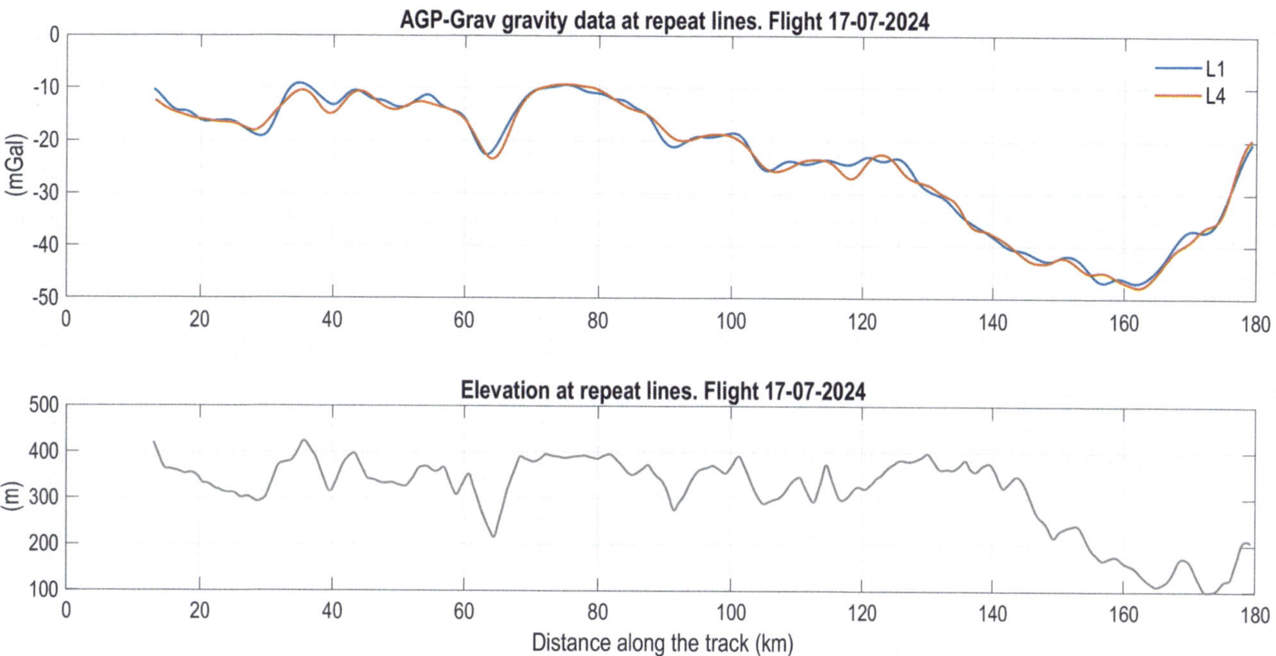

Fig. 9 Gravity disturbance estimates obtained from the AGP-Grav measurements at repeated flight lines, mGal. The gravimetric filter cutoff frequency is 0.01 Hz (above). Elevation, m (below)

Table 3 Statistics from the airborne gravity data collected by the AGP-Grav strapdown gravimeter at two repeated flight lines

	Gravimetric filter cutoff frequency (half-transmission point), Hz			
Parameter	1/80	1/100	1/120	1/140
Half-wavelength spatial resolution, m	2200	2750	3300	3850
STD, mGal	1.35	0.97	0.80	0.74

point) of 0.01 Hz (equivalent to a half-wavelength spatial resolution of 2.8 km) or less. An accuracy of 0.74 mGal (STD) corresponds to filtering with a cutoff frequency of 1/140 Hz (equivalent to a half-wavelength spatial resolution of 3.9 km).

A visual comparison of the gravity disturbance estimates and the topography elevation (Fig. 9) confirms their expected correlation. This also confirms the ability of the developed postprocessing strategy to effectively filter out carrier accelerations and instrumental errors from raw gravimeter measurements.

Figure 10 shows good visual correlation between the airborne data and topography as well as significantly higher half-wavelength spatial resolution of the airborne gravity (2.8 km) compared to the XGM2019 gravity (the stated resolution is about 10 km).

The number of cross-over points (intersections of the survey and crossing lines) from 18 flights is 504 (Fig. 11). From the cross-over analysis, the internal accuracy of the obtained airborne gravity disturbances is 1.36 mGal (the root mean square error) and −0.33 mGal (the mean value) without any crossover adjustment, which is an excellent result for strapdown airborne gravimetry considering the draped flight conditions.

6 Conclusions

Strapdown airborne gravimeters are increasingly used in aerogeophysical surveys allowing flights with small, slow-moving, multi-instrumented aircraft, which ensures cost-effectiveness and higher spatial resolution of collected gravity data.

In the paper, we evaluate the accuracy of the airborne gravimetry results obtained from measurements with a new strapdown system (AGP-Grav). The cross-over analysis yields a 1.4 mGal internal accuracy (root-mean-square error) of the results under draped flight conditions. The new strapdown gravimeter also showed a high stability of the vertical accelerometer bias, with the linear drift not exceeding 0.09 mGal/h, which is ensured by the high-precision thermal stabilization system.

Comparison of the AGP-Grav and iMAR gravity datasets from a repeated line flight showed that the results from both sensors are similar, with a root-mean-square error value of 0.87 mGal. However, AGP-Grav showed slightly better internal accuracy (root-mean-square value) based on the repeated line analysis (0.81 mGal) compared to the iMAR instrument (0.94 mGal).

The postprocessing of the raw gravimeter data was performed using the software developed at Lomonosv MSU in

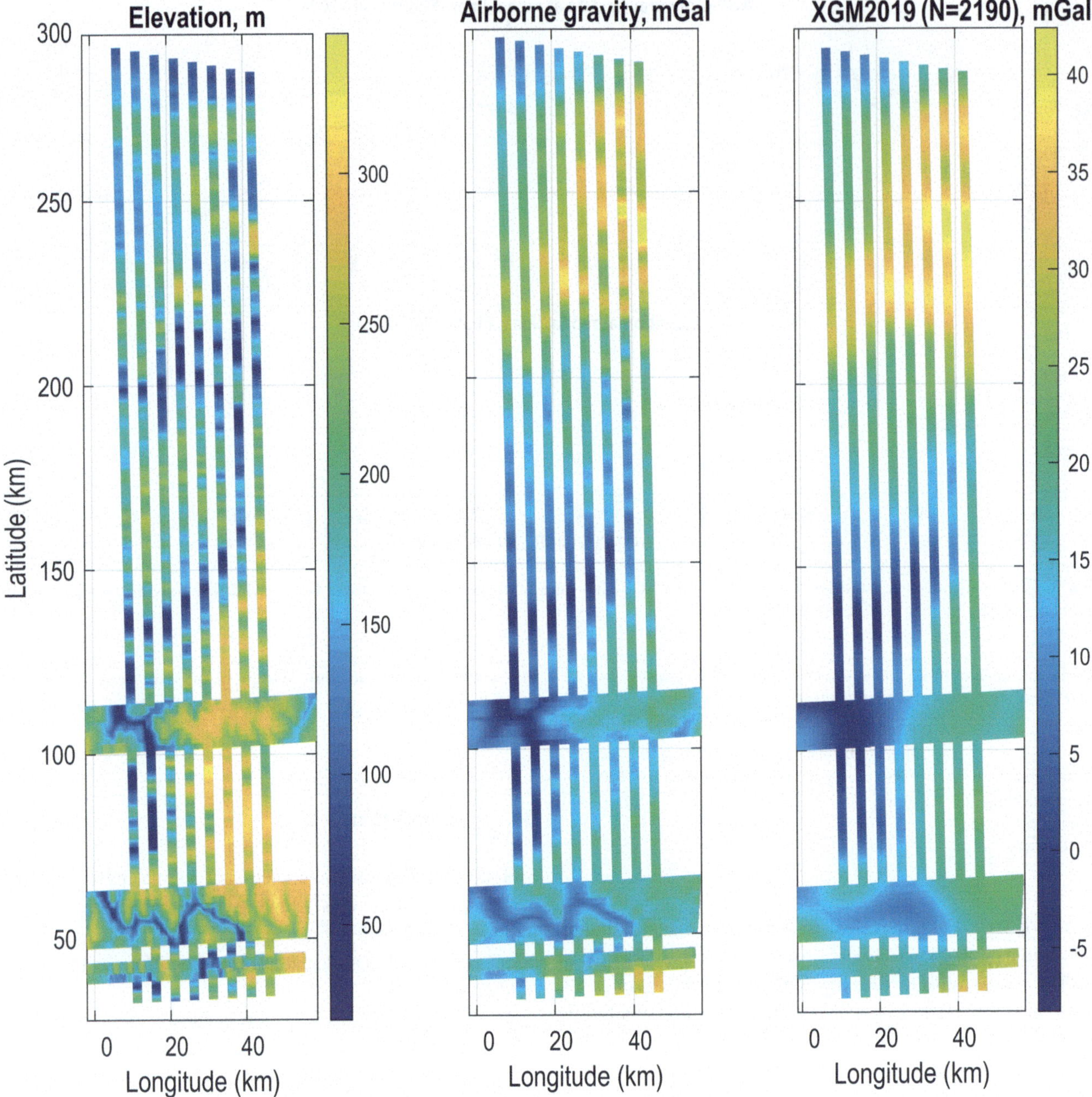

Fig. 10 Topography elevation, m (left). The airborne gravity disturbances at the flight lines obtained from the AGP-Grav data with the gravimetric filter cutoff frequency of 0.01 Hz, mGal (center). The gravity disturbances derived from the global gravity model data (XGM2019), mGal (right)

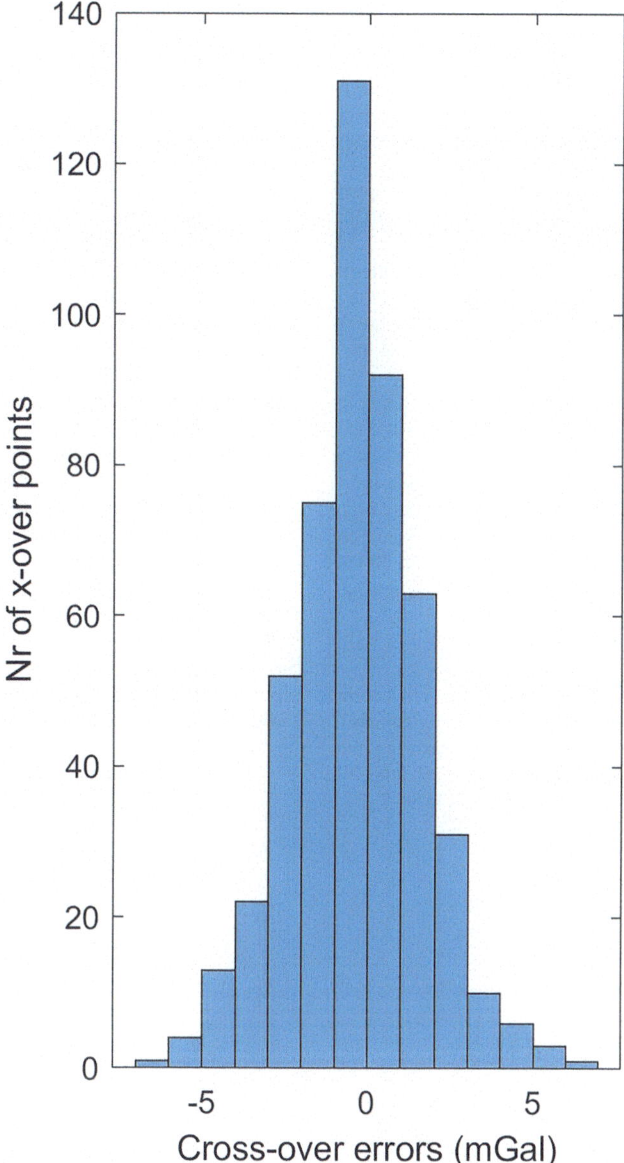

Fig. 11 Histogram derived from the cross-over errors, i.e., the differences between the airborne gravity disturbances (obtained from the AGP-Grav data) at the intersection points of survey and crossing lines

2020–2023 (the INS-GNSS and IMU-GRAV programs). The main subroutines of the software are described in the paper. The programs cover all postprocessing stages (including raw data quality control, calculation of GNSS solutions, integrated IMU/GNSS solutions, and gravimetry solutions) and can be used by researchers without expertise in the field of inertial navigation. The software programs were tested in dozens of airborne gravimetry campaigns with iMAR strapdown sensors and new AGP-Grav gravimeters.

Acknowledgements We thank the editor and two anonymous reviewers for their constructive comments and corrections, which helped us to improve the quality of the manuscript.

References

Ayres-Sampaio D, Deurloo R, Bos M, Magalhães A, Bastos L (2015) A comparison between three IMUs for strapdown airborne gravimetry. Surv Geophys 36:571–586. https://doi.org/10.1007/s10712-015-9323-5

Bolotin YV, Popelensky MY (2007) Accuracy analysis of airborne gravity when gravimeter parameters are identified in flight. J Math Sci 146:5911–5919. https://doi.org/10.1007/s10958-007-0405-x

Brovkin GI, Kontarovich OR, Golovan AA, Vyazmin VS (2021) Results of the first Russian airborne gravimetry survey with a strapdown gravimeter. In: Proceedings of the 4th international geological geophysical conference and exhibition GeoEurasia-2021. Geological exploration in modern life, vol 2. PoliPRESS, Tver, pp 107–111 de Barros Camara E, Guimarães SNP (2016) Magnetic airborne survey—geophysical flight. Geosci Instrum Method Data Syst 5:181–192. https://doi.org/10.5194/gi-5-181-2016

Foroughi I, Goli M, Ferguson S, Pagiatakis S (2024) Optimizing airborne flight line spacing for geoid determination with full gravity vectors. International Association of Geodesy Symposia, Springer, Berlin, Heidelberg. https://doi.org/10.1007/1345_2024_253

Golovan AA, Vavilova NB (2007) Satellite navigation. Raw data processing for geophysical applications. J Math Sci 146:5920–5930. https://doi.org/10.1007/s10958-007-0406-9

Golovan AA, Vyazmin VS (2023) Methodology of airborne gravimetry surveying and strapdown gravimeter data processing. Gyroscopy Navig 14(1):36–47. https://doi.org/10.1134/S2075108723010029

Jekeli C (2001) Inertial navigation systems with geodetic applications. Walter de Gruyter, Berlin

Jensen TE (2022) Spatial resolution of airborne gravity estimates in Kalman filtering. J Geod Sci 12(1):185–194. https://doi.org/10.1515/jogs-2022-0143

Johann F, Becker D, Becker M, Ince ES (2020) Multi-scenario evaluation of the direct method in strapdown airborne and shipborne gravimetry. In: Freymueller JT, Sánchez L (eds) 5th symposium on terrestrial Gravimetry: static and Mobile measurements (TG-SMM 2019), vol 153. International Association of Geodesy Symposia, Springer, Cham. https://doi.org/10.1007/1345_2020_127

Kailath T, Sayed AH, Hassibi B (2000) Linear estimation. Prentice Hall, Englewood Cliffs

Krasnov A, Sokolov A, Bolotin Y, Golovan A, Parusnikov N, Motorin A et al (2022) Data processing methods for onboard gravity anomaly measurements. In: Methods and technologies for measuring the earth's gravity field parameters. Earth systems data and models, vol 5. Springer, Cham. https://doi.org/10.1007/978-3-031-11158-7_2

Olson D (2010) GT-1A and GT-2A airborne gravimeters: improvements in design, operation, and processing from 2003 to 2010. In: Proceedings of the ASEG-PESA airborne gravity 2010 workshop, Sydney, Australia

Peshekhonov VG, Stepanov OA, Rozentsvein VG, Krasnov AA, Sokolov AV (2022) State-of-the-art strapdown airborne gravimeters: analysis of the development. Gyroscopy Navig 13(4):189–209. https://doi.org/10.1134/S2075108722040101

Simav M (2020) Results from the first strapdown airborne gravimetry campaign over the Lake District of Turkey. Surv Rev. https://doi.org/10.1080/00396265.2020.1826140

Studinger M, Bell R, Frearson N (2008) Comparison of AIRGrav and GT-1A airborne gravimeters for research appli-cations. Geophysics 73:151–161

Vavilova NB, Golovan AA, Kozlov AV, Papusha IA, Zorina OA, Izmailov EA et al (2021) INS/GNSS integration with compensated data synchronization errors and displacement of GNSS antenna. Experience of practical realization. Gyroscopy Navig 12:236–246. https://doi.org/10.1134/S207510872103007X

Vyazmin VS, Golovan AA (2023) Scalar and vector strapdown airborne gravimetry on aircraft and UAV: methodology of surveying and data processing. In: 30th Saint Petersburg international conference on integrated navigation systems (ICINS). CSRI Elektropribor, Saint Petersburg, pp 1–6. https://doi.org/10.23919/ICINS51816.2023.10168453

Zingerle P, Pail R, Gruber T, Oikonomidou X (2019) The combined global gravity field model XGM2019e. J Geod 94:1–12. https://doi.org/10.1007/s00190-020-01398-0

GOCE SGG Wavelet Multi-resolution Analysis with the Latest Level 2 GOCE Data in the Wider Hellenic Region

Eleftherios A. Pitenis and Georgios S. Vergos

Abstract

Over a decade since the official ending of the GOCE satellite mission (2009–2013), GOCE based gravity data continue to provide useful information to the study of the static gravity field of the Earth. The present study deals with the application of GOCE SGG wavelet multi-resolution analysis to the latest available Level 2 GOCE data, in order to be used for geoid and gravity field modeling applications. In total, three different methods of computing disturbing gravity gradients at satellite altitude were applied in this study, based on the latest EGG_NOM_2, SST_PSO_2 and EGG_TRF_2 products, as well as on information coming from Global Geopotential Models (GGMs). The EGG_NOM_2 and SST_PSO_2 derived gradients were then projected from the satellite altitude to a mean orbital sphere, with a special focus on the study area covering the wider Hellenic and Eastern Mediterranean regions. To the mean sphere projected disturbing gravity gradients exhibited still track-wise noise, so that a 2D Wavelet Multi-Resolution Analysis (WL MRA) using a Daubechies wavelet at 14 levels of decomposition was applied. Various reconstruction scenarios using different levels of decomposition were considered, corresponding to different spatial resolutions. The optimal one for our purposes was deemed to be the one using levels 7–14, corresponding to spatial resolutions from approximately 117 km onwards. The resulting filtered disturbing gravity gradient field are ready to be used in studies with different goals, as in the case of geodetic and geophysical works. Comparisons of the resulting mean fields using different synthesis scenarios are carried out, so as to gain insight on the spectral content of the various fields.

Keywords

2D WL-MRA · EGG_NOM_2 · EGG_TRF_2 · Global geopotential model · GOCE · Satellite gravity gradiometry · SST_PSO_2

1 Introduction

ESA's GOCE (Gravity field and steady-state Ocean Circulation Explorer) satellite mission, operationally active from 2009 to 2013, took advantage of the satellite gravity gradiometry technique to measure the static gravity field of the Earth. For this purpose, the mission was equipped with a technologically advanced Electrostatic Gravity Gradiometer (EGG) using three pairs of accelerometers placed orthogonally in the diamond configuration (Müller 2003). Satellite-to-Satellite Tracking Instrument (SSTI) was used to determine the precise orbit of the satellite and the long-wavelengths of the gravity field of the Earth through the high-low Satellite-to-Satellite Tracking technique (SST-hl) (Drinkwater et al. 2007; Frommknecht et al. 2011). Out of the six in total measured GOCE gravity gradients (V_{xx}, V_{yy},

E. A. Pitenis (✉) · G. S. Vergos
Laboratory of Gravity Field Research and Applications (GravLab), Department of Geodesy and Surveying, Aristotle University of Thessaloniki, Thessaloniki, Greece
e-mail: epitenis@topo.auth.gr; vergos@topo.auth.gr

J. T. Freymueller, L. Sànchez (eds.), *International Symposium on Gravity, Geoid and Height Systems 2024 (GGHS2024)*, International Association of Geodesy Symposia 158, https://doi.org/10.1007/1345_2025_296

V_{zz}, V_{xy}, V_{xz}, V_{yz}), the V_{xy} and V_{yz} ones are measured with a comparatively lower accuracy (Bouman et al. 2011). The processing of GOCE gravity data has been the subject of many different studies in the past such as in cases dealing with their appropriate filtering (Piretzidis and Sideris 2017), their downward continuation to the surface of Earth (Tóth et al. 2006; Šprlák et al. 2014; Sebera et al. 2015), the evaluation of their performance compared to global gravity models (Bouman and Fuchs 2012), orbit residuals (Gruber et al. 2011) and to GNSS/levelling data (Gruber et al. 2012; Vergos et al. 2018). The significant impact of GOCE data in studies over different scientific fields has been highlighted over the past years, with practical emphasis on geodesy, with GOCE data being used in regional geoid determination (Natsiopoulos et al. 2023), the computation of models of the crustal structure of the Earth like (Ebbing et al. 2013; Mariani et al. 2013; Reguzzoni and Sampietro 2015; Holzrichter and Ebbing 2016) and the determination of the ocean circulation and dynamic ocean topography (Bingham et al. 2011; Haines et al. 2011). The main objective of the present work was the processing of GOCE Satellite Gravity Gradient (SGG) observations, from the latest Level 2 GOCE data accessible through the ESA GOCE Online Dissemination service (ESA 2023). We focus on the processing of the data with classical filters, so as to reduce the spectral content of the data within the GOCE Measurement Bandwidth (MBW), while additional filtering steps are proposed to remove residual noise. The ultimate goal is to reduce the data to a mean sphere (MS), so as to be ready for downward continuation. The focus area is over the eastern Mediterranean basin and in particular the wider region around Greece.

2 GOCE SGG Processing at Satellite Altitude

In the first part of the study, three different approaches to compute disturbing gravity gradients at satellite altitude were applied. First, the latest available GOCE EGG_NOM_2 and SST_PSO_2 products were used, making for a total of 48 months of GOCE data being processed, ranging from October 2009 to October 2013. The choice to select the latest Level 2 data was dictated by the fact that the Laplace equation of the newly processed original unfiltered gravity gradients in GRF was found to be considerably enhanced contrary to that of the older ones, owing mainly to the reported improved behavior of the V_{yy} gradient (Siemes et al. 2018, 2019).

One of the main tasks during the pre-processing stage of GOCE SGG data is the detection of outliers in the original GOCE gravity gradients and their appropriate removal and/or reduction. These, in most cases, show inconsistency with the rest of the dataset, such as by appearing as measurement jumps and are normally flagged appropriately in the corresponding files according to the GOCE High-level Processing Facility (HPF) standards (ESA 2014). The outliers can also be detected, especially for the three diagonal gradients of the Gravitational Gradient Tensor (GGT), by using the Laplace equation. In general, the utilization of the Laplace equation can be a powerful blunder detection tool in the processing of gravity gradient data. Ideally it assumes that the sum of the three diagonal gravity gradients of the gravity field of the Earth outside the attracting masses is equal to zero, as (Heiskanen and Moritz 1967):

$$\Delta V = V_{xx} + V_{yy} + V_{zz} = 0 \tag{1}$$

An additional validation step employed was a blunder detection and removal, where an $M \pm k\sigma$ threshold was implemented, with M being the mean value, k a constant that can take different values and σ the standard deviation of the residual SGG data. This can be in essence the well known 3σ test, while in the present study a $k = 1.5$ deemed to be the best choice in terms of clearing spikes in the original data. Various other tests have also been implemented with different values for the k value, that provided though unsatisfactory results, because they either removed observations which were useful or retained observations which were blunders. An example of this pre-processing scheme is given for a sample GOCE arc, where Fig. 1 depicts in blue the original GOCE SGG observations, showcasing erroneous observations as the two spikes at a time of $\sim 9.4089 \times 10^8$ s. The standard deviation (std) of the original V_{zz} gravity gradients for the problematic dataset shown in Fig. 1 was found at the $\sim$32 Eötvös in the original data, to $\sim$18 Eötvös after the 3σ test, $\sim$15 Eötvös after the 2σ test, $\sim$14 Eötvös after the 1.5σ test and finally to $\sim$11 Eötvös after the application of the 1σ test. Despite the fact that the smallest std is found for k equal to 1, a value of 1.5 was chosen as ideal for k due to the fact that it retains useful information in the original signal, removing the aforementioned blunder, while keeping the original signal intact, in the rest of the time-series. Contrary to that, when a $k = 1$ value is used, not only the blunder, but also useful signal is removed (see the discrepancy between the red and blue lines in Fig. 1). The other two cases with $k = 2, 3$ do not manage to remove the blunders. The general equation of the method is given below:

$$M - 1.5\sigma < V_{ij} < M + 1.5\sigma \tag{2}$$

More analytically, in our study, SGG data originally measured in the Gradiometer's Reference Frame (GRF) were firstly reduced by a Global Geopotential Model (GGM), in a remove-compute-restore concept. The latter was based on a combination between the GO_CONS_GCF_2_TIM_R6 (to

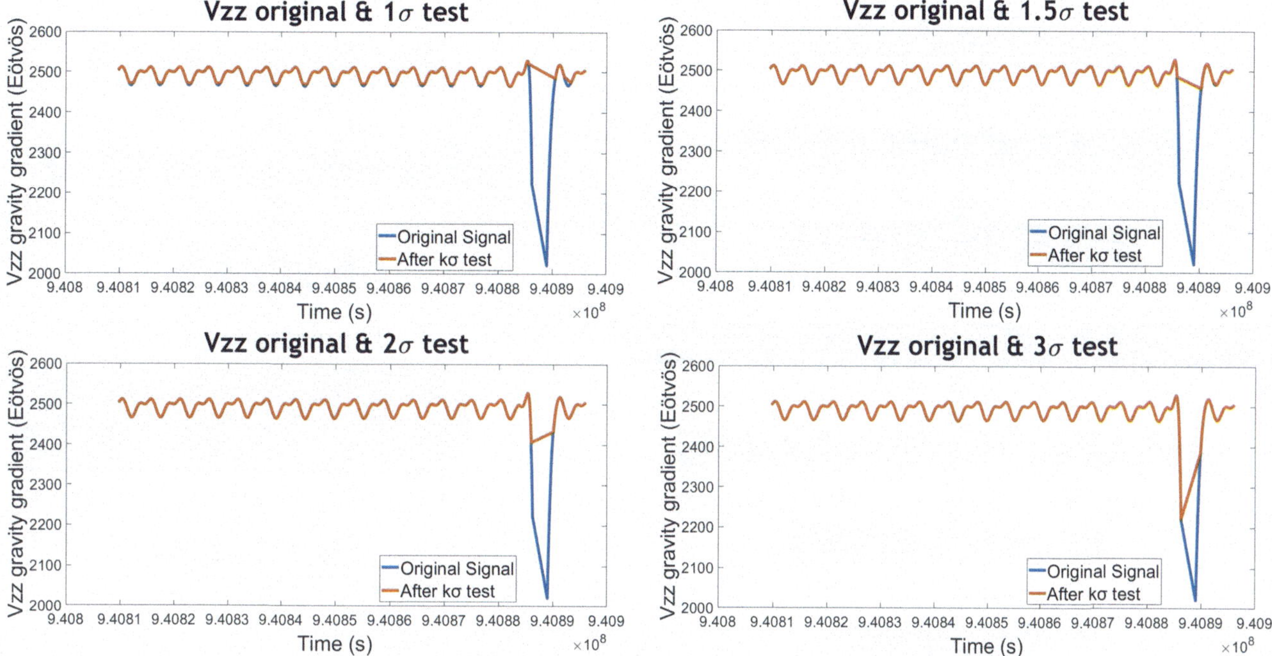

Fig. 1 Application of different $k\sigma$ tests to the original V_{zz} gravity gradients

a maximum d/o 165) (Brockmann et al. 2019) and EGM2008 (between d/o 166 to 2,190) (Pavlis et al. 2012) models as in Pitenis et al. (2022). These GGM based gradients were computed initially in the Local North Oriented Frame (LNOF) and then transformed to the GRF. The transformation process requires first their transformation from the LNOF to the Earth Fixed Reference Frame (EFRF), then from the EFRF to the Inertial Reference Frame (IRF), and finally from the IRF to the GRF. For an analytical documentation of the transformation process we refer to the detailed GOCE Level 2 Product Data Handbook (GO-MA-HPF-GS-0110) (ESA 2014). The resulting residual gravity gradient field in the GRF was then filtered using a band-pass Finite Impulse Response (FIR) filter with order equal to 1,500 aiming to retain only the useful signal inside the GOCE Measurement Bandwidth (MBW), which is limited between 5 and 100 mHz (0.005–0.1 Hz) (Pitenis et al. 2022). Considering the use of FIR filters in GOCE gravity data processing, it has been established in earlier studies such as those of Polgár et al. (2013) and Wan et al. (2012). It has also been studied as part of one of our previous works focusing on various different ways of GOCE SGG filtering to the MBW (Pitenis et al. 2022). After filtering, the GGM contribution has been restored and the filtered GOCE gradients have been transformed to the LNOF. They were then reduced to the GRS80 (Geodetic Reference System 1980) normal gravity field (Moritz 2000), resulting to the computation of disturbing gravity gradients (T_{ij}) at satellite altitude in the LNOF frame. The general equation for the computation of the disturbing gravity gradients is defined as follows:

$$T_{ij} = V_{ij} - \overline{V}_{ij} \tag{3}$$

where V_{ij} are the aforementioned processed gravity gradients transformed in the LNOF, $\overline{V}_{ij}$ the normal gravity gradients in LNOF and i, j the corresponding x, y, z axes.

In the second processing approach for computing disturbing gravity gradients at satellite altitude, all the available EGG_TRF_2 files carrying gravity gradient information in monthly format in the LNOF were processed. These files, according to the GOCE HPF standards, contain calibrated gravity gradients in the LNOF derived through the processing of EGG_NOM_2 and SST_PSO_2 files from previous product releases as well as by taking advantage of GGM information. The computation of the disturbing gravity gradient field in the present case required the subtraction of normal gravity gradients from the EGG_TRF_2 gradients in LNOF. In this case again the entire available EGG_TRF_2 dataset was used, making for a total of 45 processed months ranging from November 2009 to October 2013. It must be noted that data were missing for three months, namely July, August and September 2010.

Lastly, as a third way of evaluation, disturbing gravity gradients were computed based on GGM information by using a combination scheme of the GO_CONS_GCF_2_TIM_R6 and EGM2008 models at satellite altitude. Moreover,

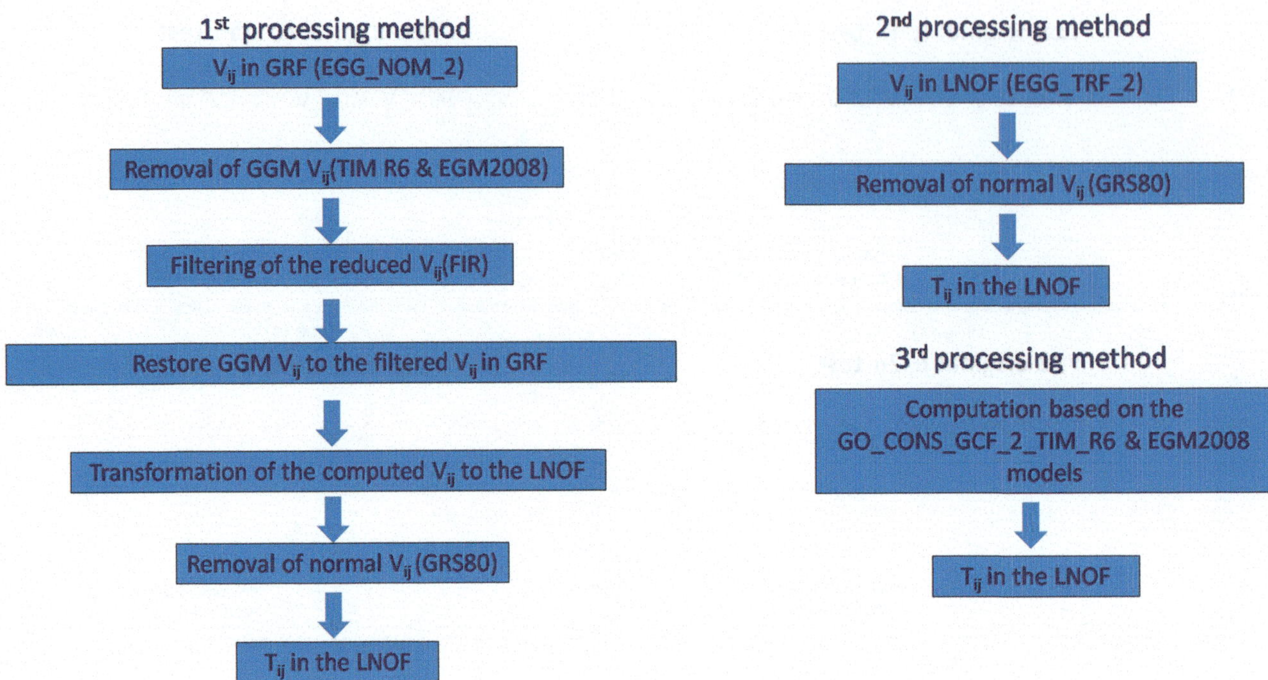

Fig. 2 The three GOCE SGG processing methods applied in the study

GO_CONS_GCF_2_TIM_R6 only and GO_CONS_GCF_2_DIR_R6 only cases were studied, which are based on the time-wise (TIM) and the direct approaches (DIR) respectively (Pail et al. 2011). The time span of the data used was the same as that of the new EGG_NOM_2 dataset, ranging from October 2009 to October 2013.

Generally, the validation of the computation procedure showed that in total the disturbing gravity gradients computed from the three aforementioned approaches (see Fig. 2) at satellite altitude presented an agreeable statistical accordance between them. In greater detail, the standard deviation values of the differences of the disturbing gravity gradients obtained from the first processing approach against those of the second and the third ones reached the 0.063 Eötvös, while their according mean values were approximately equal to zero. These results prompted for the further use of the EGG_NOM_2 and SST_PSO_2 derived gravity gradients in our work.

3 SGG Projection to MO and 2D Wavelet MRA

The next part of the study focused initially to the projection of the computed disturbing gravity gradients from the satellite altitude to a selected Mean Orbit (MO). In the past a handful of methods have been proposed in the bibliography, such as the method by Novák et al. (2013) that was reproduced by Tsoulis and Moukoulis (2019) and the method proposed by Tóth and Földváry (2005). In the present study, the Taylor expansion outlined by Tóth and Földváry (2005) and used in Natsiopoulos et al. (2023) has been employed. The resulting projected data were then cut for the chosen area of study and gridded on a $1' \times 1'$ grid using linear interpolation. The region of interest encompasses the wider Hellenic and Eastern Mediterranean regions, since it extends for latitudes (φ): 28–45° and for longitudes (λ): 10–40° (see Fig. 3).

It was found out that the projected gridded T_{ij} to the MO, still appeared to be noisy due to the appearance of satellite orbit tracks in them. To remove this noise, a 2D Wavelet MRA (WL MRA) (Mallat 1989) using a Daubechies db10 wavelet (Daubechies 1992) at 14 levels of decomposition was chosen to be applied both to the T_{zz} and ($-T_{xx} - T_{yy} = T_{zz}$) gravity gradients. The computation of the $-T_{xx} - T_{yy}$ gravity gradient is based on the fundamental principles of the Laplace equation referring to the diagonal gravity gradients of the gravity gradient tensor.

Regarding to the theory of wavelets and their basic concepts, their localization both in the time and space domains is actually their greater advantage compared to the classic Fast Fourier Transform (FFT) (Roland 2005). Wavelet MRA has been used in the past as, indicatively, for the evaluation of GOCE/GRACE based GGMs (Peidou and Vergos 2016) for modelling the gravity field of the Earth by using both wavelet frames (Panet et al. 2005) and domain decomposition methods (Panet et al. 2011), as well by applying multipole wavelets (Klees and Wittwer 2007). In WL MRA the

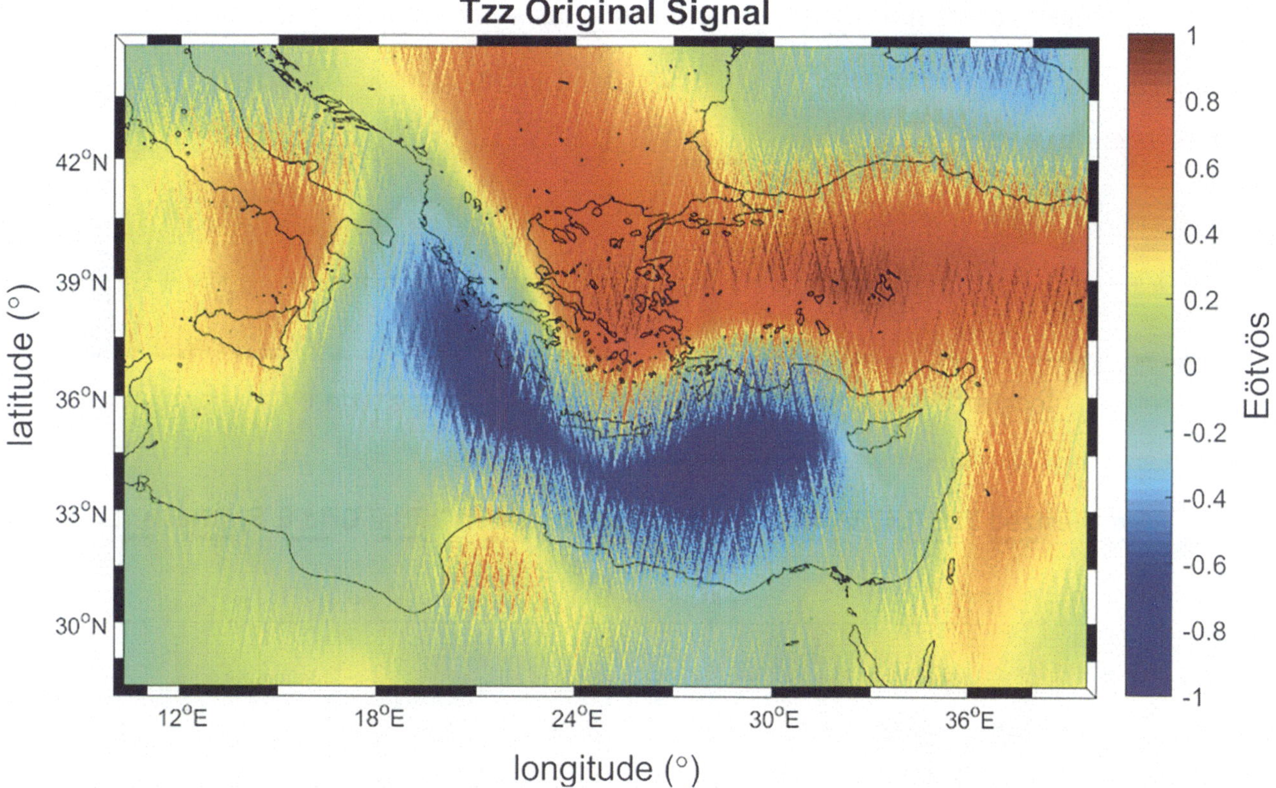

Fig. 3 The projected T_{zz} gradient in the area under study

Table 1 Spatial resolutions of the 14 levels of decomposition (units: km)

Levels	Resolution (km)		Levels	Resolution (km)	
Level 1	1.83	3.67	Level 8	234.67	469.33
Level 2	3.67	7.33	Level 9	469.33	938.67
Level 3	7.33	14.67	Level 10	938.67	1,877.33
Level 4	14.67	29.33	Level 11	1,877.33	3,754.67
Level 5	29.33	58.67	Level 12	3,754.67	7,509.33
Level 6	58.67	117.33	Level 13	7,509.33	15,018.67
Level 7	117.33	234.67	Level 14	15,018.67	30,037.33

original signal is decomposed to different levels of decomposition, with each level corresponding to a different spatial resolution. The first level of decomposition corresponds to the spatial resolution of the original signal, whereas the maximum one refers to its extent (Grebenitcharsky and Moore 2014). Generally, in WL MRA the spatial resolution of each level of decomposition is double that of the previous one, following the so-called dyadic pattern. In the case of the present study, Table 1 summarizes the spatial resolution of the various levels of decomposition given the original resolution of 1 arcmin of the available SGG data to the MO.

The reconstruction of the original decomposed signal, based on its wavelet decomposition in various levels of detailed and approximation coefficients, can be defined as follows (Daubechies 1992):

$$s = \left(d_1^{\mathrm{H}} + d_1^{\mathrm{D}} + d_1^{\mathrm{V}}\right) + \cdots + \left(a_{14} + d_{14}^{\mathrm{H}} + d_{14}^{\mathrm{D}} + d_{14}^{\mathrm{V}}\right) \tag{4}$$

In the above equation, a_n is the approximate coefficient of the last level, while d_n^{H}, d_n^{D}, d_n^{V} are the horizontal, diagonal and vertical detail coefficients of the levels of decomposition, respectively.

In case all the levels of decomposition are going to be used in the signal reconstruction, as expected, the original signal will be computed due to the orthogonality of wavelets (Peidou and Vergos 2016). The purpose of the signal reconstruction dictates which levels of decomposition are actually going to be used in the reconstruction and which are going to be neglected. In the present work, various signal reconstruction scenarios were undertaken by using different combinations of levels of decomposition. The most characteristic reconstructions were those that used: (1) levels 5–14 (L5 ... L14), (2) levels 6–14 (L6 ... L14), (3) levels 7–14 (L7 ... L14) and (4) levels 8–14 (L8 ... L14).

The L5 ... L14 reconstruction (see Fig. 4) seems to be affected by the orbit tracks as noisy features, with only few of them removed. In the L6 ... L14 reconstruction (see Fig. 5) more orbit tracks have been removed but a few of them still remain in the reconstructed signal. The L7 ... L14 (see Fig. 6) reconstruction seems to be mostly unaffected by this track-like kind of noise, while in the L8 ... L14 (see Fig. 7)

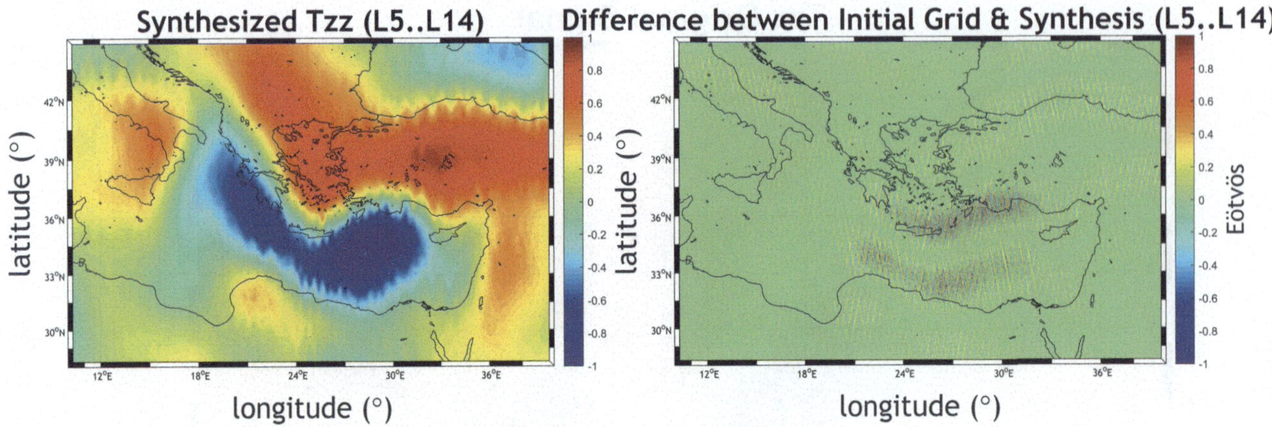

Fig. 4 The L5 . . . L14 reconstruction and its difference with the original signal

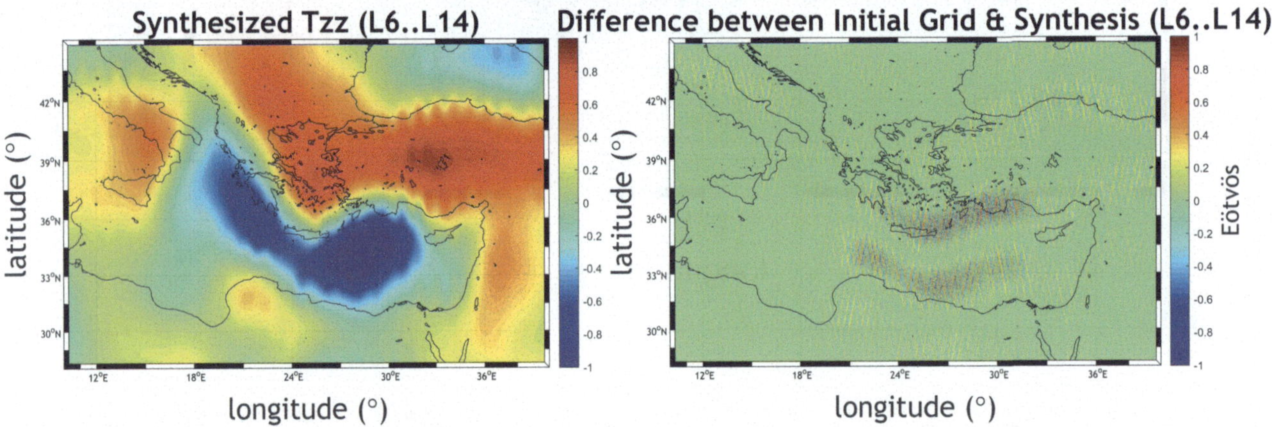

Fig. 5 The L6 . . . L14 reconstruction and its difference with the original signal

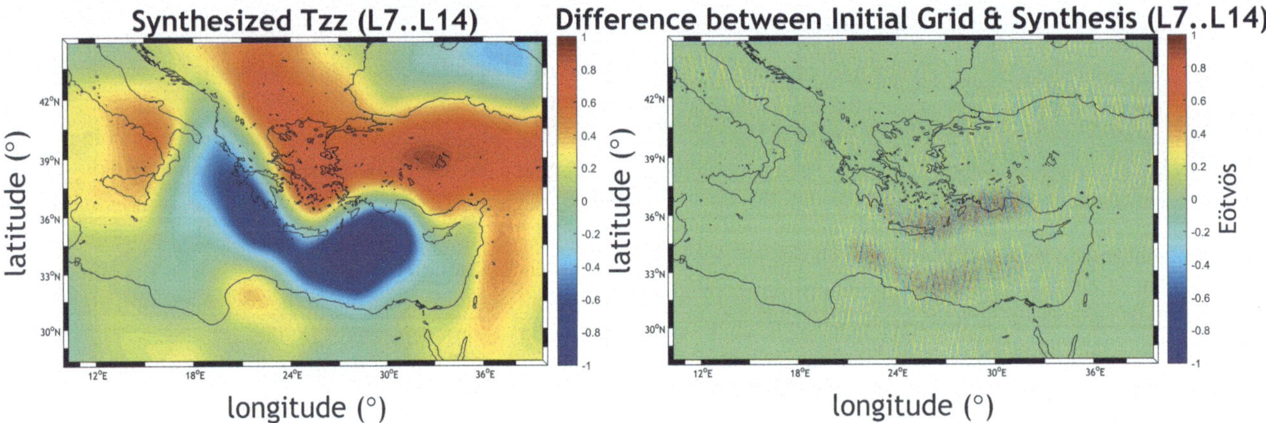

Fig. 6 The L7 . . . L14 reconstruction and its difference with the original signal

reconstruction not only noise but useful gradient signal is actually removed. The removal of this useful signal along with noise in the L8 . . . L14 reconstruction can be seen as yellow areas in Fig. 7, when the L8 . . . L14 reconstruction is compared to the original signal. The corresponding statistics of the aforementioned selective reconstructions are presented in Table 2. Taking all this into account, the selective signal reconstruction using only levels 7–14, approximately corresponding to spatial resolutions from 117 km onwards, was chosen as the ideal one for removing noise while retaining the useful signal in our study. It is worth mentioning that both the T_{zz} and the $-T_{xx} - T_{yy}$ gradients showed a similar behavior in terms of their statistics when WL MRA was applied to them, something encouraging about the general

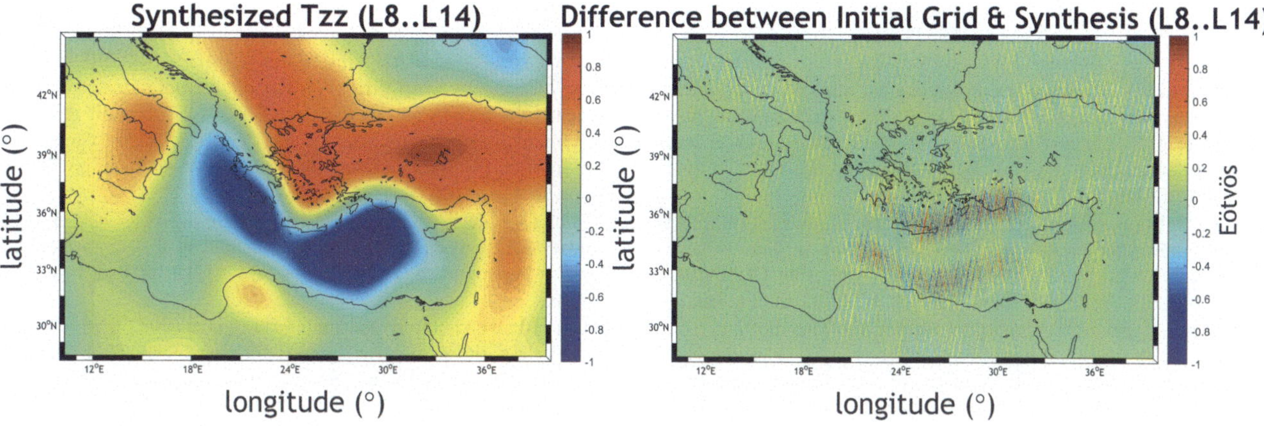

Fig. 7 The L8 ... L14 reconstruction and its difference with the original signal

Table 2 Statistics of selected T_{zz} reconstructions (units: Eötvös)

Reconstructions	Min	Max	Mean	Std	RMS
Synthesis L5 ... L14	−1.246	0.900	0.134	0.384	0.407
Synthesis L6 ... L14	−1.242	0.902	0.134	0.384	0.407
Synthesis L7 ... L14	−1.240	0.894	0.134	0.384	0.407
Synthesis L8 ... L14	−1.232	0.903	0.135	0.382	0.405

Table 3 Statistics of the Laplace equation before and after 2D WL-MRA (units: Eötvös)

	Min	Max	Mean	Std
Before WL-MRA	−0.292	0.249	0.001	0.039
Synthesis L5 ... L14	−0.020	0.018	0.000	0.002
Synthesis L6 ... L14	−0.011	0.007	0.000	0.001
Synthesis L7 ... L14	−0.004	0.003	0.000	0.001
Synthesis L8 ... L14	−0.002	0.001	0.000	0.000

accuracy of our results. To further validate the accuracy of the selective reconstructions, the according Laplace equation was computed before and after the application of WL-MRA. These statistics are presented in detail in Table 3. The greatest impact can be seen in the standard deviation values of the Laplace equation which decreases from 0.039 Eötvös in the original signal to 0.001 Eötvös in the chosen L7 ... L14 reconstruction. Significant improvement can be identified also as regards to the corresponding minimum and maximum values of the equation.

4 Conclusions

The aim of the present study was the processing of the Level 2 GOCE data to compute disturbing gravity gradients in the LNOF at satellite altitude ready to be used in studies with geodetic and geophysical goals in the wider Hellenic area. For this purpose all the methods applied took advantage of the latest processed data of the EGG_NOM_2, SST_PSO_2 and EGG_TRF_2 products for the entire mission. In total the disturbing gravity gradients computed from the original GOCE SGG data, derived from the EGG_NOM_2 and SST_PSO_2 files, showed good statistical accordance, as discussed before, with those computed from the EGG_TRF_2 and GGM ones. Then they were chosen to be projected to the selected MO of 250 km and focused over the wider Hellenic area. Finally they were filtered using 2D Wavelet Multi-Resolution Analysis to remove remaining tracks appearing as noise in the projected data. This resulted to the computation of the final T_{zz} and $-T_{xx} - T_{yy}$ gravity gradient fields of the study. Regarding to the statistics of the Laplace equation before and after the application of WL-MRA, in the later case they showcased a significant improvement considering both the range of their minimum and maximum values and also their standard deviation ones.

References

Bingham RJ, Knudsen P, Andersen O, Pail R (2011) An initial estimate of the North Atlantic steady-state geostrophic circulation from GOCE. Geophys Res Lett 38:L01606. https://doi.org/10.1029/2010GL045633

Bouman J, Fuchs MJ (2012) GOCE gravity gradients versus global gravity field models. Geophys J Int 189:846–850. https://doi.org/10.1111/j.1365-246X.2012.05428.x

Bouman J, Fiorot S, Fuchs M et al (2011) GOCE gravitational gradients along the orbit. J Geod 85:791–805. https://doi.org/10.1007/s00190-011-0464-0

Brockmann JM, Schubert T, Mayer-Gürr T, Schuh W-D (2019) The Earth's gravity field as seen by the GOCE satellite – an improved sixth release derived with the time-wise approach (GO_CONS_GCF_2_TIM_R6). GFZ Data Serv 3:1–9. https://doi.org/10.5880/ICGEM.2019.003

Daubechies I (1992) Ten lectures on wavelets. Soc Ind Appl Math. https://doi.org/10.1137/1.9781611970104

Drinkwater MR, Haagmans R, Muzi D et al (2007) The GOCE gravity mission: ESA'S first core earth explorer. In: Proceedings of the 3rd GOCE user workshop, ESA SP-627, Frascati, Italy

Ebbing J, Bouman J, Fuchs M et al (2013) Advancements in satellite gravity gradient data for crustal studies. Lead Edge 32:900–906. https://doi.org/10.1190/tle32080900.1

ESA (2014) GOCE high level processing facility GOCE level 2 product data handbook (No. GO-MA-HPF-GS-0110, Issue 5.0), EGG-C report

ESA (2023) ESA GOCE online dissemination. https://goce-ds.eo.esa.int/oads/access/collection/GOCE_Level_2/tree. Accessed Feb 2023

Frommknecht B, Lamarre D, Meloni M et al (2011) GOCE level 1b data processing. J Geod 85:759–775. https://doi.org/10.1007/s00190-011-0497-4

Grebenitcharsky R, Moore P (2014) Application of wavelets for along-track multiresolution analysis of GOCE SGG data. In: Gravity, geoid and height systems: proceedings of the IAG symposium GGHS2012, Oct 9–12, 2012, Venice, Italy. Springer International Publishing, Cham, pp 41–50. https://doi.org/10.1007/978-3-319-10837-7_6

Gruber T, Visser PNAM, Ackermann C, Hosse M (2011) Validation of GOCE gravity field models by means of orbit residuals and geoid comparisons. J Geod 85:845–860. https://doi.org/10.1007/s00190-011-0486-7

Gruber T, Gerlach C, Haagmans R (2012) Intercontinental height datum connection with GOCE and GPS-levelling data. J Geod Sci 2(4):270–280. https://doi.org/10.2478/v10156-012-0001-y

Haines K, Johannessen JA, Knudsen P et al (2011) An ocean modelling and assimilation guide to using GOCE geoid products. Ocean Sci 7(1):151–164. https://doi.org/10.5194/os-7-151-2011

Heiskanen WA, Moritz H (1967) Physical geodesy. W.H. Freeman, San Francisco, CA

Holzrichter N, Ebbing J (2016) A regional background model for the Arabian Peninsula from modeling satellite gravity gradients and their invariants. Tectonophysics 692. https://doi.org/10.1016/j.tecto.2016.06.002

Klees R, Wittwer T (2007) Local gravity field modelling with multi-pole wavelets. In: Dynamic planet: monitoring and understanding a dynamic planet with geodetic and oceanographic tools IAG symposium, Cairns, Australia, Aug 22–26, 2005. Springer, Berlin, pp 303–308

Mallat SG (1989) A theory for multiresolution signal decomposition: the wavelet representation. IEEE Trans Pattern Anal Mach Intell. https://doi.org/10.1109/34.192463

Mariani P, Braitenberg C, Ussami N (2013) Explaining the thick crust in Paraná basin, Brazil, with satellite GOCE gravity observations. J South Am Earth Sci 45. https://doi.org/10.1016/j.jsames.2013.03.008

Moritz H (2000) Geodetic reference system 1980. J Geod 74:128–133. https://doi.org/10.1007/s001900050278

Müller J (2003) GOCE gradients in various reference frames and their accuracies. Adv Geosci. https://doi.org/10.5194/adgeo-1-33-2003

Natsiopoulos DA, Mamagiannou EG, Pitenis EA et al (2023) GOCE downward continuation to the Earth's surface and improvements to local geoid modeling by FFT and LSC. Remote Sens 15. https://doi.org/10.3390/rs15040991

Novák P, Sebera J, Val'ko M et al (2013) Towards a better understanding of the Earth's interior and geophysical exploration research. Final report: ESA contract no. 4000103566; ESTEC project 4000103566/11/NL/FvO/ef

Pail R, Bruinsma S, Migliaccio F et al (2011) First GOCE gravity field models derived by three different approaches. J Geod. https://doi.org/10.1007/s00190-011-0467-x

Panet I, Jamet O, Diament M, Chambodut A (2005) Modelling the earth's gravity field using wavelet frames. Int Assoc Geod Symp 129. https://doi.org/10.1007/3-540-26932-0_9

Panet I, Kuroishi Y, Holschneider M (2011) Wavelet modelling of the gravity field by domain decomposition methods: an example over Japan. Geophys J Int 184. https://doi.org/10.1111/j.1365-246X.2010.04840.x

Pavlis NK, Holmes SA, Kenyon SC, Factor JK (2012) The development and evaluation of the Earth Gravitational Model 2008 (EGM2008). J Geophys Res Solid Earth. https://doi.org/10.1029/2011JB008916

Peidou AC, Vergos GS (2016) Wavelet multi-resolution analysis of recent GOCE/GRACE GGMs. In: IGFS 2014: proceedings of the 3rd International Gravity Field Service (IGFS), Shanghai, China, Jun 30–Jul 6, 2014. Springer International Publishing, Cham, pp 53–61. https://doi.org/10.1007/1345_2015_44

Piretzidis D, Sideris MG (2017) Adaptive filtering of GOCE-derived gravity gradients of the disturbing potential in the context of the space-wise approach. J Geod. https://doi.org/10.1007/s00190-017-1010-5

Pitenis E, Mamagiannou E, Natsiopoulos DA et al (2022) FIR, IIR and wavelet algorithms for the rigorous filtering of GOCE SGG data to the GOCE MBW. Remote Sens 14:3024. https://doi.org/10.3390/rs14133024

Polgár Z, Sujbert L, Földváry L et al (2013) Filter design for GOCE gravity gradients. Geocarto Int 28. https://doi.org/10.1080/10106049.2012.687401

Reguzzoni M, Sampietro D (2015) GEMMA: an Earth crustal model based on GOCE satellite data. Int J Appl Earth Obs Geoinf 35:31–43. https://doi.org/10.1016/j.jag.2014.04.002

Roland M (2005) Untersuchungen zur Kombination terrestrischer Schweredaten und aktueller globaler Schwerefeldmodelle. Doctoral dissertation, Fachrichtung Vermessungswesen der Universität Hannover

Sebera J, Pitoňák M, Hamáčková E, Novák P (2015) Comparative study of the spherical downward continuation. Surv Geophys 36:253–267. https://doi.org/10.1007/s10712-014-9312-0

Siemes C, Rexer M, Schlicht A et al (2018) On the reprocessing of GOCE gravity gradients. In: EGU General Assembly 2018, Apr 4–13, Vienna, Austria

Siemes C, Rexer M, Schlicht A, Haagmans R (2019) GOCE gradiometer data calibration. J Geod 93. https://doi.org/10.1007/s00190-019-01271-9

Šprlák M, Sebera J, Val'ko M, Novák P (2014) Spherical integral formulas for upward/downward continuation of gravitational gradients onto gravitational gradients. J Geod 88:179–197. https://doi.org/10.1007/s00190-013-0676-6

Tóth G, Földváry L (2005) Effect of geopotential model errors on the projection of GOCE gradiometer observables. Int Assoc Geod Symp 129. https://doi.org/10.1007/3-540-26932-0_13

Tóth G, Földváry L, Tziavos IN, Ádám J (2006) Upward/downward continuation of gravity gradients for precise geoid determination. Acta Geod Geophys Hungarica 41:21–30. https://doi.org/10.1556/AGeod.41.2006.1.3

Tsoulis D, Moukoulis C (2019) Processing aspects of level 2 GOCE gradiometer data for regional applications. Geophys J Int. https://doi.org/10.1093/gji/ggy485

Vergos GS, Erol B, Natsiopoulos DA et al (2018) Preliminary results of GOCE-based height system unification between Greece and Turkey over marine and land areas. Acta Geod Geophys 53. https://doi.org/10.1007/s40328-017-0204-x

Wan XY, Yu JH, Zeng YY (2012) Frequency analysis and filtering processing of gravity gradient data from GOCE. Chin J Geophys 55:530–538

Part IV

Methods

Modelling Gravity and Geoid by Least Squares Collocation with Planar Covariance Models

R. Barzaghi, D. Carrion, and Ö. Koç

Abstract

Least Squares Collocation is a method frequently used in modelling the Earth gravity field and the geoid. Either global or local approaches of this method have been studied and applied. The collocation estimator for gravity and the geoid depends on selecting appropriate auto- and cross-covariance functions for both the data and the functional of the anomalous potential to be estimated. Usually, these models are given in terms of series expansions depending on the Legendre Polynomials. In this paper, a different class of auto- and cross-covariance functions are studied which can be applied in planar approximation. These functions have been used in estimating the geoid from gravity anomalies using the data of the Colorado test. Results proved that this approach could give estimates that are equivalent to those obtained using the well-known covariance models by Tscherning and Rapp.

Keywords

Colorado test · Hankel transform · Least squares collocation · Planar approximation

1 Introduction

Least Squares Collocation (LSC) is a well-known method applied in geodesy in estimating any functional of the anomalous potential $T(P)$ given a combination of other functionals of $T(P)$ as input data (Moritz 1980). As an example, one can use gravity anomaly data and compute the geoid undulation as done by using the Stokes's integral approach (Heiskanen and Moritz 1967). Also, contrary to the Stokes's integral approach where only gravity data is used as input, LSC allows the combination of all the heterogeneous data available within a given computation area. Furthermore, LSC allows the use of input data at their original measurement locations, eliminating the need for gridding. The critical point of this method is that it depends on the inversion of a covariance matrix having a dimension equal to the amount of data. This limits seriously the LSC application when a large dataset is available. The covariance matrix of the data and the cross-covariances are typically computed based on the covariance functional models that have been proposed by Tschering and Rapp (1974). The auto- and cross-covariance models of any functional of the anomalous potential $T(P)$ are derived, through linear functional covariance propagation, from the auto-covariance of $T(P)$ (Moritz 1980). They were originally defined for global LSC applications and then modified for local or regional computations by Knudsen (1987). These models are fitted to the empirical covariance function of the data. This allows defining the parameters that enter in the covariance models that are then used in the LSC estimation formulas. In local applications of LSC, such models are fitted to empirical covariances of data reduced by the Global Geopotential Models (GGMs) and the Residual Terrain Component (RTC) effect. Since the recent GGMs are complete to high degree and order (e.g. to d/o 2,190) and the terrain effect can be computed based on high-resolution digital terrain model (DTM) (like SRTM, Farr et al. 2007), the reduced data

R. Barzaghi (✉) · D. Carrion · Ö. Koç
DICA-Politecnico di Milano, Milano, Italy
e-mail: riccardo.barzaghi@polimi.it

J. T. Freymueller, L. Sànchez (eds.), *International Symposium on Gravity, Geoid and Height Systems 2024 (GGHS2024)*, International Association of Geodesy Symposia 158, https://doi.org/10.1007/1345_2025_302

display very short correlation length (the distance at which the correlation function is half its value at the origin). The standard models mentioned above can be used with these reduced data, but it is sometimes difficult to find a proper fit with the empirical covariance values. In this paper, a planar geometry approach to gravity field approximation based on LSC is proposed. In this context, a new covariance model of the anomalous potential is defined and studied based on the zero order Bessel function of first kind (Watson 1948). Since reduced data have, as mentioned, very short correlation lengths, the planar approximation should not be so coarse, and reasonable results can be expected in a windowed LSC approach (Grigoriadis et al. 2021). In this approach, only data up to the distance of the order of the correlation length from the computation point will be considered. The computation test performed using the Colorado area data (Wang et al. 2021) proved that, at least for this case, the results obtained using the planar LSC approach are comparable with those obtained with the standard LSC.

2 The Collocation Approach in Planar Approximation

The planar geometry that we adopted is represented in Fig. 1.

If we assume to remove the topography down to the plane (X, Y), the anomalous potential $T(P)$ will be harmonic in the semi-space $Z > 0$. By solving the Dirichlet problem for this geometry, we obtain the formula expressing the anomalous potential $T(P)$ in its harmonicity domain given $T(Q_0)$ on the plane $S_0 = (X, Y)$

$$T(P) = \frac{1}{2\pi} \int_{S_0} T(Q_0) \frac{z}{\left(z^2 + |P_0 - Q_0|^2\right)^{3/2}} dS_0 \qquad (1)$$

Assuming that $T(Q_0)$ is a weakly stationary process, ergodic in the mean and the covariance (Moritz 1980), by the covariance propagation law, we get that the covariance $C_{TT}(\cdot)$ as a function of the covariance $C^0_{TT}(\cdot)$ of $T(Q_0)$ is given by (see Fig. 1 for the definition of the point positions)

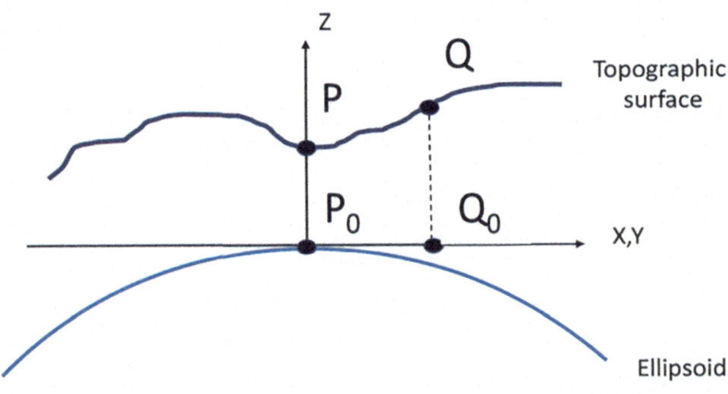

Fig. 1 The geometry of the planar approximation

$$C_{TT}(P_1, P_2) = \frac{1}{4\pi^2} \int_{S_0} dS_{01} \int_{S_0} dS_{02}\, C^0_{TT}(Q_{01}, Q_{02}) \cdot \frac{z_1}{\left(z_1^2+|P_{01}-Q_{01}|^2\right)^{3/2}} \cdot \frac{z_2}{\left(z_2^2+|P_{02}-Q_{02}|^2\right)^{3/2}} \qquad (2)$$

Equation (2) can be then rewritten in a more convenient form as it follows. We set

$$\underline{X}_1 = P_{01} \quad \underline{X}_2 = P_{02} \quad \underline{Y}_1 = Q_{01} \quad \underline{Y}_2 = Q_{02} \qquad (3)$$

and we assume that

$$C^0_{TT}(Q_{01}, Q_{02}) = C^0_{TT}(|\underline{Y}_1 - \underline{Y}_2|) \qquad (4)$$

Equation (2) can be formally written as a convolution

$$C_{TT}(P_1, P_2) = g(z_2, |\underline{X}_2 - \underline{Y}_2|) * C^0_{TT}(|\underline{Y}_1 - \underline{Y}_2|) * g(z_1, |\underline{X}_1 - \underline{Y}_1|) \qquad (5)$$

with

$$g(z, |\underline{X} - \underline{Y}|) = \frac{z}{\left(z^2 + |\underline{X} - \underline{Y}|^2\right)^{3/2}} \qquad (6)$$

By Fourier Transform it can be proved that (see also Forsberg 1987)

$$C_{TT}(|\underline{X}_1 - \underline{X}_2|, z_1, z_2) = \int_0^{+\infty} k\, e^{-k(z_1+z_2)}\, S^0_{TT}(k)\, J_0(k|\underline{X}_2 - \underline{X}_1|)\, dk \qquad (7)$$

where $S^0_{TT}(k)$ is the Hankel spectrum of $C^0_{TT}(|\underline{X}_1 - \underline{X}_2|)$

$$C^0_{TT}(|\underline{X}_1 - \underline{X}_2|) = \int_0^{+\infty} k\, S^0_{TT}(k)\, J_0(k|\underline{X}_2 - \underline{X}_1|)\, dk \qquad (8)$$

and $J_0(\cdot)$ is the zero order Bessel function of the first kind (Bracewell 2000).

Then, selecting the Hankel spectrum of $C^0_{TT}(\cdot)$ as

$$S^0_{TT}(k) = \frac{A}{\alpha^2} e^{-\frac{k\lambda}{\alpha}d}\, \delta\left(\frac{K}{\alpha} - 1\right) \quad \forall \lambda, \alpha, A > 0 \qquad (9)$$

where $d = |\underline{X}_2 - \underline{X}_1|$ and $\delta(\cdot)$ is the impulse symbol (Bracewell 2000), we can express the covariances (7) and (8) in closed forms as

$$\begin{aligned} C^0_{TT}(|\underline{X}_1 - \underline{X}_2|) &= \\ &= \int_0^{+\infty} k\, S^0_{TT}(k)\, J_0(k|\underline{X}_1 - \underline{X}_2|)\, dk = \\ &= A e^{-\lambda|\underline{X}_1 - \underline{X}_2|}\, J_0(\alpha|\underline{X}_1 - \underline{X}_2|) \end{aligned} \qquad (10)$$

and

$$\begin{aligned} &C_{TT}\left(\left|\underline{X}_1-\underline{X}_2\right|,z_1,z_2\right)= \\ &=\int_0^{+\infty} k\, e^{-k(z_1+z_2)}\, S_{TT}^0(k)\, J_0\left(k\left|\underline{X}_2-\underline{X}_1\right|\right) dk \\ &=Ae^{-\lambda\left|\underline{X}_1-\underline{X}_2\right|}\, e^{-\alpha(z_1+z_2)}\, J_0\left(\alpha\left|\underline{X}_1-\underline{X}_2\right|\right)= \\ &=C_{TT}^0\left(\left|\underline{X}_1-\underline{X}_2\right|\right) e^{-\alpha(z_1+z_2)} \end{aligned} \tag{11}$$

In these covariance models, the parameter A is the value of $C_{TT}^0\left(\left|\underline{X}_1-\underline{X}_2\right|\right)$ in the origin which is, by definition, the variance of the signal $T(Q_0)$. The λ parameter controls the way covariance is tending to zero while $\left|\underline{X}_1-\underline{X}_2\right|$ is increasing. This is an essential parameter for conveniently decreasing the correlation at large distances, an expected feature from the statistical and physical point of view. The α parameter allows tuning another important feature of the covariance function, i.e. the correlation length of $C_{TT}^0\left(\left|\underline{X}_1-\underline{X}_2\right|\right)$, the length at which the covariance is half its value in the origin. Also, by (9) α can be interpreted as the spatial frequency that we assume to have in $T(Q_0)$ (Flury 2006). Furthermore, this parameter is related to the standard upward continuation operator in the semi-space $Z > 0$ (Forsberg 1987).

Then, by applying covariance propagation to gravity anomaly that, in planar approximation, can be written as

$$\Delta g \cong -\frac{\partial T}{\partial z} \tag{12}$$

we can get explicitly the auto-coraviance of Δg and the cross-covariance between $\Delta g(P)$ and $t(P)$

$$\begin{aligned} &C_{\Delta g\Delta g}\left(P_1,P_2\right)=-\tfrac{\partial}{\partial z_2}\left[-\tfrac{\partial}{\partial z_1}C_{TT}\left(P_1,P_2\right)\right]= \\ &=C_{TT}^0\left(\left|\left|\underline{X}_1-\underline{X}_2\right|\right|\right)\left\{-\tfrac{\partial}{\partial z_2}\left[-\tfrac{\partial}{\partial z_1}e^{-\alpha(z_1+z_2)}\right]\right\}= \\ &=C_{TT}^0\left(\left|\underline{X}_1-\underline{X}_2\right|\right)\alpha^2 e^{-\alpha(z_1+z_2)}= \\ &=A\alpha^2 e^{-\lambda\left|\underline{X}_1-\underline{X}_2\right|}\, J_0\left(\alpha\left|\underline{X}_1-\underline{X}_2\right|\right) e^{-\alpha(z_1+z_2)} \end{aligned} \tag{13}$$

$$\begin{aligned} &C_{T\Delta g}\left(P_1,P_2\right)=-\tfrac{\partial}{\partial z}\left[C_{TT}\left(P_1,P_2\right)\right]= \\ &=C_{TT}^0\left(\left|\underline{X}_1-\underline{X}_2\right|\right)\alpha e^{-\alpha(z_1+z_2)}= \\ &=A\alpha e^{-\lambda\left|\underline{X}_1-\underline{X}_2\right|}\, J_0\left(\alpha\left|\underline{X}_1-\underline{X}_2\right|\right) e^{-\alpha(z_1+z_2)} \end{aligned} \tag{14}$$

As currently done in LSC, by properly tuning the parameters contained in (13) and (14), these models can suitably represent the empirical covariance values of residual gravity data. In this way, we will be able to compute the LSC estimation of T having Δg as observation. In the next section, these formulas will be applied in the LSC computation of the height anomaly from gravity using the data of the Colorado test.

3 An Application to the Colorado Test Area

In the frame of an international test aiming at comparing different methods in regional geoid computation, terrestrial and airborne gravity datasets and GPS/leveling data were made available in an area of Colorado, USA. Particularly, a highly precise GPS/levelling line (Geoid Slope Validation Survey 2017-GSVS17, van Westrum et al. 2021) was assumed as the benchmark for comparing all the applied geoid computation methods (Wang et al. 2021). In Fig. 2 the available data are displayed.

In the computations of this paper, only terrestrial gravity data have been considered. The processing for the computation of the residual gravity values has been performed as described in Grigoriadis et al. (2021). Gravity anomaly data have been reduced for the long wavelength components using the XGM2016 GGM to d/o 719 (Pail et al. 2018).

The Residual Terrain Component effect has been computed by applying the spectral method (Rexer et al. 2018) based on the CGIAR-CSI SRTM digital elevation model (Farr et al. 2007).

The obtained residual free air gravity anomalies

$$\Delta g_r = \Delta g - \Delta g_{GGM} - \Delta g_{RTC} \tag{15}$$

are shown in Fig. 3.

The empirical covariance values of Δg_r and the best-fit spherical model by Tscherning (1976, Model 1) and Knudsen (1987) with degree variances

$$\sigma_n^2 = \frac{A}{(n-2)} \tag{16}$$

are plotted in Fig. 4 (COVFIT of the GRAVSOFT package by Forsberg and Tscherning (2008) has been used in the fitting procedure).

The same empirical covariance values and the best fit covariance function (13) are presented in Fig. 5. The fitting parameters, estimated by Least Squares fitting, considering $z_1 = z_2 = 2000$ m which is the mean altitude of the area containing the data, are:

$$A = 3.25 \times 10^7\ \frac{\text{cm}^4}{\text{s}^4},\quad \lambda = 0.02\ \text{km}^{-1},\quad \alpha = 0.126\ \text{km}^{-1}$$

The two model functions are very close to the empirical values till its minimum, while a better agreement with the empirical values is shown by the $J_0(\cdot)$ based model at larger distances. Based on these covariances, two LSC estimates have been computed. The GEOCOL program of the GRAV-

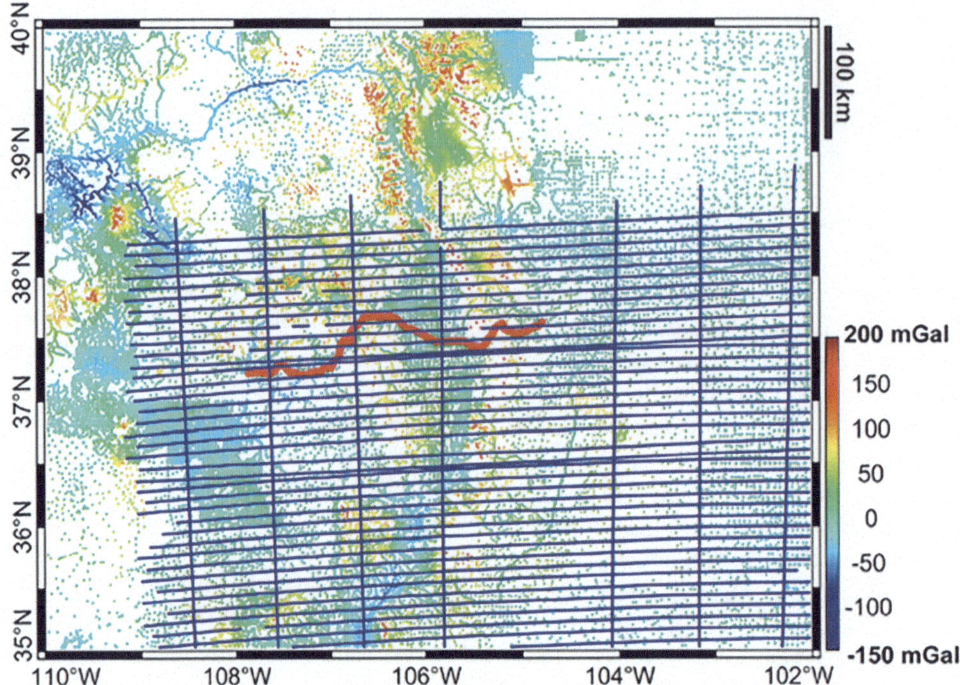

Fig. 2 The gravity data of the Colorado test: in red the GSVS17 line; blue lines are airborne gravity data

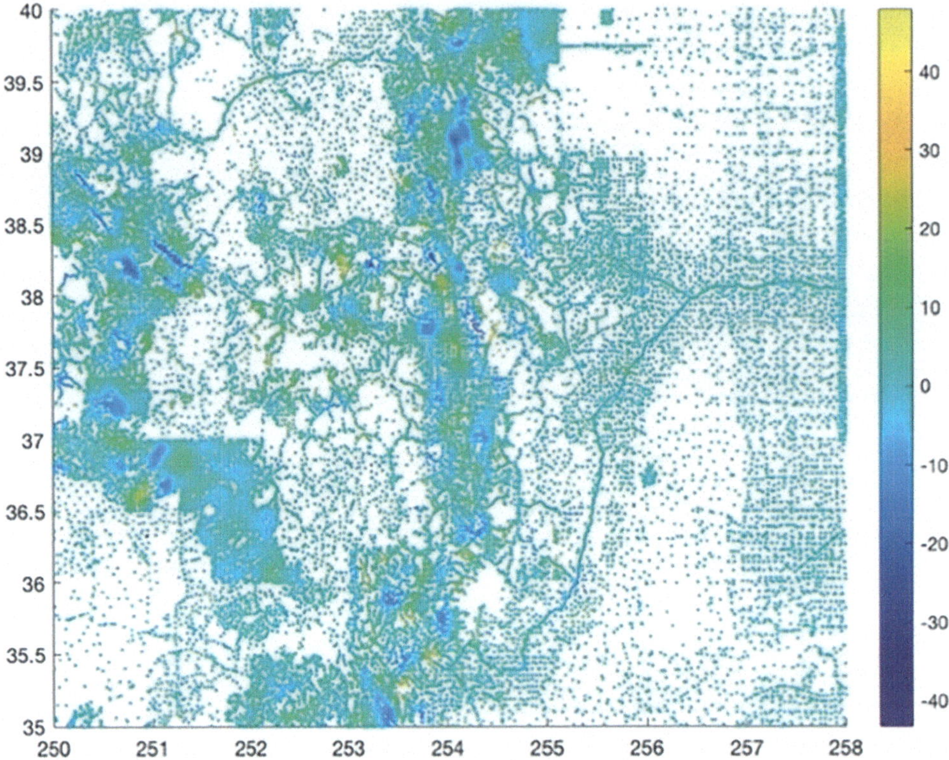

Fig. 3 Terrestrial residual free-air gravity data (mGal) in the Colorado area

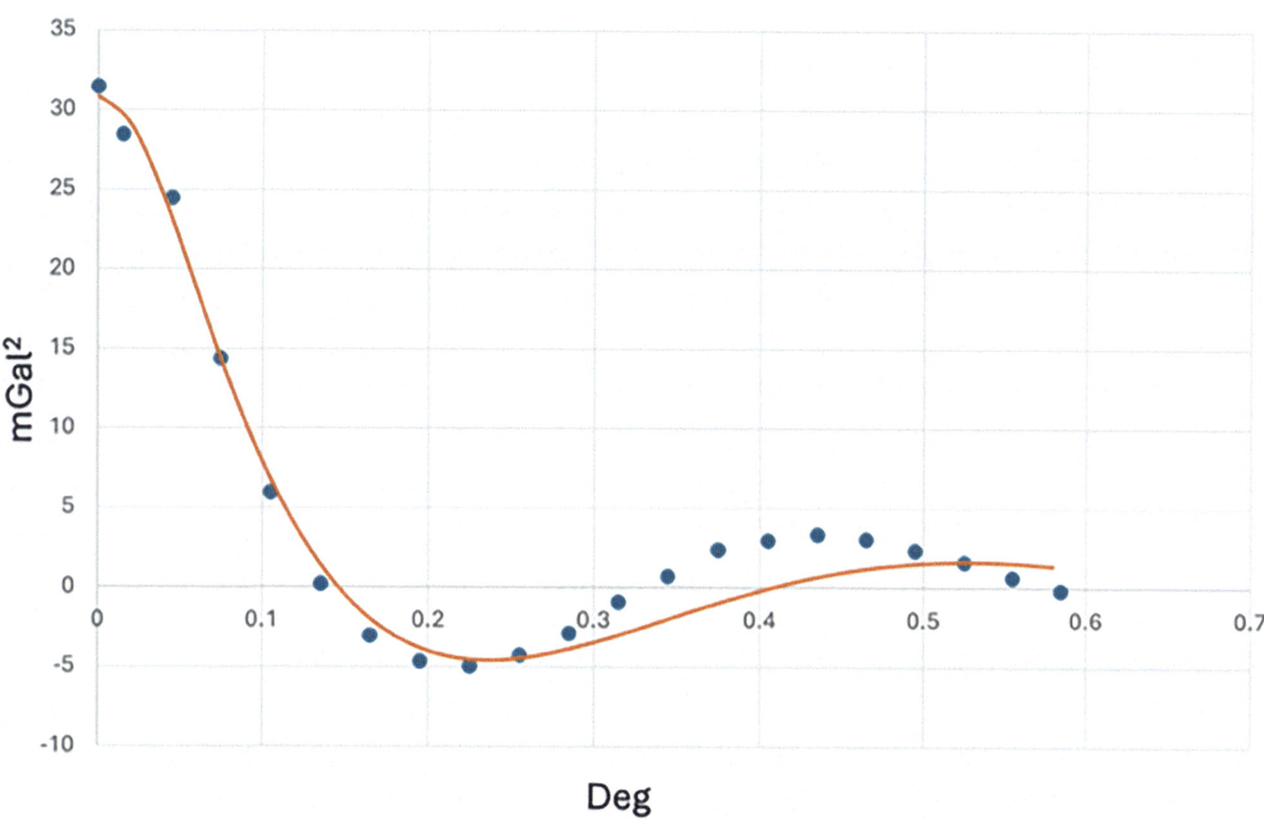

Fig. 4 The empirical covariance values (blue dots) and the best fitting covariance model according to Knudsen (1987) (solid orange line)

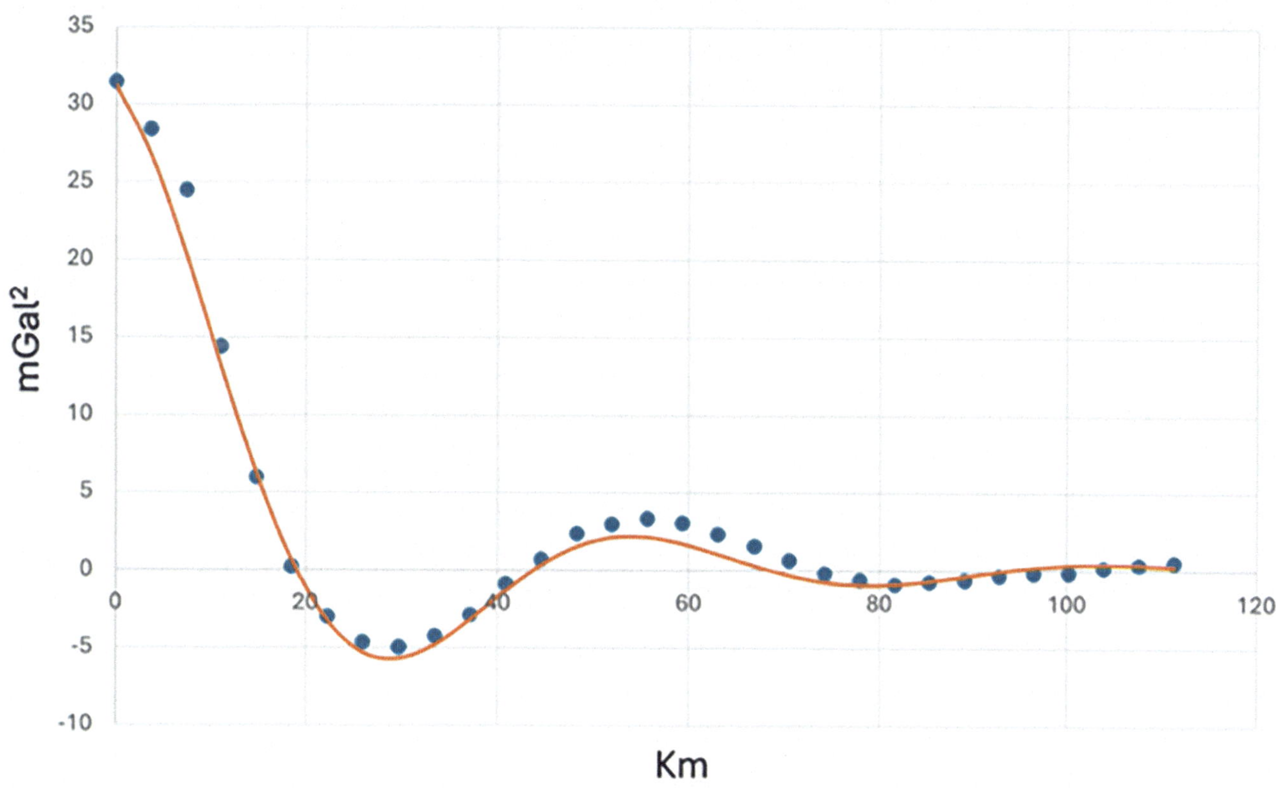

Fig. 5 The empirical covariance values (blue dots) and the best fitting covariance model of formula (13) (solid orange line)

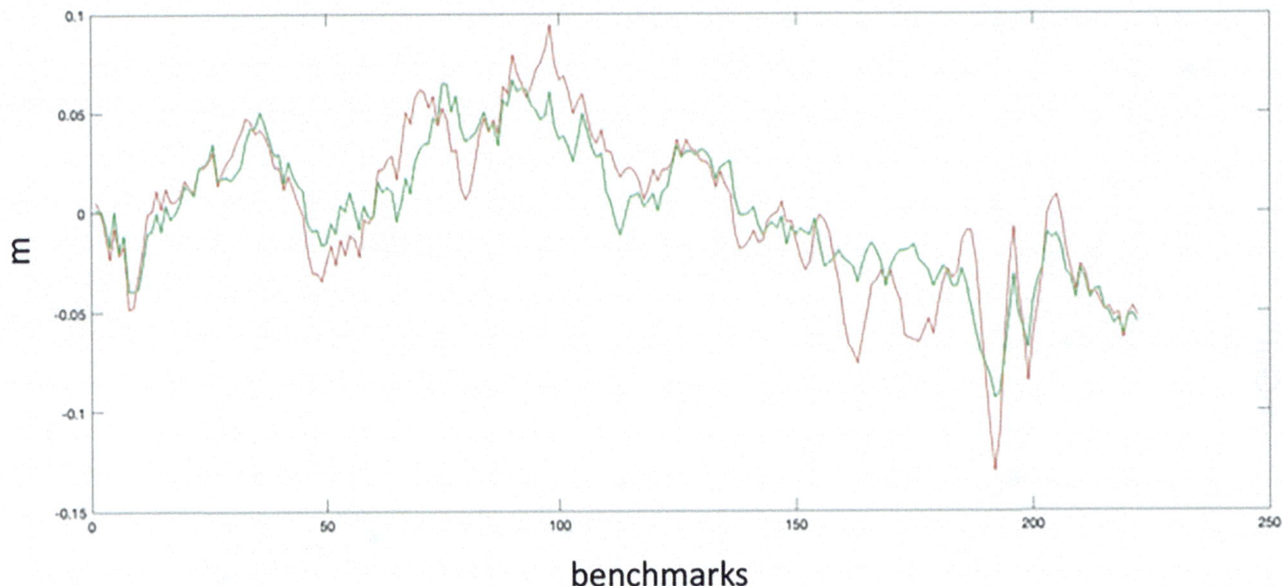

Fig. 6 The height anomaly difference between spherical GRAVSOFT (GEOCOL) estimates and GSVS17 values (green line), and between planar approach estimates and GSVS17 values (red line)

SOFT package (Forsberg and Tscherning 2008) has been used to get residual height anomalies ζ_r in the points of the GSVS17 line. The same has been done using the planar approximation approach devised in this paper. Both solutions were based on windowed data to 10 km (~0.09°) from each computation point[1].

Finally, the GGM and the RTC height anomaly components have been added to the two estimated residual values to get the total height anomaly to be compared with the GSVS17 values. In Fig. 6, the differences between the estimated height anomalies and the corresponding GSVS17 values are shown.

Statistics of this comparison are listed in Table 1.

As one can see from Fig. 6 and the values in Table 1, the two LSC estimates are practically equivalent. The slightly better agreement between empirical and model covariance in Fig. 5 has no significant impact since only windowed data up to 10 km have been used in both computations. However, what is relevant in our opinion is that the planar approach detailed in this paper proved to be feasible giving results that are in line with the standard LSC approach used in the GRAVSOFT package.

[1]The CPU time for inverting a $n \times n$ matrix is $O(n^3)$ (Teukolsky et al. 2007). As in the Colorado database the mean density of gravity data is around 2′, each height anomaly estimated point by windowed collocation is bases on around 40 gravity points. This implies a very remarkable reduction of the CPU time as compared to the solution based on the whole gavity dataset (containing 58,121 gravity values).

Table 1 The statistics of the difference between the LSC estimated height anomalies and the GSVS17 line values

	$\hat{\zeta}_{GRAVSOFT} - \zeta_{GSVS17}$	$\hat{\zeta}_{PLANAR} - \zeta_{GSVS17}$
E [m]	1.077	1.075
σ [m]	0.034	0.038
Min [m]	0.983	0.962
Max [m]	1.143	1.156

4 Conclusions

In this paper, LSC as applied in planar approximation is presented. A new class of covariance models has been derived that are based on the $J_0(\cdot)$ Bessel function of the first kind. A first result obtained in the context of the Colorado test, proved that LSC in planar approach can be applied. As a matter of fact, the LSC estimate computed via the planar approximation is statistically equivalent to the one based on the GEOCOL program of the GRAVSOFT package. However, further investigations are needed to prove the full applicability of this method, particularly for defining the maximum spatial extent to which this approach can be properly used.

References

Bracewell RN (2000) The Fourier transform and its applications. McGraw-Hill Book Company

Farr TG et al (2007) The shuttle radar topography mission. Rev Geophys 45:RG2004. https://doi.org/10.1029/2005RG000183

Flury J (2006) Short-wavelength spectral properties of the gravity field from a range of regional data sets. J Geod 79:624–640

Forsberg R (1987) A new covariance model for inertial gravimetry and gradiometry. J Geophys Res 92(B2):1305–1310

Forsberg R, Tscherning CC (2008) An overview manual for the GRAVSOFT geodetic gravity field modelling programs, 2nd edn. Department of Geodetic Science and Surveying, No. 341, The Ohio State University, Columbus

Grigoriadis VN, Vergos GS, Barzaghi R, Carrion D, Koç Ö (2021) Collocation and FFT-based geoid estimation within the Colorado 1 cm geoid experiment. J Geod 95:52. https://doi.org/10.1007/s00190-021-01507-7

Heiskanen W, Moritz H (1967) Physical geodesy. Freeman and Company, San Francisco

Knudsen P (1987) Estimation and modelling of the local empirical covariance function using gravity and satellite altimeter data. Bull Geod 61:145–160

Moritz H (1980) Advanced physical geodesy. Wichmann, Karlsruhe

Pail R, Fecher T, Barnes D et al (2018) Short note: the experimental geopotential model XGM2016. J Geod 92:443–451. https://doi.org/10.1007/s00190-017-1070-6

Rexer M, Hirt C, Blazej B, Holmes S (2018) Solution to the spectral filter problem of residual terrain modelling (RTM). J Geod 92:675–690

Teukolsky SA, Vetterling WT, Flannery BP (2007) Numerical recipes. The art of scientific computing, 3rd edn. Cambridge University Press. ISBN: 0-521-88068-8

Tschering CC, Rapp RH (1974) Closed covariance expressions for gravity anomalies, geoid undulations, and deflections of the vertical implied by anomaly degree variance models. Department of Geodetic Science and Surveying, No. 208, The Ohio State University, Columbus

Tscherning CC (1976) Covariance expressions for second and lower order derivatives of the anomalous potential. OSU Report No. 225

van Westrum D, Ahlgren K, Hirt C, Guillaume S (2021) A Geoid Slope Validation Survey (2017) in the rugged terrain of Colorado, USA. J Geod 95:9. https://doi.org/10.1007/s00190-020-01463-8

Wang YM et al (2021) Colorado geoid computation experiment: overview and summary. J Geod 95:127. https://doi.org/10.1007/s00190-021-01567-9

Watson GN (1948) Theory of Bessel functions, 2nd edn. Cambridge University Press

Orthogonality Properties of Spherical and Spheroidal Harmonic Functions: A Short Overview

Michal Šprlák

Abstract

Boundary-value problems (BVPs) play a crucial role in determining the external gravitational field. The Laplace differential equation, together with scalar, vector, and tensor boundary conditions, has been used to formulate and solve numerous spherical BVPs. The close proximity of planetary bodies to the rotational ellipsoid has led theoreticians to deal with more complex spheroidal BVPs.

Systematic treatment of BVPs is allowed by the orthogonality of respective basis functions. This property is taken for granted for the spherical case, as it exactly holds for the scalar spherical harmonics and can be extended for the vector and higher-order tensor analogues. Equivalent paradigms could theoretically be introduced for the spheroidal geometry. In the first attempt, however, we end up with the lack of this peculiar property for the spheroidal basis functions. Thus, the spheroidal BVPs may seem extremely demanding or are impossible to be solved analytically. Fortunately, the orthogonality can be restored in a more general weighted sense that has been examined for the scalar and vector spheroidal harmonics.

In this contribution, we summarize orthogonality properties of spherical and spheroidal harmonics. For the spherical case, orthogonality of two scalar basis functions is the simplest one and can naturally be extended for the vector, second-order, and third-order tensor basis functions. For the spheroidal case, the concept of orthogonality has been generalized in a weighted sense for the scalar, vector, and even (though incorrectly) second-order tensor basis functions. The most recent attempts of extending the weighted orthogonality for the second- and third-order tensor basis functions have failed.

Keywords

Basis functions · Boundary-value problem · Gravitational tensor · Sphere · Spheroid

1 Introduction

Generally, we describe gravitational field by different quantities (e.g., Novák et al. 2017). The simplest is the potential V (zeroth-order gravitational tensor) being a single number in every point. The gravitational gradient (first-order gravitational tensor) is a vector defined by the three components (V_x, V_y, V_z) at a given place. The second-order gravitational tensor is more complex and specified by the six components $(V_{xx}, V_{xy}, V_{xz}, V_{yy}, V_{yz}, V_{zz})$ at one location. The most demanding is the third-order gravitational tensor represented by the ten components $(V_{xxx}, V_{xxy}, V_{xxz}, V_{xyy}, V_{xyz}, V_{xzz}, V_{yyy}, V_{yyz}, V_{yzz}, V_{zzz})$ at a position. The components of each quantity refer to the local north-oriented reference frame and can systematically be grouped by the number of vertical and horizontal indices.

M. Šprlák (✉)
NTIS—New Technologies for the Information Society, Faculty of Applied Sciences, University of West Bohemia in Pilsen, Plzeň, Czech Republic
e-mail: michal.sprlak@gmail.com

J. T. Freymueller, L. Sànchez (eds.), *International Symposium on Gravity, Geoid and Height Systems 2024 (GGHS2024)*, International Association of Geodesy Symposia 158, https://doi.org/10.1007/1345_2025_285

Several mathematical parametrizations of the gravitational field and its quantities exist, but harmonic expansions are perhaps the most popular (e.g., Moritz 2010). This apparatus allows efficient manipulation with various quantities, for example, when deriving spectral characteristics or correlation functions, solving boundary-value problems, or designing filters (e.g., Bölling and Grafarend 2005; Jekeli 2013; Yang et al. 2024). In these nontrivial tasks, we heavily rely on one peculiar property of harmonic basis functions called orthogonality (e.g., Jackson 2004).

In this contribution, we provide a brief pragmatic summary of orthogonality properties in the context of different gravitational field quantities for two distinct geometries:

1. For the sphere, by examining orthogonality of the scalar, first-order horizontal, and second-order horizontal tensor basis functions. A transparent extension of orthogonality for the third-order horizontal tensor basis functions is possible by using polar coordinates on the sphere.
2. For the oblate spheroid, by introducing weighted orthogonality for the scalar and first-order horizontal basis functions. We also discuss a recent orthogonalization effort for the second- and third-order spheroidal tensor basis functions.

2 Orthogonality on the Sphere

We firstly examine orthogonality on the sphere. For this geometry, see Fig. 1, we naturally introduce the geocentric spherical coordinates (r, φ, λ), where r is the geocentric radius, φ is the spherical latitude, and λ is the spherical longitude, and use them as variables of a position. The components of the gravitational field quantities, i.e., those of the gravitational vector, second-order gravitational tensor, and third-order gravitational tensor, refer to the spherical north-oriented reference frame. Vectors $(\boldsymbol{e}_x, \boldsymbol{e}_y, \boldsymbol{e}_z)$ define the frame's basis point to the north, west, and radially outward, respectively; see Fig. 1. Integrations will be performed over the unit sphere and abbreviated as $\int_S (\cdot)\, \mathrm{d}S = \int_{\lambda=0}^{2\pi} \int_{\varphi=-\frac{\pi}{2}}^{\frac{\pi}{2}} (\cdot) \cos\varphi \, \mathrm{d}\varphi \, \mathrm{d}\lambda$.

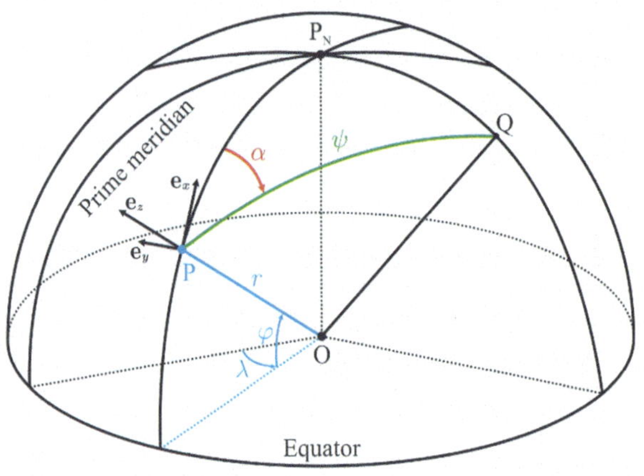

Fig. 1 Reference frames on the sphere. P_N is the north pole, and O is the center of the sphere

2.1 Gravitational Potential

The gravitational potential can be expressed by the external spherical harmonic series, which converges above the minimum Brillouin sphere (e.g., Heiskanen and Moritz 1967):

$$V(r, \varphi, \lambda) = \frac{GM}{R} \sum_{n=0}^{\infty} \sum_{m=-n}^{n} \left(\frac{R}{r}\right)^{n+1} \overline{C}_{n,m}\, \overline{Y}_{n,m}(\varphi, \lambda). \tag{1}$$

The symbol GM is the geocentric gravitational constant, R is a metric scale factor (equal to the radius of the geocentric sphere, where the orthogonality of the basis functions is being investigated), $\overline{C}_{n,m}$ is the spherical harmonic coefficient of degree n and order m, and $\overline{Y}_{n,m}$ is the spherical harmonic of degree n and order m. Both spherical harmonic coefficients and spherical harmonics are normalized for numerical reasons and indicated by an overbar.

Importantly, the functions $\overline{Y}_{n,m}$ fulfil the orthogonality relationship (e.g., Heiskanen and Moritz 1967):

$$\frac{1}{4\pi} \int_S \overline{Y}_{n_1,m_1}(\varphi, \lambda)\, \overline{Y}_{n_2,m_2}(\varphi, \lambda)\, \mathrm{d}S = \delta_{n_1,n_2}\, \delta_{m_1,m_2}. \tag{2}$$

Namely, the integral over the unit sphere S (divided by 4π representing the surface area of the unit sphere) from the product of the two spherical harmonics equals one, when degrees and orders of those two functions are identical. Otherwise, the integral vanishes. This is indicated by the multiplications of the two Kronecker deltas $\delta_{n,m} = \begin{cases} 1, \text{if } n = m \\ 0, \text{if } n \neq m \end{cases}$ on the right-hand side of Eq. (2).

The related mathematical framework is known as scalar spherical harmonics (e.g., Freeden and Schreiner 2009).

2.2 Horizontal Components of the Gravitational Vector

We continue with the horizontal components V_x and V_y of the gravitational vector grad V. The external spherical harmonic series result from the application of the differential operators $D^x = \frac{1}{r}\frac{\partial}{\partial\varphi}$ and $D^y = -\frac{1}{r\cos\varphi}\frac{\partial}{\partial\lambda}$ to Eq. (1). After this operation, the functions $\overline{Y}_{n,m}$ are differentiated by φ or λ and multiplied by the inverse of $\cos\varphi$:

$$V_x(r,\varphi,\lambda) = \frac{GM}{R^2}\sum_{n=0}^{\infty}\sum_{m=-n}^{n}\left(\frac{R}{r}\right)^{n+2} \times \overline{C}_{n,m}\frac{\partial}{\partial\varphi}\overline{Y}_{n,m}(\varphi,\lambda), \tag{3}$$

$$V_y(r,\varphi,\lambda) = -\frac{GM}{R^2}\sum_{n=0}^{\infty}\sum_{m=-n}^{n}\left(\frac{R}{r}\right)^{n+2} \times \overline{C}_{n,m}\frac{1}{\cos\varphi}\frac{\partial}{\partial\lambda}\overline{Y}_{n,m}(\varphi,\lambda). \tag{4}$$

These horizontal derivatives destroy orthogonality. In other words, the integral over the unit sphere from the product of two spherical harmonics both differentiated by φ is not zero for all different degrees and orders. The same is true for the differentiation by λ factorized by one over $\cos^2\varphi$, i.e.:

$$\frac{1}{4\pi}\int_S \frac{\partial}{\partial\varphi}\overline{Y}_{n_1,m_1}(\varphi,\lambda)\frac{\partial}{\partial\varphi}\overline{Y}_{n_2,m_2}(\varphi,\lambda)\,dS \neq \delta_{n_1,n_2}\delta_{m_1,m_2}, \tag{5}$$

$$\frac{1}{4\pi}\int_S \frac{1}{\cos^2\varphi}\frac{\partial}{\partial\lambda}\overline{Y}_{n_1,m_1}(\varphi,\lambda)\frac{\partial}{\partial\lambda}\overline{Y}_{n_2,m_2}(\varphi,\lambda)\,\mathrm{d}S \neq \delta_{n_1,n_2}\delta_{m_1,m_2}. \tag{6}$$

Nevertheless, the orthogonality holds, but for the combination of the two horizontal differentiations as follows (e.g., Meissl 1971):

$$\frac{1}{4\pi}\int_S \left[\frac{\partial}{\partial\varphi}\overline{Y}_{n_1,m_1}(\varphi,\lambda)\frac{\partial}{\partial\varphi}\overline{Y}_{n_2,m_2}(\varphi,\lambda) + \frac{1}{\cos^2\varphi}\frac{\partial}{\partial\lambda}\overline{Y}_{n_1,m_1}(\varphi,\lambda)\frac{\partial}{\partial\lambda}\overline{Y}_{n_2,m_2}(\varphi,\lambda)\right]\mathrm{d}S = n_1(n_1+1)\,\delta_{n_1,n_2}\delta_{m_1,m_2}. \tag{7}$$

This expression is an increment of a general framework called vector spherical harmonics (e.g., Freeden and Schreiner 2009).

2.3 Horizontal Components of the Second-Order Gravitational Tensor

We proceed to the horizontal components of the second-order gravitational tensor grad $\otimes$ grad V. V_{xx}, V_{xy}, and V_{yy} are defined by the action of the three independent differential operators $D^{xx} = \frac{1}{r}\left(\frac{\partial}{\partial r} + \frac{1}{r}\frac{\partial^2}{\partial\varphi^2}\right)$, $D^{xy} = -\frac{1}{r^2\cos\varphi}\left(\tan\varphi\frac{\partial}{\partial\lambda} + \frac{\partial^2}{\partial\varphi\partial\lambda}\right)$, and $D^{yy} = \frac{1}{r}\left(\frac{\partial}{\partial r} - \frac{\tan\varphi}{r}\frac{\partial}{\partial\varphi} + \frac{1}{r\cos^2\varphi}\frac{\partial^2}{\partial\lambda^2}\right)$. This operation leads to the external spherical harmonic series:

$$V_{xx}(r,\varphi,\lambda) = -\frac{GM}{R^3}\sum_{n=0}^{\infty}\sum_{m=-n}^{n}\left(\frac{R}{r}\right)^{n+3}\overline{C}_{n,m} \times \left[(n+1) - \frac{\partial^2}{\partial\varphi^2}\right]\overline{Y}_{n,m}(\varphi,\lambda), \tag{8}$$

$$V_{xy}(r,\varphi,\lambda) = -\frac{GM}{R^3}\sum_{n=0}^{\infty}\sum_{m=-n}^{n}\left(\frac{R}{r}\right)^{n+3}\overline{C}_{n,m} \times \frac{1}{\cos\varphi}\left[\tan\varphi\frac{\partial}{\partial\lambda} + \frac{\partial^2}{\partial\varphi\partial\lambda}\right]\overline{Y}_{n,m}(\varphi,\lambda), \tag{9}$$

$$V_{yy}(r,\varphi,\lambda) = -\frac{GM}{R^3}\sum_{n=0}^{\infty}\sum_{m=-n}^{n}\left(\frac{R}{r}\right)^{n+3}\overline{C}_{n,m} \times \left[(n+1) + \tan\varphi\frac{\partial}{\partial\varphi} - \frac{1}{\cos^2\varphi}\frac{\partial^2}{\partial\lambda^2}\right] \times \overline{Y}_{n,m}(\varphi,\lambda). \tag{10}$$

Similar to V_x and V_y, these second-order differentiations are non-orthogonal when considered individually. We can again restore this property, if we combine spherical harmonics of $(V_{xx} - V_{yy})$ and $2V_{xy}$ (e.g., Rummel 1997):

$$\frac{1}{4\pi}\int_S \left\{\left[n_1(n_1+1) + 2\frac{\partial^2}{\partial\varphi^2}\right]\overline{Y}_{n_1,m_1}(\varphi,\lambda) \times \left[n_2(n_2+1) + 2\frac{\partial^2}{\partial\varphi^2}\right]\overline{Y}_{n_2,m_2}(\varphi,\lambda) + \frac{4}{\cos^2\varphi}\left[\tan\varphi + \frac{\partial^2}{\partial\varphi\partial\lambda}\right]\overline{Y}_{n_1,m_1}(\varphi,\lambda) \times \left[\tan\varphi\frac{\partial}{\partial\lambda} + \frac{\partial^2}{\partial\varphi\partial\lambda}\right]\overline{Y}_{n_2,m_2}(\varphi,\lambda)\right\}dS = (n_1-1)\,n_1(n_1+1)(n_1+2)\,\delta_{n_1,n_2}\delta_{m_1,m_2}. \tag{11}$$

Other details can be found in the theory of tensor spherical harmonics (e.g., Freeden and Schreiner 2009).

Note that vertical derivatives of the quantities examined up to now follow the rules of their original equivalents and are not presented separately.

2.4 Orthogonal Combinations in Polar Coordinates

The orthogonal horizontal combinations in Eqs. (7) and (11) are simplified when expressed in the polar coordinates α (being the direct azimuth) and ψ (being the spherical distance); see Fig. 1.

The first- and second-order horizontal combinations are simply decomposed into the azimuthal part (depending on α) and the isotropic part (depending on ψ):

$$V_x(r,\psi,\alpha) = \cos\alpha\mathcal{D}_1\, V(r,\psi,\alpha)\,, \tag{12}$$

$$V_y(r,\psi,\alpha) = -\sin\alpha\mathcal{D}_1\, V(r,\psi,\alpha)\,, \tag{13}$$

$$V_{xx}(r,\psi,\alpha) - V_{yy}(r,\psi,\alpha) = 2\,\cos 2\alpha\mathcal{D}_2\, V(r,\psi,\alpha)\,, \tag{14}$$

$$2\,V_{xy}(r,\psi,\alpha) = -2\,\sin 2\alpha\mathcal{D}_2\, V(r,\psi,\alpha)\,, \tag{15}$$

where $\mathcal{D}_1 = \frac{1}{r}\frac{\partial}{\partial\cos\psi}$ and $\mathcal{D}_2 = \frac{\left(1-\cos^2\psi\right)}{2r^2}\frac{\partial^2}{\partial\cos^2\psi}$. The multiples of α and the order of the differentiation by $\cos\psi$ are identical with the order of the components themselves.

This remarkable symmetry was also employed to find an orthogonal combination among the horizontal components of the third-order gravitational tensor grad $\otimes$ grad $\otimes$ grad V, namely:

$$V_{xxx}(r,\psi,\alpha) - 3\,V_{xyy}(r,\psi,\alpha) = 4\cos 3\alpha\mathcal{D}_3\, V(r,\psi,\alpha)\,, \tag{16}$$

$$V_{yyy}(r,\psi,\alpha) - 3\,V_{xxy}(r,\psi,\alpha) = 4\sin 3\alpha\mathcal{D}_3\, V(r,\psi,\alpha)\,, \tag{17}$$

where $\mathcal{D}_3 = \frac{\left(1-\cos^2\psi\right)^{3/2}}{4r^3}\frac{\partial^3}{\partial\cos^3\psi}$.

The last two expressions allowed solution of the related external gravitational curvature boundary-value problem with purely horizontal components (e.g., Šprlák and Novák 2016; Šprlák 2022).

3 Orthogonality on the Spheroid

We next consider orthogonality on the reference spheroid defined by major and minor semiaxes a and b; see Fig. 2. Besides other options, we employ the Jacobi spheroidal coordinates (u,β,λ) as position variables, where u is the minor semiaxis of the spheroid, β is the reduced latitude, and λ is the spheroidal longitude. Use of these curvilinear coordinates is essential in conjunction with harmonic series, as it allows solution of the Laplace equation by separation of variables.

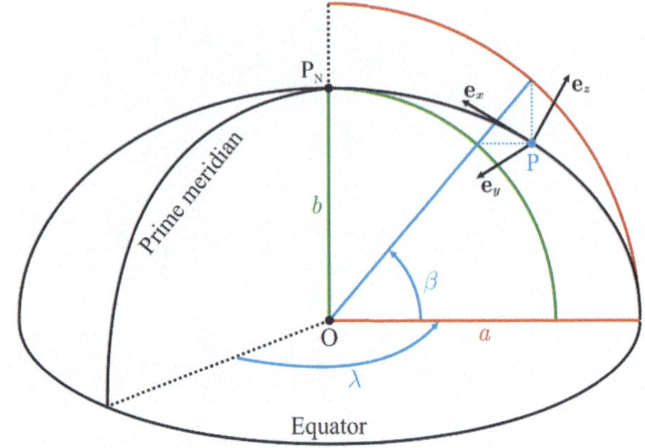

Fig. 2 Reference frames on the spheroid. P_N is the north pole, and O is the center of the spheroid

Components of the gravitational vector, second-order gravitational tensor, and third-order gravitational tensor are related to the spheroidal north-oriented reference frame. This frame is given by the basis $(\boldsymbol{e}_x,\boldsymbol{e}_y,\boldsymbol{e}_z)$, vectors of which are directed to the north, west, and normally outward, respectively; see Fig. 2. Integrals over the reference spheroid will be designated by the short-hand notation $\int_S(\cdot)\,\mathrm{d}S = \int_{\lambda=0}^{2\pi}\int_{\beta=-\frac{\pi}{2}}^{\frac{\pi}{2}}(\cdot)\,a\,\sqrt{b^2+\varepsilon^2\sin^2\beta}\cos\beta\,\mathrm{d}\beta\,\mathrm{d}\lambda$, where $\varepsilon = \sqrt{a^2-b^2}$ is the linear eccentricity.

3.1 Gravitational Potential

The gravitational potential can alternatively be expanded by the external spheroidal harmonic series with the convergence domain above the minimum Brillouin spheroid (e.g., Heiskanen and Moritz 1967):

$$V(u,\beta,\lambda) = \frac{GM}{a}\sum_{n=0}^{\infty}\sum_{m=-n}^{n}\frac{Q_{n,|m|}\left(i\frac{u}{\varepsilon}\right)}{Q_{n,|m|}\left(i\frac{b}{\varepsilon}\right)}\,\overline{C}_{n,m}\,\overline{Y}_{n,m}(\beta,\lambda)\,. \tag{18}$$

$Q_{n,m}$ is the associated Legendre function of the second kind of degree n and order m, $\overline{C}_{n,m}$ is the spheroidal harmonic coefficient of degree n and order m, and $\overline{Y}_{n,m}$ represents the spheroidal harmonic of degree n and order m.

The functions $\overline{Y}_{n,m}$ are not orthogonal. In other words, the integral over the surface of the reference spheroid S (divided by the corresponding surface area, which can be calculated for given major and minor semiaxes) from the product of two spheroidal harmonics does not vanish for all different degrees and orders:

$$\frac{1}{\int_S \mathrm{d}S} \int_S \overline{Y}_{n_1,m_1}(\beta,\lambda)\, \overline{Y}_{n_2,m_2}(\beta,\lambda)\, \mathrm{d}S \neq \delta_{n_1,n_2}\, \delta_{m_1,m_2}. \tag{19}$$

This is caused by the square-root term of the surface element $\mathrm{d}S = a\sqrt{b^2+\varepsilon^2\sin^2\beta}\cos\beta\, \mathrm{d}\beta\, \mathrm{d}\lambda$.

To restore orthogonality, we introduce the weighting factor $w = \frac{a}{\sqrt{b^2+\varepsilon^2\sin^2\beta}}$, which cancels out the troublesome square root of dS (e.g., Ardalan 1999):

$$\frac{1}{\int_S w(a,b,\beta)\, \mathrm{d}S} \int_S \overline{Y}_{n_1,m_1}(\beta,\lambda)\, \overline{Y}_{n_2,m_2}(\beta,\lambda) \times w(a,b,\beta)\, \mathrm{d}S = \delta_{n_1,n_2}\, \delta_{m_1,m_2}. \tag{20}$$

The concept of weighted orthogonality may appear artificial, but it is well established in the theory of orthogonal series (e.g., Jackson 2004).

3.2 Horizontal Components of the Gravitational Vector

We now continue with orthogonality principles for the horizontal components V_x and V_y of the gravitational vector. The respective external spheroidal harmonic series originate from the application of the differential operators $D^x = \frac{1}{\sqrt{u^2+\varepsilon^2\sin^2\beta}}\frac{\partial}{\partial\beta}$ and $D^y = \frac{1}{\sqrt{u^2+\varepsilon^2}\cos\beta}\frac{\partial}{\partial\lambda}$ to Eq. (18). This operation leads to the spheroidal harmonic expansions:

$$V_x(u,\beta,\lambda) = \frac{GM}{a}\sum_{n=0}^{\infty}\sum_{m=-n}^{n}\frac{Q_{n,|m|}\left(i\frac{u}{\varepsilon}\right)}{Q_{n,|m|}\left(i\frac{b}{\varepsilon}\right)}\overline{C}_{n,m} \times \frac{1}{\sqrt{u^2+\varepsilon^2\sin^2\beta}}\frac{\partial}{\partial\beta}\overline{Y}_{n,m}(\beta,\lambda), \tag{21}$$

$$V_y(u,\beta,\lambda) = -\frac{GM}{a}\sum_{n=0}^{\infty}\sum_{m=-n}^{n}\frac{Q_{n,|m|}\left(i\frac{u}{\varepsilon}\right)}{Q_{n,|m|}\left(i\frac{b}{\varepsilon}\right)}\overline{C}_{n,m} \times \frac{1}{\sqrt{u^2+\varepsilon^2}\cos\beta}\frac{\partial}{\partial\lambda}\overline{Y}_{n,m}(\beta,\lambda). \tag{22}$$

Identical to the spherical case, the horizontal differentiations destroy orthogonality. To reach the desired orthogonality, the horizontal differentiations are combined and the weighting factor has to be introduced (e.g., Lowes and Winch 2012; Šprlák and Tangdamrongsub 2022):

$$\frac{1}{\int_S w(a,b,\beta)\, \mathrm{d}S} \times \int_S \left[\frac{1}{b^2+\varepsilon^2\sin^2\beta}\frac{\partial}{\partial\beta}\overline{Y}_{n_1,m_1}(\beta,\lambda)\frac{\partial}{\partial\beta}\overline{Y}_{n_2,m_2}(\beta,\lambda) + \frac{1}{(b^2+\varepsilon^2)\cos^2\beta}\frac{\partial}{\partial\lambda}\overline{Y}_{n_1,m_1}(\beta,\lambda)\frac{\partial}{\partial\lambda}\overline{Y}_{n_2,m_2}(\beta,\lambda)\right] \times w(a,b,\beta)\, \mathrm{d}S = \frac{3}{2b^2+a^2}\left[n_1(n_1+1) - \frac{\varepsilon^2}{b^2+\varepsilon^2}m_1^2\right] \times \delta_{n_1,n_2}\,\delta_{m_1,m_2}. \tag{23}$$

The weighting factor $w = \frac{\sqrt{b^2+\varepsilon^2\sin^2\beta}}{a}$ differs from the one for the potential in Eq. (20).

Grafarend et al. (2006) suggested another orthogonalization, namely:

$$\frac{1}{\int_S \mathrm{d}S}\int_S \left[\frac{w_x(a,b,\beta)}{b^2+\varepsilon^2\sin^2\beta}\frac{\partial}{\partial\beta}\overline{Y}_{n_1,m_1}(\beta,\lambda)\frac{\partial}{\partial\beta}\overline{Y}_{n_2,m_2}(\beta,\lambda) + \frac{w_y(a,b,\beta)}{(b^2+\varepsilon^2)\cos^2\beta}\frac{\partial}{\partial\lambda}\overline{Y}_{n_1,m_1}(\beta,\lambda)\frac{\partial}{\partial\lambda}\overline{Y}_{n_2,m_2}(\beta,\lambda)\right]\mathrm{d}S = \delta_{n_1,n_2}\,\delta_{m_1,m_2}. \tag{24}$$

The two distinct weighting factors for each horizontal component are $w_x = \sqrt{b^2+\varepsilon^2\sin^2\beta}\left(\frac{b^2}{4\varepsilon}\ln\frac{a+\varepsilon}{a-\varepsilon} + \frac{a}{2}\right)$ and $w_y = \frac{a^2}{\sqrt{b^2+\varepsilon^2\sin^2\beta}}\left(\frac{b^2}{4\varepsilon}\ln\frac{a+\varepsilon}{a-\varepsilon} + \frac{a}{2}\right)$.

3.3 Components of Higher-Order Gravitational Tensors

We now consider components of higher-order gravitational tensors. For brevity, we exemplify only the second-order vertical component. Its differential operator is of the form $D^{zz} = -\frac{u\varepsilon^2\cos^2\beta}{\left(u^2+\varepsilon^2\sin^2\beta\right)^2}\frac{\partial}{\partial u} + \frac{\varepsilon^2\sin\beta\cos\beta}{\left(u^2+\varepsilon^2\sin^2\beta\right)^2}\frac{\partial}{\partial\beta} + \frac{u^2+\varepsilon^2}{u^2+\varepsilon^2\sin^2\beta}\frac{\partial^2}{\partial u^2}$. This operator includes the first and the second derivatives by u, and the differentiation by β, and is thus more complex than its spherical counterpart. Its application to Eq. (18) leads to the spheroidal harmonic series:

$$
\begin{aligned}
&V_{zz}(u,\beta,\lambda)\\
&=\frac{GM}{a}\sum_{n=0}^{\infty}\sum_{m=-n}^{n}\left[-\frac{u\varepsilon^2\cos^2\beta}{\left(u^2+\varepsilon^2\sin^2\beta\right)^2}\frac{\frac{\partial}{\partial u}Q_{n,|m|}\left(i\frac{u}{\varepsilon}\right)}{Q_{n,|m|}\left(i\frac{b}{\varepsilon}\right)}\right.\\
&+\frac{\varepsilon^2\sin\beta\cos\beta}{\left(u^2+\varepsilon^2\sin^2\beta\right)^2}\frac{Q_{n,|m|}\left(i\frac{u}{\varepsilon}\right)}{Q_{n,|m|}\left(i\frac{b}{\varepsilon}\right)}\frac{\partial}{\partial\beta}\\
&\left.+\frac{u^2+\varepsilon^2}{u^2+\varepsilon^2\sin^2\beta}\frac{\frac{\partial^2}{\partial u^2}Q_{n,|m|}\left(i\frac{u}{\varepsilon}\right)}{Q_{n,|m|}\left(i\frac{b}{\varepsilon}\right)}\right]\overline{C}_{n,m}\,\overline{Y}_{n,m}(\beta,\lambda)\,. \qquad (25)
\end{aligned}
$$

To our best knowledge, there are only two investigations on orthogonality properties of the corresponding spheroidal harmonic basis in Eq. (25):

1. The comprehensive work by Bölling and Grafarend (2005) asserted orthogonality in the weighted sense, but missed any proof of this property.
2. Šprlák (2023) identified fundamental errors in the study by Bölling and Grafarend (2005). In particular, it was found that the integral over the reference spheroid from the product of the two weighted second-order vertical spheroidal harmonics was not zero for all different degrees and orders.

A significant effort has recently been made by the author of this study to restore orthogonality of the spheroidal harmonic basis in Eq. (25). The following manipulations, of either analytical or numerical nature, were performed:

- Expressing the differential operator D^{zz} in different spheroidal coordinate systems
- Expressing the differential operator D^{zz} by equivalent decompositions
- Combining the differential operators of various components of the second-order gravitational tensor
- Introducing distinct weighting factors

None of these operations led to orthogonality.

Equivalent tasks were carried out for the other components of the second- and third-order gravitational tensors, but with identical conclusions.

4 Conclusions

In this contribution, we summarized orthogonality properties of the spherical and oblate spheroidal harmonic basis functions. The concept of orthogonality is well understood and routinely used on the sphere even for higher-order gravitational tensors. The notion of weighted orthogonality has been introduced on the spheroid and worked out for the gravitational potential and gravitational gradient. The latest results, however, show that weighted orthogonality fails for higher-order gravitational tensors on the spheroid. More insight into this issue may be revealed by a rigorous investigation following the Sturm-Liouville theory (e.g., Arfken and Weber 2005, Chap. 8). Alternatively, other orthogonalization procedures will have to be employed to routinely manipulate with these quantities, for example, when expressing their spectral characteristics or correlation functions, solving respective boundary-value problems, or designing filters.

Acknowledgements This research was supported financially by the Czech Science Foundation through the Standard Project no. 23-07031S. Thoughtful and constructive comments of two anonymous reviewers are gratefully acknowledged. Thanks are also extended to the editors for handling our manuscript.

References

Ardalan AA (1999) High resolution regional geoid computation in the World Geodetic Datum 2000 based upon collocation of linearized observational functionals of the type GPS, gravity potential and gravity intensity. PhD. Thesis, Department of Geodesy and Geoinformatics, Stuttgart University, Stuttgart, Germany

Arfken GB, Weber HJ (2005) Mathematical methods for physicists, 6th edn. Elsevier Academic Press, New York

Bölling C, Grafarend EW (2005) Ellipsoidal spectral properties of the earth's gravitational potential and its first and second derivatives. J Geod 79:300–330

Freeden W, Schreiner M (2009) Spherical functions of mathematical geosciences. A scalar, vectorial, and tensorial setup. In: Advances in geophysical and environmental mechanics and mathematics. Springer, Berlin Heidelberg, Germany

Grafarend EW, Finn G, Ardalan AA (2006) Ellipsoidal vertical deflections and ellipsoidal gravity disturbance: case studies. Stud Geophys Geod 50:1–57

Heiskanen WA, Moritz H (1967) Physical geodesy. Freeman and Co., San Francisco

Jackson D (2004) Fourier series and orthogonal polynomials. Dover Publications, Inc., New York

Jekeli C (2013) Correlation modeling of the gravity field in classical geodesy. In: Freeden W, Nashed ZM, Sonar T (eds) Handbook of geomathematics. Springer, Berlin, Germany, pp 1–34

Lowes FJ, Winch DE (2012) Orthogonality of harmonic potentials and fields in spheroidal and ellipsoidal coordinates: application to geomagnetism and geodesy. Geophys J Int 191:491–507

Meissl P (1971) A study of covariance functions related to the Earth's disturbing potential. Report no. 151, Department of Geodetic Science, The Ohio State University, Columbus

Moritz H (2010) Classical physical geodesy. In: Freeden W, Nashed ZM, Sonar T (eds) Handbook of geomathematics. Springer, Berlin, Germany, pp 127–158

Novák P, Šprlák M, Tenzer R, Pitoňák M (2017) Integral formulas for transformation of potential field parameters in geosciences. Earth-Sci Rev 164:208–231

Rummel R (1997) Spherical spectral properties of the earth's gravitational potential and its first and second derivatives. In: Sansò F, Rummel R (eds) Geodetic boundary value problems in view of the one centimeter geoid, Lecture notes in earth sciences, vol 65. Springer, Berlin, Germany, pp 359–404

Šprlák M (2022) Gravitational curvature boundary value problem. In: Sideris MG (ed) Encyclopedia of geodesy, encyclopedia of earth sciences series. Springer, Cham, Switzerland

Šprlák M (2023) Comments and corrections to: "Ellipsoidal spectral properties of the earth's gravitational potential and its first and second derivatives" by Bölling and Grafarend (2005) in J.Geod. 79(6–7):300–330. J Geod 97:101

Šprlák M, Novák P (2016) Spherical gravitational curvature boundary-value problem. J Geod 90:727–737

Šprlák M, Tangdamrongsub N (2022) Vertical and horizontal boundary-value problems on a spheroidal boundary. In: Sideris MG (ed) Encyclopedia of geodesy, encyclopedia of earth sciences series. Springer, Cham, Switzerland

Yang F, Liu S, Forootan E (2024) A spatial-varying non-isotropic Gaussian-based convolution filter for smoothing GRACE-like temporal gravity fields. J Geod 98:66

Spectral Combination of Vertical and Horizontal Spheroidal Boundary-Value Problems: A Theoretical Study

Martin Pitoňák, Jiří Belinger, Pavel Novák, and Michal Šprlák

Abstract

The spectral combination method or technique encompasses all procedures to combine heterogeneous datasets by spectral weights, which depend on spherical harmonic degree n. It was initially developed to combine terrestrial gravity data and a global geopotential model optimally to calculate the geoid or quasigeoid. Later on, this technique was extended to combine solutions of spherical geodetic boundary-value problems. It is well-known that the Earth is considerably flattened at the poles, and its shape is closer to a rotational ellipsoid rather than a sphere. Spheroidal formulation provides higher accuracy despite being more complex. This contribution applies the spectral combination method to solutions of vertical and horizontal spheroidal boundary-value problems. For this purpose, we derive the corresponding spectral weights for each solution of vertical and horizontal spheroidal boundary-value problems, as well as for their combination.

Keywords

Gravity field modelling · Spectral combination · Spectral weights · Spheroidal boundary-value problem

1 Introduction

From space, the Earth appears to be a perfect sphere. However, in the seventeenth century, it became evident that the Earth was not a perfect sphere, as was noted in Newton's Principia, published in 1687. Since 1957, the launch of artificial satellites has provided a powerful new tool for studying the Earth's shape. By analyzing satellite orbits, scientists have calculated the Earth's flattening with far greater accuracy. Additionally, satellites have revealed a wealth of data on subtle irregularities in the Earth's shape, which are much less noticeable than the equatorial bulge. Artificial satellites helped to determine the shapes of other telluric planets and asteroids as well. The shape of Earth and telluric planets is closer to the rotational ellipsoid than to the sphere.

The spectral combination technique includes all methods for merging heterogeneous gravity datasets using spectral weights that vary based on the spherical harmonic degree n. Initially designed for the optimal integration of terrestrial gravity data with a global geopotential model (GGM) to determine the geoid or quasi-geoid (e.g., Sjoberg 1980; Wenzel 1982). This approach was expanded to combine solutions to spherical geodetic boundary-value problems (e.g., Eshagh 2011, 2012; Pitoňák et al. 2018, 2023a,b).

So far, this method has been applied only for spherical approximation. In this contribution, we go further, motivated by the high-accuracy requirements for gravitational field modelling. We decided to apply the spectral combination technique to solutions of vertical and horizontal spheroidal boundary-value problems (Šprlák and Tangdamrongsub 2018). To do so, we formulate and derive the corresponding spectral weights. These spectral weights modify

M. Pitoňák (✉) · J. Belinger · P. Novák · M. Šprlák
NTIS—New Technologies for the Information Society, Faculty of Applied Sciences, University of West Bohemia, Plzeň, Czech Republic
e-mail: pitonakm@ntis.zcu.cz; belinger@ntis.zcu.cz; panovak@ntis.zcu.cz; michal.sprlak@gmail.com

J. T. Freymueller, L. Sànchez (eds.), *International Symposium on Gravity, Geoid and Height Systems 2024 (GGHS2024)*,
International Association of Geodesy Symposia 158, https://doi.org/10.1007/1345_2025_304

the corresponding integral kernels. The spheroidal integral transformations, modified in this way, are then suitable for the continuation and transformation of the gravitational vector components, as the spectral weights stabilise the downward continuation process.

The paper is theoretical and has the following structure. The solution to the vertical and horizontal spheroidal boundary-value problems in the spatial domain is summarized in Sect. 2, followed by deriving the spectral weights for these solutions in Sect. 3. The importance of the spectral weights is proved in Sect. 4. The study is concluded in Sect. 5.

2 Solutions of the Vertical and Horizontal Spheroidal Boundary-Value Problems in Spatial Domain

The gravitational vector **V** is composed of three components: one vertical V^u and two horizontal V^β and V^λ. Integral transform of the gravitational vector component V^u onto the gravitational potential V is defined as (see, Šprlák and Tangdamrongsub 2018):

$$V(u,\Omega) = \frac{1}{4\pi}\int_{\Omega'} V^u(b_0,\Omega')\mathcal{K}^u\left(u,\Omega,b_0,\Omega'\right)\mathrm{d}\Omega', \quad (1)$$

where the corresponding integral kernel is:

$$\mathcal{K}^u\left(u,\Omega,b_0,\Omega'\right) = \frac{L(b_0,\beta')}{a_0} \times \sum_{n=0}^{\infty}\sum_{m=-n}^{n} \frac{Q_{n,|m|}\left(i\frac{u}{\varepsilon}\right)}{\frac{\partial}{\partial u}Q_{n,|m|}\left(i\frac{u}{\varepsilon}\right)|_{b_0}} \overline{Y}_{n,m}\left(\Omega'\right)\overline{Y}_{n,m}\left(\Omega\right). \quad (2)$$

where $L(u,\beta) = \sqrt{u^2+\varepsilon^2\sin^2\beta}$. The infinitesimal area is defined as $\mathrm{d}\Omega' = \cos\beta'\mathrm{d}\beta'\mathrm{d}\lambda'$. The one-parametric spheroidal coordinates (Hofmann-Wellenhof and Moritz 2006, Sect. 1.15) are defined by the semi-minor axis u of the biaxial ellipsoid passing through a point with the linear eccentricity ε equal to that of the reference spheroid, the reduced latitude β and the spheroidal longitude λ. The linear eccentricity $\varepsilon = \sqrt{a_0^2-b_0^2}$ is determined by the semi-major and semi-minor axes of the reference spheroid a_0 and b_0. For any point on the reference spheroid, the semi-minor axis $u = b_0$. $Q_{n,|m|}$ are complex functions with $i = \sqrt{-1}$ which are known as the associated Legendre functions of the second kind. The symbol $\overline{Y}_{n,m}$ stands for the fully normalized spheroidal harmonics and are defined as follows:

$$\overline{Y}_{n,m}\left(\Omega\right) = \overline{P}_{n,|m|}\left(\sin\beta\right)\begin{cases}\cos m\lambda, & \forall m \geq 0,\\ \sin|m|\lambda, & \forall m < 0,\end{cases} \quad (3)$$

note that $\overline{P}_{n,|m|}$ are the fully normalized associated Legendre functions of the first kind, and $\Omega = (\beta,\lambda)$. The relation between the gravitational potential and the gravitational vector components V^β and V^λ is described by the following integral transform (ibid):

$$V(u,\Omega) = \frac{1}{4\pi}\int_{\Omega'}\left[V^\beta(b_0,\Omega')\mathcal{K}^{\beta,o}\left(u,\Omega,b_0,\Omega'\right) - V^\lambda(b_0,\Omega')\mathcal{K}^{\lambda,o}\left(u,\Omega,b_0,\Omega'\right)\right]\mathrm{d}\Omega'. \quad (4)$$

Equation (1) represents the solution of the vertical spheroidal boundary-value problem while Eq. (4) describes the horizontal boundary-value problem. The index o in Eq. (4) distinguishes between the two orthogonalization methods. The first method ($o = 1$) was proposed by Grafarend et al. (2006), and the second approach ($o = 2$) was suggested by Lowes and Winch (2012). The corresponding sub-integral kernels have the following form:

$$\mathcal{K}^{\beta,1}\left(u,\Omega,b_0,\Omega'\right) = L(b_0,\beta')\sum_{n=1}^{\infty}\sum_{m=-n}^{n}\frac{1}{n(n+1)}\frac{Q_{n,|m|}\left(i\frac{u}{\varepsilon}\right)}{Q_{n,|m|}\left(i\frac{b_0}{\varepsilon}\right)}\overline{Y}_{n,m}\left(\Omega\right) \times\frac{\partial}{\partial\beta'}\overline{Y}_{n,m}\left(\Omega'\right), \quad (5)$$

$$\mathcal{K}^{\lambda,1}\left(u,\Omega,b_0,\Omega'\right) = \frac{a_0}{\cos\beta'}\sum_{n=1}^{\infty}\sum_{m=-n}^{n}\frac{1}{n(n+1)}\frac{Q_{n,|m|}\left(i\frac{u}{\varepsilon}\right)}{Q_{n,|m|}\left(i\frac{b_0}{\varepsilon}\right)}\overline{Y}_{n,m}\left(\Omega\right) \times\frac{\partial}{\partial\lambda'}\overline{Y}_{n,m}\left(\Omega'\right), \quad (6)$$

$$\mathcal{K}^{\beta,2}\left(u,\Omega,b_0,\Omega'\right) = a_0^2L(b_0,\beta')\sum_{n=1}^{\infty}\sum_{m=-n}^{n}\frac{1}{a_0^2n(n+1)-\varepsilon^2m^2} \times\frac{Q_{n,|m|}\left(i\frac{u}{\varepsilon}\right)}{Q_{n,|m|}\left(i\frac{b_0}{\varepsilon}\right)}\overline{Y}_{n,m}\left(\Omega\right)\frac{\partial}{\partial\beta'}\overline{Y}_{n,m}\left(\Omega'\right), \quad (7)$$

$$\mathcal{K}^{\lambda,2}\left(u,\Omega,b_0,\Omega'\right) = \frac{a_0 L^2\left(b_0,\beta'\right)}{\cos\beta'} \sum_{n=1}^{\infty} \sum_{m=-n}^{n} \frac{1}{a_0^2 n(n+1) - \varepsilon^2 m^2} \times \frac{Q_{n,|m|}\left(i\frac{u}{\varepsilon}\right)}{Q_{n,|m|}\left(i\frac{b_0}{\varepsilon}\right)} \overline{Y}_{n,m}\left(\Omega\right) \frac{\partial}{\partial\lambda'} \overline{Y}_{n,m}\left(\Omega'\right). \tag{8}$$

The position of the calculation point is defined by spheroidal coordinates (u, Ω) while the spheroidal coordinates of the integration element are represented by (b_0, Ω'). It is clear from Eqs. (5) and (7), and from Eqs. (6) and (8) that the corresponding pairs of sub-integral kernels have different analytical solutions. The reason is the different types of orthogonalization that change the formal structure of corresponding integral kernels. The infinite summations in Eqs. (2), (5), (6), (7) and (8) are in a practical calculation truncated in a certain degree.

3 Spectral Weighting and Combination

Now, we will explain how spectral weights for the solution of the vertical and horizontal spheroidal BVPs and their combination were derived. We aim to apply Eqs. (1) and (4) for downward continuation (DWC), i.e., to estimate the gravitational potential on the surface of the reference spheroid ($u = b_0$), see Fig. 1. Downward continuation is a classical inverse problem in potential theory. Because the gravitational potential satisfies Laplace's equation in a source-free region, it is a harmonic and thus analytic function. While the analytic nature of gravitational potential guarantees that continuation is mathematically possible, it is inherently ill-posed because high-degree spherical harmonic terms grow rapidly toward the Earth's surface. The spectral weighting approach presented here serves as a regularization mechanism, attenuating noise-dominated components and stabilizing the inversion. In the first step, we change the boundaries of the computation point and the integration element. We rewrite Eqs. (1) and (4) into the following forms:

$$V\left(b_0,\Omega\right) = \frac{1}{4\pi} \int_{\Omega'} V^u(u,\Omega')\mathcal{K}^u\left(u,\Omega,b_0,\Omega'\right)\mathrm{d}\Omega', \tag{9}$$

$$V\left(b_0,\Omega\right) = \frac{1}{4\pi} \int_{\Omega'} \left[V^\beta(u,\Omega')\mathcal{K}^{\beta,o}\left(u,\Omega,b_0,\Omega'\right) - V^\lambda(u,\Omega')\mathcal{K}^{\lambda,o}\left(u,\Omega,b_0,\Omega'\right) \right] \mathrm{d}\Omega', \tag{10}$$

or into the spectral form:

$$V^{(j)}(b_0,\Omega) = \sum_{n=1}^{\infty} \sum_{m=-n}^{n} \left(q_{n,m}^j\right)^{-1} V_{n,m}^j(u,\Omega'), \tag{11}$$

where the index j represents vertical (u) and horizontal ($\beta\lambda$) solutions. The remaining coefficients used in the previous equation are defined in Table 1. Note that we change the ratio of the associated Legendre functions of the second kind in sub-integral kernels defined in Eqs. (2), (5), (7), (6) and (8) according to Table 1. The prove that equations presented above can be derive by relations provided in Šprlák and Tangdamrongsub (2018).

In what follows, we describe how the spectral weights for the one-component and two-component estimators are derived. The term one-component estimator means we estimate the gravitational potential from one integral transformation, either from the vertical or the horizontal solution of the spheroidal boundary-value problem. The two-component estimator represents the gravitational potential calculated from a combination of vertical and horizontal solutions of the spheroidal boundary-value problem. The following five steps summarize the derivation of the spectral weights:

(i) Modification of the Eq. (11) by the spectral weights $a_{n,m}^j$;
(ii) Definition of the difference between the estimated value and its theoretical counterpart;
(iii) Definition of the global root mean square error (GRMSE);
(iv) Minimisation of GRMSE with respect to $a_{n,m}^j$;
(v) Determination of $a_{n,m}^j$.

3.1 Spectral DWC by One-component Estimator

Following the steps defined above, we modify firstly Eq. (11) by spectral weights. The modified equation is defined as follows:

$$V^{(j)}(b_0,\Omega) = \sum_{n=1}^{\infty} \sum_{m=-n}^{n} a_{n,m}^j V_{n,m}^j(u,\Omega'). \tag{12}$$

Secondly, we express the difference between the gravitational potential obtained by either vertical or horizontal components of the gravitational vector and its theoretical value. We assume that input values are affected only by random errors $\varepsilon^{(j)}$. The difference is:

$$\delta V^{(j)}(b_0,\Omega) = V^{(j)}(b_0,\Omega) + \varepsilon^{(j)}(b_0,\Omega) - V(b_0,\Omega)$$

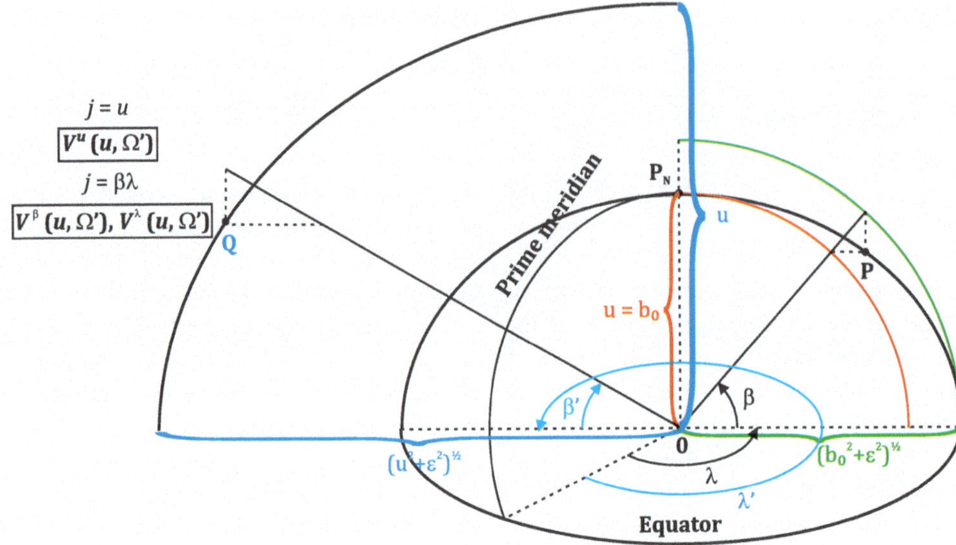

Fig. 1 The geometry of the downward continuation to the surface of the reference ellipsoid

Table 1 Definition of the coefficients used in Eq. (11)

j	u	$\beta\lambda$
$\left(q^j_{n,m}\right)^{-1}$	$\dfrac{Q_{n,\lvert m\rvert}\left(i\frac{b_0}{\varepsilon}\right)}{\frac{\partial}{\partial u}Q_{n,\lvert m\rvert}\left(i\frac{u}{\varepsilon}\right)}$	$\dfrac{Q_{n,\lvert m\rvert}\left(i\frac{b_0}{\varepsilon}\right)}{Q_{n,\lvert m\rvert}\left(i\frac{u}{\varepsilon}\right)}$
$V^j_{n,m}(u,\Omega')$	$\dfrac{1}{4\pi}\displaystyle\int_{\Omega'}\left[L(u,\beta')\;\; V^u(u,\Omega')\overline{Y}_{n,\lvert m\rvert}(\Omega')\right]\mathrm{d}\Omega'$	$o=1:$ $\frac{1}{4\pi n(n+1)}\int_{\Omega'}\Big[L(u,\beta')V^{\beta}(u,\Omega')\frac{\partial}{\partial\beta'}\overline{Y}_{n,\lvert m\rvert}(\Omega')$ $-\frac{a_0}{\cos\beta'}V^{\lambda}(u,\Omega')\frac{\partial}{\partial\lambda'}\overline{Y}_{n,\lvert m\rvert}(\Omega')\Big]\,\mathrm{d}\Omega'$ $o=2:$ $\frac{1}{4\pi(a_0^2 n(n+1)-\varepsilon^2 m^2)}$ $\times\int_{\Omega'}\Big[a_0^2(u,\beta')L(u,\beta')V^{\beta}(u,\Omega')\frac{\partial}{\partial\beta'}\overline{Y}_{n,\lvert m\rvert}(\Omega')$ $-\frac{a_0 L^2(u,\beta')}{\cos\beta'}V^{\lambda}(u,\Omega')\frac{\partial}{\partial\lambda'}\overline{Y}_{n,\lvert m\rvert}(\Omega')\Big]\,\mathrm{d}\Omega'$

$$
\begin{aligned}
&=\sum_{n=1}^{\infty}\sum_{m=-n}^{n} a^j_{n,m}V^j_{n,m}(u,\Omega)+\sum_{n=1}^{\infty}\sum_{m=-n}^{n} a^j_{n,m}\varepsilon^j_{n,m}(u,\Omega)\\
&-\sum_{n=1}^{\infty}\sum_{m=-n}^{n} V_{n,m}(b_0,\Omega).
\end{aligned}
\tag{13}
$$

If we consider the following relation $q^j_{n,m}V_{n,m}=V^j_{n,m}$, then we can rewrite the previous equation as follows:

$$
\begin{aligned}
&\delta V^{(j)}(b_0,\Omega)\\
&=\sum_{n=1}^{\infty}\sum_{m=-n}^{n} a^j_{n,m}q^j_{n,m}V_{n,m}(b_0,\Omega)\\
&\quad+\sum_{n=1}^{\infty}\sum_{m=-n}^{n} a^j_{n,m}\varepsilon^j_{n,m}(u,\Omega)-\sum_{n=0}^{\infty}\sum_{m=-n}^{n} V_{n,m}(b_0,\Omega)\\
&=\sum_{n=1}^{\infty}\sum_{m=-n}^{n}\left(a^j_{n,m}q^j_{n,m}-1\right)V_{n,m}(b_0,\Omega)\\
&\quad+\sum_{n=1}^{\infty}\sum_{m=-n}^{n} a^j_{n,m}\varepsilon^j_{n,m}(u,\Omega).
\end{aligned}
\tag{14}
$$

Thirdly, we form the GRMSE of the obtained gravitational potential. Its expression is:

$$
\begin{aligned}
&M\left\{(\delta V^{(j)}(b_0,\Omega))^2\right\}\\
&=E\left\{\int_{S'}(\delta V^{(j)}(b_0,\Omega))^2\mathrm{d}S'\right\}\\
&=E\left\{\int_{S'}\left[\sum_{n=1}^{\infty}\sum_{m=-n}^{n}\left(a^j_{n,m}q^j_{n,m}-1\right)V_{n,m}(b_0,\Omega)\right]\right.\\
&\quad\left.\times\left[\sum_{n=1}^{\infty}\sum_{m=-n}^{n}\left(a^j_{n,m}q^j_{n,m}-1\right)V_{n,m}(b_0,\Omega)\right]\mathrm{d}S'\right\}\\
&+2E\left\{\int_{S'}\left[\sum_{n=1}^{\infty}\sum_{m=-n}^{n}\left(a^j_{n,m}q^j_{n,m}-1\right)V_{n,m}(b_0,\Omega)\right]\right.
\end{aligned}
\tag{15}
$$

$$\times\left[\sum_{n=1}^{\infty}\sum_{m=-n}^{n} a^j_{n,m}\varepsilon^j_{n,m}(u,\Omega)\right]\mathrm{d}S'\Bigg\}$$
$$+E\left\{\int_{S'}\left[\sum_{n=1}^{\infty}\sum_{m=-n}^{n} a^j_{n,m}\varepsilon^j_{n,m}(u,\Omega)\right]\right.$$
$$\left.\times\left[\sum_{n=1}^{\infty}\sum_{m=-n}^{n} a^j_{n,m}\varepsilon^j_{n,m}(u,\Omega)\right]\mathrm{d}S'\right\}.$$

The symbol S' denotes the spheroidal surface and its infinitesimal element is $\mathrm{d}S' = L(u',\beta')\upsilon(u')\cos\beta'\mathrm{d}\beta'\mathrm{d}\lambda'$. In order to simplify the previous equation, we consider that the vertical $V^u_{n,m}$ and horizontal $V^{\beta\lambda}_{n,m}$ components of gravitational vector and their errors $\varepsilon^u_{n,m}$, $\varepsilon^{\beta\lambda}_{n,m}$ are not mutually correlated. Then, the second term of Eq. (15) $2E\left\{\int_{S'}\left[\sum_{n,m}\left(a^j_{n,m}q^j_{n,m}-1\right)V_{n,m}(b_0,\Omega)\right]\left[\sum_{n,m}a^j_{n,m}\varepsilon^j_{n,m}(u,\Omega)\right]\mathrm{d}S'\right\} = 0$. Further, we can prove that $E\left\{\int_{S'}(\delta V^{(j)}(b_0,\Omega'))^2\mathrm{d}S'\right\} = E\left\{\frac{1}{4\pi}\int_{\Omega'}(\delta V^{(j)}(b_0,\Omega'))^2\mathrm{d}\Omega'\right\}$. The last assumption is that the signal degree-order variances and error degree-order variances are defined as $c_n = E\left\{\frac{1}{4\pi}\int_{\Omega'}[V_{n,m}(b_0,\Omega')]^2\,\mathrm{d}\Omega'\right\}$, $\left(\sigma^j_n\right)^2 = E\left\{\frac{1}{4\pi}\int_{\Omega'}\left[\varepsilon^j_{n,m}(u,\Omega')\right]^2\mathrm{d}\Omega'\right\}$, respectively. Finally, we get for the GRMSE:

$$M\left\{(\delta V^{(j)}(b_0,\Omega))^2\right\} = \sum_{n=1}^{\infty}\sum_{m=-n}^{n}\left(a^j_{n,m}q^j_{n,m}-1\right)^2 c_n + \sum_{n=1}^{\infty}\sum_{m=-n}^{n} a^j_{n,m}\left(\sigma^j_n\right)^2. \tag{16}$$

Fourthly, we differentiate Eq. (16) with respect to $a^j_{n,m}$ and find its minimum by equating to zero:

$$\frac{\partial M\left\{(\delta V(b_0,\Omega))^2\right\}}{\partial a^j_{n,m}} = 2\sum_{n=1}^{\infty}\sum_{m=-n}^{n} a^j_{n,m}\left(\sigma^j_n\right)^2 + 2\sum_{n=1}^{\infty}\sum_{m=-n}^{n} q^j_{n,m}\left(a^j_{n,m}q^j_{n,m}-1\right)c_n. \tag{17}$$

Finally, we get Eq. (18), that is, the least-squares solution for the spectral weights for the one-component estimator:

$$a^j_{n,m} = \frac{c_n q^j_{n,m}}{c_n\left(q^j_{n,m}\right)^2 + \left(\sigma^j_n\right)^2}. \tag{18}$$

3.2 Spectral DWC by Two-Component Estimator

Firstly, the two-component estimator suitable for DWC is defined as:

$$V^{(u,\beta\lambda,o)}(b_0,\Omega) = \sum_{n=1}^{\infty}\sum_{m=-n}^{n} a^{u,\beta\lambda,o}_{n,m} V^u_{n,m}(u,\Omega') + \sum_{n=1}^{\infty}\sum_{m=-n}^{n} a^{\beta\lambda,u,o}_{n,m} V^{\beta\lambda}_{n,m}(u,\Omega'). \tag{19}$$

Secondly, we define the difference between gravitational potential obtained from two-component estimator and its true value as follows:

$$\begin{aligned}
&\delta V^{(u,\beta\lambda,o)}(b_0,\Omega)\\
&= V^{(u)}(u,\Omega) + \varepsilon^{(u)}(u,\Omega) + V^{(\beta\lambda,o)}(u,\Omega)\\
&\quad + \varepsilon^{(\beta\lambda,o)}(u,\Omega) - V(b_0,\Omega)\\
&= \sum_{n=1}^{\infty}\sum_{m=-n}^{n} a^{u,\beta\lambda,o}_{n,m}V^u_{n,m}(u,\Omega) + \sum_{n=1}^{\infty}\sum_{m=-n}^{n} a^{u,\beta\lambda,o}_{n,m}\varepsilon^u_{n,m}(u,\Omega)\\
&\quad + \sum_{n=1}^{\infty}\sum_{m=-n}^{n} a^{\beta\lambda,u,o}_{n,m}V^{\beta\lambda}_{n,m}(u,\Omega)\\
&\quad + \sum_{n=1}^{\infty}\sum_{m=-n}^{n} a^{\beta\lambda,u,o}_{n,m}\varepsilon^{\beta\lambda,o}_{n,m}(u,\Omega) - \sum_{n=1}^{\infty}\sum_{m=-n}^{n} V_{n,m}(b_0,\Omega)\\
&= \sum_{n=1}^{\infty}\sum_{m=-n}^{n} a^{u,\beta\lambda,o}_{n,m}q^u_{n,m}V_{n,m}(b_0,\Omega)\\
&\quad + \sum_{n=1}^{\infty}\sum_{m=-n}^{n} a^{u,\beta\lambda,o}_{n,m}\varepsilon^u_{n,m}(u,\Omega)\\
&\quad + \sum_{n=1}^{\infty}\sum_{m=-n}^{n} a^{\beta\lambda,u,o}_{n,m}q^{\beta\lambda}_{n,m}V_{n,m}(b_0,\Omega)\\
&\quad + \sum_{n=1}^{\infty}\sum_{m=-n}^{n} a^{\beta\lambda,u,o}_{n,m}\varepsilon^{\beta\lambda,o}_{n,m}(u,\Omega)\\
&\quad - \sum_{n=1}^{\infty}\sum_{m=-n}^{n} V_{n,m}(b_0,\Omega)\\
&= \sum_{n=1}^{\infty}\sum_{m=-n}^{n}\left(a^{u,\beta\lambda,o}_{n,m}q^u_{n,m} + a^{\beta\lambda,u,o}_{n,m}q^{\beta\lambda}_{n,m} - 1\right)V_{n,m}(b_0,\Omega)\\
&\quad + \sum_{n=1}^{\infty}\sum_{m=-n}^{n} a^{u,\beta\lambda,o}_{n,m}\varepsilon^u_{n,m}(u,\Omega)\\
&\quad + \sum_{n=1}^{\infty}\sum_{m=-n}^{n} a^{\beta\lambda,u,o}_{n,m}\varepsilon^{\beta\lambda,o}_{n,m}(u,\Omega)
\end{aligned} \tag{20}$$

As the third step, we express the GRMSE of the two-component estimator as:

$$
\begin{aligned}
M\left\{(\delta V^{(u,\beta\lambda,o)}(b_0,\Omega))^2\right\} &= E\left\{\int_{S'}(\delta V^{(u,\beta\lambda,o)}(b_0,\Omega'))^2 \mathrm{d}S'\right\}\\
&= E\left\{\int_{S'}\left[\sum_{n=1}^{\infty}\sum_{m=-n}^{n}\left(a_{n,m}^{u,\beta\lambda,o}q_{n,m}^{u}+a_{n,m}^{\beta\lambda,u,o}q_{n,m}^{\beta\lambda}-1\right)V_{n,m}(b_0,\Omega')\right]^2\mathrm{d}S'\right\}\\
&+2E\left\{\int_{S'}\left[\sum_{n=1}^{\infty}\sum_{m=-n}^{n}\left(a_{n,m}^{u,\beta\lambda,o}q_{n,m}^{u}+a_{n,m}^{\beta\lambda,u,o}q_{n,m}^{\beta\lambda}-1\right)V_{n,m}(b_0,\Omega')\right]\left[\sum_{n=1}^{\infty}\sum_{m=-n}^{n}a_{n,m}^{u,\beta\lambda,o}\varepsilon_{n,m}^{u}(u,\Omega')\right]\mathrm{d}S'\right\}\\
&+2E\left\{\int_{S'}\left[\sum_{n=1}^{\infty}\sum_{m=-n}^{n}\left(a_{n,m}^{u,\beta\lambda,o}q_{n,m}^{u}+a_{n,m}^{\beta\lambda,u,o}q_{n,m}^{\beta\lambda}-1\right)V_{n,m}(b_0,\Omega')\right]\left[\sum_{n=1}^{\infty}\sum_{m=-n}^{n}a_{n,m}^{\beta\lambda,u,o}\varepsilon_{n,m}^{\beta\lambda,o}(u,\Omega')\right]\mathrm{d}S'\right\}\\
&+E\left\{\int_{S'}\left[\sum_{n=1}^{\infty}\sum_{m=-n}^{n}a_{n,m}^{u,\beta\lambda,o}\varepsilon_{n,m}^{u}(u,\Omega')\right]^2\mathrm{d}S'\right\}+2E\left\{\int_{S'}\left[\sum_{n=1}^{\infty}\sum_{m=-n}^{n}a_{n,m}^{u,\beta\lambda,o}\varepsilon_{n,m}^{u}(u,\Omega')\right]\right.\\
&\left.\times\left[\sum_{n=1}^{\infty}\sum_{m=-n}^{n}a_{n,m}^{\beta\lambda,u,o}\varepsilon_{n,m}^{\beta\lambda,o}(u,\Omega')\right]\mathrm{d}S'\right\}+E\left\{\int_{S'}\left[\sum_{n=1}^{\infty}\sum_{m=-n}^{n}a_{n,m}^{\beta\lambda,u,o}\varepsilon_{n,m}^{\beta\lambda,o}(u,\Omega')\right]^2\mathrm{d}S'\right\}.
\end{aligned}
\tag{21}
$$

If we consider the same assumptions as for the one-component estimator, we get the following form for the GRMSE:

$$
\begin{aligned}
M\left\{(\delta V^{(u,\beta\lambda,o)}(b_0,\Omega))^2\right\} &= \sum_{n=1}^{\infty}\sum_{m=-n}^{n}\left(a_{n,m}^{u,\beta\lambda,o}q_{n,m}^{u}+a_{n,m}^{\beta\lambda,u,o}q_{n,m}^{\beta\lambda,o}-1\right)^2 c_n(b_0,\Omega)\\
&+\sum_{n=1}^{\infty}\sum_{m=-n}^{n}a_{n,m}^{u,\beta\lambda,o}\left(\sigma_{n,m}^{u}\right)^2\\
&+\sum_{n=1}^{\infty}\sum_{m=-n}^{n}a_{n,m}^{\beta\lambda,u,o}\left(\sigma_{n,m}^{\beta\lambda,o}\right)^2,
\end{aligned}
\tag{22}
$$

Fourthly, we take the derivative of Eq. (22) with respect to $a_{n,m}^{u,\beta\lambda,o}$ and $a_{n,m}^{\beta\lambda,u,o}$. Then, we equate the result to zero, and we get the system of equations:

$$
\begin{aligned}
&\sum_{n=1}^{\infty}\sum_{m=-n}^{n}a_{n,m}^{u,\beta\lambda,o}\left(\sigma_{n,m}^{u}\right)^2+\sum_{n=1}^{\infty}\sum_{m=-n}^{n}q_{n,m}^{u}\left(a_{n,m}^{\beta\lambda,u,o}q_{n,m}^{\beta\lambda,o}+a_{n,m}^{u,\beta\lambda,o}q_{n,m}^{u}-1\right)c_n\\
&\sum_{n=1}^{\infty}\sum_{m=-n}^{n}a_{n,m}^{\beta\lambda,u,o}\left(\sigma_{n,m}^{\beta\lambda,o}\right)^2+\sum_{n=1}^{\infty}\sum_{m=-n}^{n}q_{n,m}^{\beta\lambda}\left(a_{n,m}^{\beta\lambda,u,o}q_{n,m}^{\beta\lambda,o}+a_{n,m}^{u,\beta\lambda,o}q_{n,m}^{u}-1\right)c_n.
\end{aligned}
\tag{23}
$$

The solution of the previous system of equations is the least-squares estimate of the spectral weights for the two-component estimator. The spectral weights for the two-component estimator have the following form:

$$
\begin{aligned}
a_{n,m}^{u,\beta\lambda,o} &= \frac{c_n q_{n,m}^{u} \left(\sigma_{n,m}^{\beta\lambda,o}\right)^2}{\left(\sigma_{n,m}^{u}\right)^2 \left(\sigma_{n,m}^{\beta\lambda,o}\right)^2 + c_n q_{n,m}^{u} \left(\sigma_{n,m}^{\beta\lambda,o}\right)^2 + c_n q_{n,m}^{\beta\lambda} \left(\sigma_{n,m}^{u}\right)^2}, \\
a_{n,m}^{\beta\lambda,u,o} &= \frac{c_n q_{n,m}^{\beta\lambda,o} \left(\sigma_{n,m}^{u}\right)^2}{\left(\sigma_{n,m}^{u}\right)^2 \left(\sigma_{n,m}^{\beta\lambda,o}\right)^2 + c_n q_{n,m}^{u} \left(\sigma_{n,m}^{\beta\lambda,o}\right)^2 + c_n q_{n,m}^{\beta\lambda} \left(\sigma_{n,m}^{u}\right)^2}.
\end{aligned}
\tag{24}
$$

4 Numerical Experiment

We analyzed benefits of the spectral weights derived in the previous section over Greenland (Fig. 2). Unknown values of the disturbing potential were computed on an equiangular grid with a 0.5° spacing, bounded by the parallels $\varphi \in [60°, 86°]$ and by the meridians $\lambda \in [-76°, -10°]$ on the surface of the reference ellipsoid.

In the first step, we transform EGM2008 (Pavlis et al. 2012) spherical harmonics up to degree and order 90 into their spheroidal counterparts using Hotine-Jekeli's transformation (Jekeli 1988). To guarantee that the spherical and spheroidal spectra contain the same total power, the spheroidal harmonic coefficients are computed up to a maximum degree that is 40 degrees higher than the maximum degree of the spherical harmonic coefficients.

Secondly, we generated the components of the disturbing gravitational vector from this model globally on an equiangular 0.5° grid at the satellite altitude 400 km above the reference ellipsoid. The spectral combination method assumes that the noise characteristics of the observed disturbing gravitational data are known. We polluted components of the disturbing gravitational vector by Gaussian noise with standard deviation of 10 μGal. Further, we calculated the signal and error degree-order variances.

In the third step, we demonstrated the importance of proper spectral weighting of the observed disturbing gravitational vector data. The upper part of Table 2 summarizes

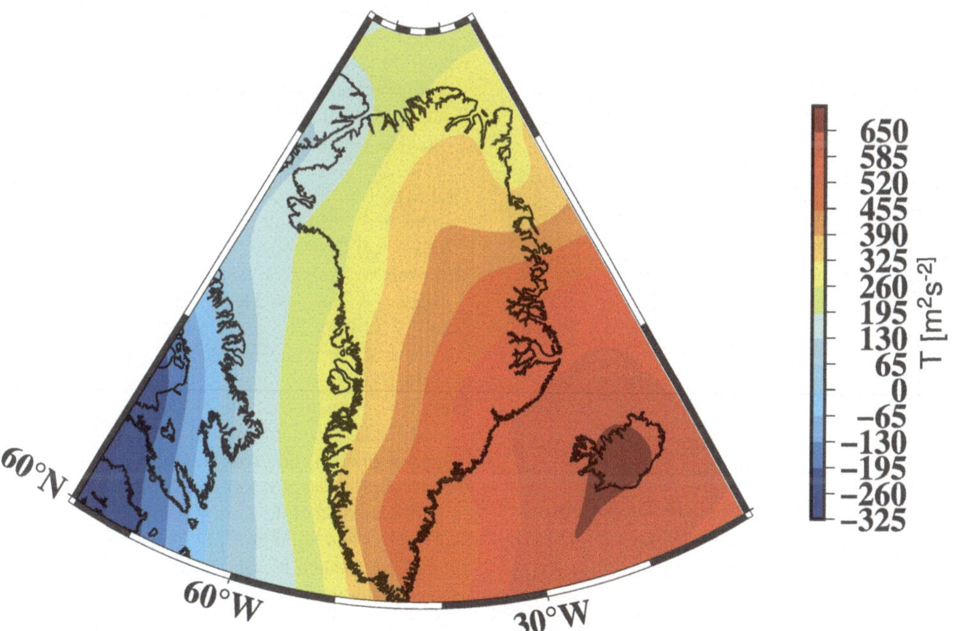

Fig. 2 The disturbing potential on the surface of the reference ellipsoid over the test area

Table 2 Statistics of the differences between the values of the disturbing gravitational potential by the numerical integration method w/o and w/ spectral weights and the reference values synthesized from EGM2008 ($m^2 s^{-2}$)

	u	$\beta\lambda,o = 1$	$\beta\lambda,o = 2$
DWC			
std	3.551	12.506	12.170
max	18.972	26.981	29.956
min	−3.548	−32.501	−19.009
mean	9.003	−9.196	11.093
DWC with SW			
std	0.384	0.435	0.429
max	1.494	1.348	1.515
min	−0.941	−1.942	−1.757
mean	0.000	−0.014	−0.040

Table 3 Statistics of the differences between the values of the disturbing gravitational potential by the numerical integration method w/o spectral weights and the reference values synthesized from EGM2008 ($m^2 s^{-2}$)

	$u,\beta\lambda,o = 1$	$u+\beta\lambda,o = 2$
std	0.435	0.429
max	1.348	1.515
min	−1.942	−1.757
mean	−0.014	−0.040

the differences between the disturbing potential obtained via Eq. (11) and the reference values. The unweighted spectral DWC results are clearly insufficient. In contrast, the lower part of the same table shows that including spectral weights, Eq. (12), significantly improves accuracy based on spectral weights. When combining all two components of disturbing gravitational vector, the standard deviation of differences reaches same value as from only horizontal solutions (see Table 3). The explanation is that the spectral weighting and combination method works as the weighted mean and horizontal solution of spheroidal boundary-value problem.

Finally, the results confirm that the application of spectral weights substantially enhances the stability and accuracy of the downward continuation process. These findings validate the theoretical derivations presented in the previous section and demonstrate the practical potential for precise gravitational field modelling.

5 Conclusions

The paper investigates spectral weighting and the combination of solutions to vertical and horizontal spheroidal boundary-value problems. The proposed method continues gravitational data measured above the reference ellipsoid downward and transforms them into the gravitational potential without the need to invert matrix operators. For this purpose, spectral weights for one- and two-component estimators were derived. The geodetic theory has been extended by applying the spectral weighting approach to bodies whose shape more closely resembles an ellipsoid than a sphere. The significance of the spectral weights was demonstrated in a numerical experiment, which confirmed that the weights were derived correctly and effectively suppress noise during the downward continuation of the data. The method can be also applied to the precise modelling of planetary gravitational fields. Another potential application lies in modelling ice mass variations in polar regions using satellite gravimetric data, as the Earth exhibits noticeable flattening in these areas. The proposed method is applicable only under the condition of a constant u and when observations are distributed on a regular grid. Nevertheless, due to the distant-zone effects, global grid coverage is required, which in practice necessitates the use of satellite-derived data that are spectrally limited to approximately degree 300 for the Earth and up to about degree 600 for atmosphere-free bodies such as the Moon.

Acknowledgements Authors were supported by the project No. 23-07031S of the Czech Science Foundation. We would like to thank two anonymous reviewers for their thoughtful and constructive comments, and to prof. Riccardo Barzaghi for handling our manuscript.

References

Ardalan AA (1999) High resolution regional geoid computation in the World Geodetic Datum 2000 based upon collocation of linearized observational functionals of the type GPS, gravity potential and gravity intensity. PhD Thesis, Department of Geodesy and Geoinformatics, Stuttgart University, Stuttgart, Germany, 239 pp.

Bölling C, Grafarend EW (2005) Ellipsoidal spectral properties of the Earth's gravitational potential and its first and second derivatives. J Geodesy 79:300–330

Eshagh M (2011) Spectral combination of vector gravimetric boundary value problems. J Geospat Inf Technol 1(3):33–50 (in Persian)

Eshagh M (2012) Spectral combination of spherical gradiometric boundary-value problems: a theoretical study. Pure Appl Geophys 169:2201–2215

Grafarend EW, Finn G, Ardalan AA (2006) Ellipsoidal vertical deflections and ellipsoidal gravity disturbance: case studies. Studia Geophysica et Geodaetica 50:1–57

Hofmann-Wellenhof B, Moritz H (2006) Physical geodesy. Springer Science & Business Media, Austria

Jekeli C (1988) The exact transformation between ellipsoidal and spherical harmonic expansions. Manuscripta Geodaetica 13:106–113

Lowes FJ, Winch DE (2012) Orthogonality of harmonic potentials and fields in spheroidal and ellipsoidal coordinates: application to geomagnetism and geodesy. Geophys J Int 191:491–507

Pavlis NK, Holmes SA, Kenyon SC, Factor JK (2012) The development and evaluation of the Earth Gravitational Model 2008 (EGM2008). J Geophys Res Solid Earth 117:B04406

Pitoňák M, Eshagh M, Šprlák M, Tenzer R, Novák P (2018) Spectral combination of spherical gravitational curvature boundary-value problems. Geophys J Int 214(2):773–791

Pitoňák M, Šprlák M, Ophaug V, Omang OC, Novák P (2023a) Validation of space-wise GOCE gravitational gradient grids using the spectral combination method and GNSS/levelling data. Surv Geophys 44:739–782

Pitoňák M, Šprlák M, Novák P (2023b) Estimation of Height Anomalies from Gradients of the Gravitational Potential Using a Spectral Combination Method. In: Freymueller JT, Sánchez L (eds) X Hotine-Marussi Symposium on Mathematical Geodesy. HMS 2022. International Association of Geodesy Symposia, vol 155. Springer, Cham

Sjöberg LE (1980) Least squares combination of satellite harmonics and integral formulas in physical geodesy. Gerlands Beiträge zur Geophysik 89:371–377

Šprlák M, Tangdamrongsub N (2018) Vertical and horizontal spheroidal boundary-value problems. J Geodesy 92:811–826

Wenzel HG (1982) Geoid computation by least squares spectral combination using integral kernels. In: Proceedings of the IAG General Assembly, Tokyo, pp. 438–453. Springer, Berlin

Global Gravitational Field Modelling for Spheroidal Planetary Bodies: Theory and Numerical Aspects

Jiří Belinger, Veronika Dohnalová, Martin Pitoňák, Pavel Novák, and Michal Šprlák

Abstract

Determination of gravitational fields generated by various planetary bodies (including Earth, Earth's Moon or neighboring planets) represents a crucial task in modern geodesy and planetodesy. In most cases, calculation of gravitational field quantities is based on the spherical approximation of planetary bodies. Spherical approximation is valid for bodies with shapes close to sphere or in cases when lower gravitational field quantities estimation accuracy is acceptable. On the other hand, shapes of planetary bodies are often significantly closer to spheroids. The spheroidal approximation is essential as it significantly extends region near the surface of flattened planetary bodies allowing safe computation without any divergence issues.

This contribution describes proposed theoretical solution for the gravitational field modelling of planetary bodies using spheroidal harmonic functions and selected numerical aspects of this solution. Theory described in this contribution provides background for calculation of gravitational field quantities using spheroidal harmonic synthesis. Special attention is paid to the theoretical solution for calculation of associated Legendre functions based on hypergeometric functions and recursions.

To support and validate these theoretical assumptions, we have implemented derived equations into scripts and conducted various tests to evaluate correctness, numerical properties and efficiency of the proposed approach.

Keywords

Associated Legendre functions · Gravitational field modelling · Hypergeometric functions · Recursive expressions · Spheroidal harmonic functions

1 Introduction

The gravitational field research concerning spheroidal approximation of reference surfaces (used in harmonic series representations of gravitational field quantities) is becoming essential with increasing knowledge of irregularly shaped planetary bodies and higher demands with respect to the field calculation accuracy. Properties of methods using spheroidal approach have already been studied in Novák and Šprlák (2018). The importance of spheroidal approximation consideration is depicted in Fig. 1, where both (spherical and spheroidal) approximations are represented for dwarf planet Ceres and asteroid Vesta. These planetary bodies were selected because of their significant flattening ($f \approx 0.0747$ for Ceres and $f \approx 0.2204$ for Vesta, see Marchi et al. 2022). Note that only outer parts of the spherical (blue) and spheroidal (red) surfaces are demonstrably suitable for harmonic series representation, as there are known divergence issues in the inner parts (Hu and Jekeli 2015).

J. Belinger (✉) · V. Dohnalová · M. Pitoňák · P. Novák · M. Šprlák
NTIS – New Technologies for the Information Society, Faculty of Applied Sciences, University of West Bohemia in Pilsen, Plzeň, Czech Republic
e-mail: belinger@ntis.zcu.cz; dohnalv@ntis.zcu.cz; pitonakm@kgm.zcu.cz; panovak@ntis.zcu.cz; sprlakm@kgm.zcu.cz

J. T. Freymueller, L. Sànchez (eds.), *International Symposium on Gravity, Geoid and Height Systems 2024 (GGHS2024)*, International Association of Geodesy Symposia 158, https://doi.org/10.1007/1345_2025_297

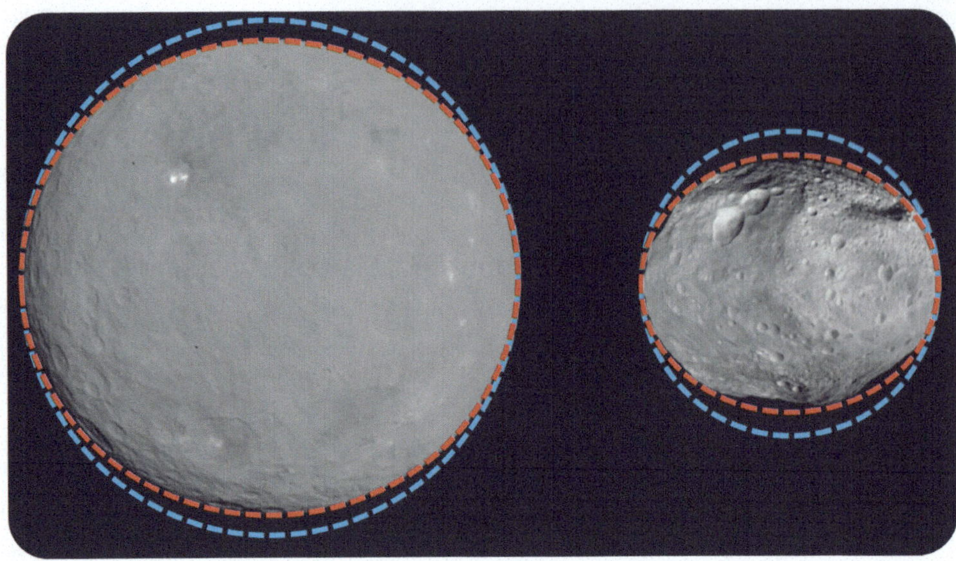

Fig. 1 Difference between spherical (blue) and spheroidal (red) approximation for the dwarf planet Ceres (left) and asteroid Vesta (right)

There are various methods for determination of gravitational field, e.g., least squares collocation (Ophaug and Gerlach 2017), radial basis functions (Freeden and Michel 2004), integral transformations (Novák et al. 2017) and spherical/spheroidal harmonic functions (Hu and Jekeli 2015; Šprlák and Han 2021). We focus on the modelling of gravitational field using spheroidal harmonic functions, as they represent a global method and more general alternative to spherical harmonic functions, which are widely used for the Earth by the geodetic community.

As stated earlier, key advantage of spheroidal approach is its potential for extension of region suitable for calculation using spheroidal harmonic series as depicted in Fig. 1 (space between blue and red surface). However, there are several challenges to overcome while evaluating spheroidal harmonic series numerically.

Firstly, there is the problem of overflow/underflow of values in calculation of associated Legendre functions (ALFs) of both first and second kind, and their partial derivatives with respect to minor semi-axis of confocal spheroid u (Fukushima 2012). In this contribution, we discuss possible options of dealing with this issue using hypergeometric (HG) functions.

Secondly, the complexity of derived equations for spheroidal harmonic synthesis is much higher compared to the spherical approach.

Thirdly, new topography calculations depend on partial derivatives with respect to reduced spheroidal latitude β and trigonometric functions dependent on β, which may lead to singularities at computational points close to the poles (Hamáčková et al. 2016).

In the following sections, we describe theoretical basis of above-mentioned solution, numerical tests of selected HG functions with respect to their value range variability and computational efficiency, numerical experiments testing correctness of implementation using recursions and finally we evaluate current results and plans for future development.

2 Theoretical Background

To enable and also to simplify calculations using spheroidal harmonic functions, we use one-parametric Jacobi ellipsoidal coordinates (u, Ω) closer described in Fig. 2 and in Heiskanen and Moritz (1967).

Our newly derived theory contains equations for calculation of gravitational field quantities for oblate and prolate spheroidal bodies using spheroidal harmonic functions including spheroidal harmonic analysis (calculation of spheroidal harmonic coefficients) and spheroidal harmonic synthesis (calculation of gradients of the gravitational potential up to the third order). In this contribution, we will focus on theory concerning spheroidal harmonic synthesis equations and their components for oblate spheroidal bodies.

Principles of deriving the relevant formulas can be described using Eq. (1) that represents the first-order partial derivative of gravitational potential V with respect to β (V_β), as it is the most concise equation allowing to explain all important parts of the above-mentioned process solving

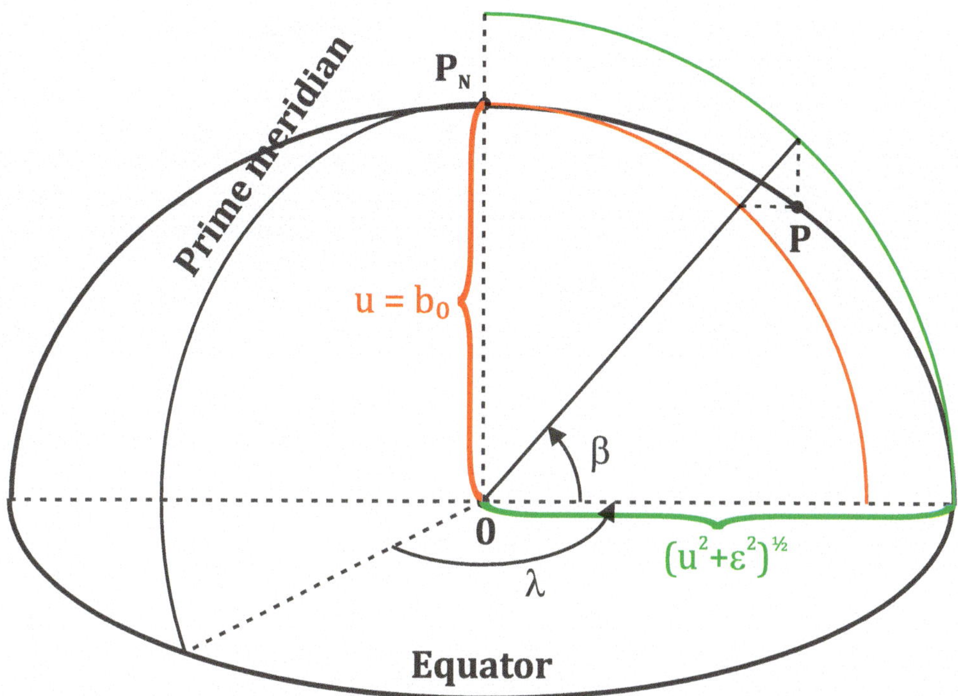

Fig. 2 Spheroidal (one-parametric Jacobi ellipsoidal) coordinates of the point P, where $u \in [0, \infty)$ denotes minor semi-axis of confocal spheroid and $\beta \in [-90°, 90°]$, $\lambda \in [0°, 360°)$ (often coupled and labelled as Ω) stand for the reduced spheroidal latitude and spheroidal longitude

singularities and value range variability. The specific form of the equation is as follows:

$$V_\beta(u, \Omega) = \frac{GM}{R_o} \sum_{n=0}^{\infty} \sum_{m=-n}^{n} \bar{C}^{o}_{n,m}(u) \frac{Q_{n,|m|}\left(i\frac{u}{\varepsilon}\right)}{Q_{n,|m|}\left(i\frac{b}{\varepsilon}\right)} \frac{1}{L} \frac{\partial}{\partial \beta} \bar{P}_{n,|m|}(\sin\beta)\; K_m(\lambda)$$

$$+ \frac{GM}{R_i} \sum_{n=0}^{\infty} \sum_{m=-n}^{n} \bar{C}^{i}_{n,m}(u) \frac{P_{n,|m|}\left(i\frac{u}{\varepsilon}\right)}{P_{n,|m|}\left(i\frac{b}{\varepsilon}\right)} \frac{1}{L} \frac{\partial}{\partial \beta} \bar{P}_{n,|m|}(\sin\beta)\; K_m(\lambda), \qquad (1)$$

where (u, Ω) are the Jacobi one-parametric coordinates of a computational point, see Fig. 2. GM is the geocentric gravitational constant, R_o and R_i are suitable scale factors (usually set equal to the minor semi-axis of the minimal Brillouin spheroid), n and m stand for degree and order of the spheroidal harmonic expansion. $\bar{C}^{o}_{n,m}$ and $\bar{C}^{i}_{n,m}$ denote the normalized (geodetic norm) spheroidal harmonic coefficients. $P_{n,|m|}$ and $Q_{n,|m|}$ define the un-normalized ($\bar{P}_{n,|m|}$ for normalized, respectively) ALFs of the first and the second kind. ε stands for the linear eccentricity calculated from major and minor semi-axes of the reference spheroid a and b, $L = \sqrt{u^2 + \varepsilon^2 \sin^2\beta}$ and:

$$K_m(\lambda) = \begin{cases} \cos m\lambda, & m \geq 0, \\ \sin |m|\lambda, & m < 0. \end{cases} \qquad (2)$$

Equation (1) consists of 2 main parts:

- Calculation outside the confocal spheroid with minor semi-axis u (upper part),
- Calculation inside the confocal spheroid with minor semi-axis u, but outside gravitating masses (lower part).

Equivalent equations were derived for all other derivatives of the gravitational potential up to the third order (3 for the gravitational gradient vector, 6 for the second-order gravitational tensor and 10 for the third-order gravitational tensor).

2.1 Hypergeometric Functions

In this section, we focus on the behaviour of ALFs when calculated using selected HG functions. More details about HG functions can be found in Hobson (1965). The tests were carried out for both kinds of ALFs and for 6 types of HG functions for each kind of ALF. The example of mentioned calculation for the ALFs of the first kind is represented by the following equation:

$$P_{n,m}(u) = \left(\frac{b}{\varepsilon}\right)^n \frac{(2n)!}{2^n n!\,(n-m)!}\, p_{n,m}(u), \tag{3}$$

where $p_{n,m}(u)$ can be obtained by the Gauss HG function ${}_2F_1$. The calculation was tested for 6 HG functions listed in Table 1.

2.2 Non-singular Representation

Non-singular representation section describes implementation of recursions substituting higher-order derivatives of ALFs with respect to β and also trigonometric functions dependent on β and thus eliminating singularities in original expressions. Derivation and implementation of spherical formulas are discussed in Hamáčková et al. (2016). In case of Eq. (1), the specific recursions substituting $\frac{\partial}{\partial\beta}\bar{P}_{n,|m|}(\sin\beta)$ are:

$$\frac{\partial}{\partial\beta}\bar{P}_{0,0}(\sin\beta) = 0, \tag{4}$$

$$\frac{\partial}{\partial\beta}\bar{P}_{n,0}(\sin\beta) = \frac{1}{2}\sqrt{2n(n+1)}\,\bar{P}_{n,1}(\sin\beta), \quad n>0, \tag{5}$$

$$\begin{aligned}\frac{\partial}{\partial\beta}\bar{P}_{n,|m|}(\sin\beta) = {} & \frac{1}{2}\sqrt{(n-|m|)(n+|m|+1)}\,\bar{P}_{n,|m|+1}(\sin\beta) \\ & - \frac{1}{2}\sqrt{(1+\delta_{|m|-1,0})(n+|m|)(n-|m|+1)}\,\bar{P}_{n,|m|-1}(\sin\beta), \quad |m|>0,\end{aligned} \tag{6}$$

where $\delta_{|m|-1,0}$ is the Kronecker delta and the final form of Eq. (1) after the implementation of recursions is:

$$\begin{aligned} V_\beta(u,\Omega) = {} & \frac{GM}{2R_oL}\sum_{n=1}^{\infty}\bar{C}^o_{n,0}(u)\,\frac{Q_{n,0}\left(i\frac{u}{\varepsilon}\right)}{Q_{n,0}\left(i\frac{b}{\varepsilon}\right)}\sqrt{2n(n+1)}\,\bar{P}_{n,1}(\sin\beta) \\ & + \frac{GM}{2R_oL}\sum_{n=1}^{\infty}\sum_{\substack{m=-n\\ m\neq 0}}^{+n}\bar{C}^o_{n,m}(u)\,\frac{Q_{n,|m|}\left(i\frac{u}{\varepsilon}\right)}{Q_{n,|m|}\left(i\frac{b}{\varepsilon}\right)}\Bigg[\sqrt{(n-|m|)(n+|m|+1)}\,\bar{P}_{n,|m|+1}(\sin\beta) \\ & - \sqrt{(1+\delta_{|m|-1,0})(n+|m|)(n-|m|+1)}\,\bar{P}_{n,|m|-1}(\sin\beta)\Bigg]K_m(\lambda) \\ & + \frac{GM}{2R_iL}\sum_{n=1}^{\infty}\bar{C}^i_{n,0}(u)\,\frac{P_{n,0}\left(i\frac{u}{\varepsilon}\right)}{P_{n,0}\left(i\frac{b}{\varepsilon}\right)}\sqrt{2n(n+1)}\,\bar{P}_{n,1}(\sin\beta) \\ & + \frac{GM}{2R_iL}\sum_{n=1}^{\infty}\sum_{\substack{m=-n\\ m\neq 0}}^{+n}\bar{C}^i_{n,m}(u)\,\frac{P_{n,|m|}\left(i\frac{u}{\varepsilon}\right)}{P_{n,|m|}\left(i\frac{b}{\varepsilon}\right)}\Bigg[\sqrt{(n-|m|)(n+|m|+1)}\,\bar{P}_{n,|m|+1}(\sin\beta) \\ & - \sqrt{(1+\delta_{|m|-1,0})(n+|m|)(n-|m|+1)}\,\bar{P}_{n,|m|-1}(\sin\beta)\Bigg]K_m(\lambda). \end{aligned} \tag{7}$$

Table 1 HG functions selected for the calculation of $P_{n,m}(u)$

Type	Equation
1	$p_{n,m}(u) = \left(\frac{\sqrt{u^2+\varepsilon^2}}{b}\right)^n {}_2F_1\left(-\frac{n-m}{2}, -\frac{n+m}{2}, -\frac{2n-1}{2}, \frac{\varepsilon^2}{u^2+\varepsilon^2}\right)$
2	$p_{n,m}(u) = \frac{u}{b} \left(\frac{\sqrt{u^2+\varepsilon^2}}{b}\right)^{n-1} {}_2F_1\left(-\frac{n+m-1}{2}, -\frac{n-m-1}{2}, -\frac{2n-1}{2}, \frac{\varepsilon^2}{u^2+\varepsilon^2}\right)$
3	$p_{n,m}(u) = \left(\frac{u}{b}\right)^n \left(\frac{\sqrt{u^2+\varepsilon^2}}{u}\right)^m {}_2F_1\left(-\frac{n-m}{2}, -\frac{n-m-1}{2}, -\frac{2n-1}{2}, -\frac{\varepsilon^2}{u^2}\right)$
4	$p_{n,m}(u) = \left(\frac{u}{b}\right)^n \left(\frac{u}{\sqrt{u^2+\varepsilon^2}}\right)^m {}_2F_1\left(-\frac{n+m}{2}, -\frac{n+m-1}{2}, -\frac{2n-1}{2}, -\frac{\varepsilon^2}{u^2}\right)$
5	$p_{n,m}(u) = \left(\frac{\sqrt{u^2+\varepsilon^2}}{b}\right)^n {}_2F_1\left[-(n-m), -(n+m), -\frac{2n-1}{2}, -\frac{1}{2}\left(\frac{u-\sqrt{u^2+\varepsilon^2}}{\sqrt{u^2+\varepsilon^2}}\right)\right]$
6	$p_{n,m}(u) = \left(\frac{u+\sqrt{u^2+\varepsilon^2}}{2b}\right)^n \left(\frac{2\sqrt{u^2+\varepsilon^2}}{u+\sqrt{u^2+\varepsilon^2}}\right)^m {}_2F_1\left[-(n-m), \frac{2m+1}{2}, -\frac{2n-1}{2}, \frac{u-\sqrt{u^2+\varepsilon^2}}{u+\sqrt{u^2+\varepsilon^2}}\right]$

Analogical non-singular expressions were derived for all other derivatives of the gravitational potential up to the third order.

3 Numerical Experiment

Equations described in Sect. 2, together with all analogical equations concerning other derivatives of the gravitational potential, were implemented into C language scripts. Using these scripts, it was possible to carry out an experiment testing correctness of both theory and its implementation.

The experiment, similarly to the theory, was divided into 2 parts, namely numerical testing of individual HG functions behaviour and testing of implemented recursions.

3.1 Properties of Hypergeometric Functions

The main purpose of the first part of experiment was to examine numerical behaviour of various types of HG functions. The emphasis was on range and variability of order of magnitude and computational efficiency. To support validity of results, tested parameters (namely u and f) were set to extreme values to cover possible extremes in case of real planetary bodies. For this part of the experiment, the major semi-axis of reference spheroid was set to 1. Maximal degree of expansion was set to 2160. We tested 6 HG functions (see Table 1), each for 3 different positions of the computational point:

- Outside the reference spheroid ($b = 0.99$, $a = 1$): $u = 10b$,
- On the reference spheroid: $u = b$,
- Inside the reference spheroid: $u = 0.1b$.

Order of magnitude range of all functions was in limits of double-precision arithmetics only for points lying on the reference spheroid, as can be seen on example (HG function of type 4) in the upper part (a) of Fig. 3. Possible solution to overflow at points outside the spheroid would be to calculate the attenuation factor (e.g., $\frac{P_{n,|m|}\left(i\frac{u}{\varepsilon}\right)}{P_{n,|m|}\left(i\frac{b}{\varepsilon}\right)}$), instead of computing functions separately. However, this solution cannot be applied for points inside the reference spheroid. For these points, different approach is sought currently.

The next parameter affecting the order of magnitude range is flattening. As shown in the middle part (b) of Fig. 3, higher flattening also causes value overflow.

In order to evaluate numerical efficiency of individual HG functions, we analysed the number of iterations needed to get the final result. The number of iretations required to reach sufficient accuracy is displayed in the lower part (c) of Fig. 3 depending on degree and order of ALFs. According to Fig. 3, computational efficiency of individual HG functions differs in both number of iterations (j) and its dependency on degree and order of ALFs.

Described results show overflow/underflow of all functions beyond the limits of double precision for points lying inside and outside the reference spheroid. Overflow/underflow of values was observed even for higher values of flattening. Experiment also confirmed dependence of the computational efficiency on degree, order and selected type of HG function. These properties must be considered before further utilization of HG functions.

3.2 Implementation of Non-singular Expressions

To test correctness of the non-singular solution using recursions, it was necessary to compare results with original singular solution, e.g., in form of Eq. (1). Since the second (inner) part of both singular and non-singular equations was not yet implemented, results were compared only for the first (outer) part.

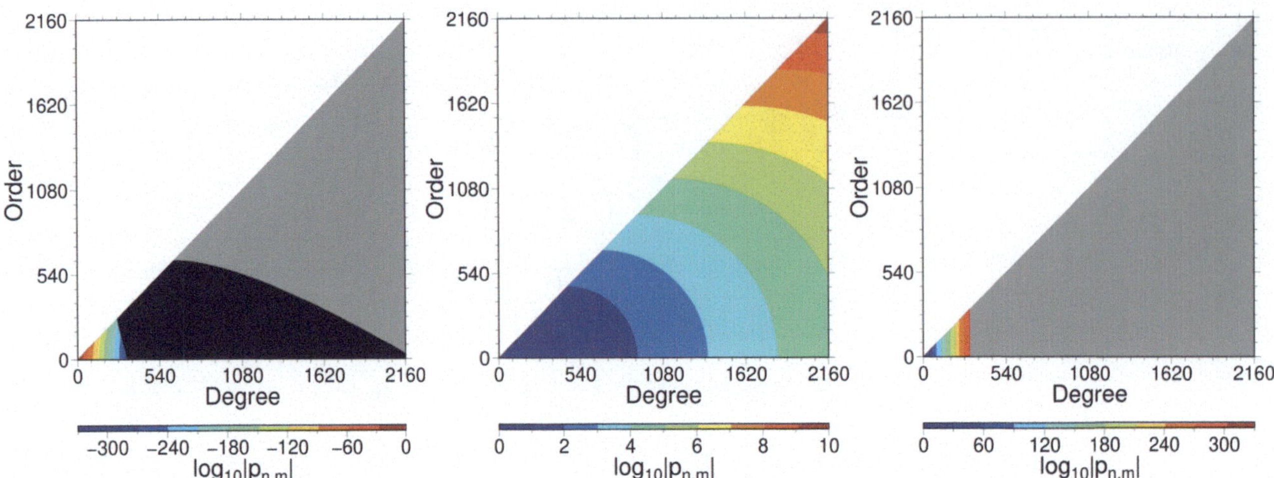

a) Range of value order of magnitude for points inside (left), on (center) and outside (right) the reference spheroid (black and grey zones denote underflow or overflow values).

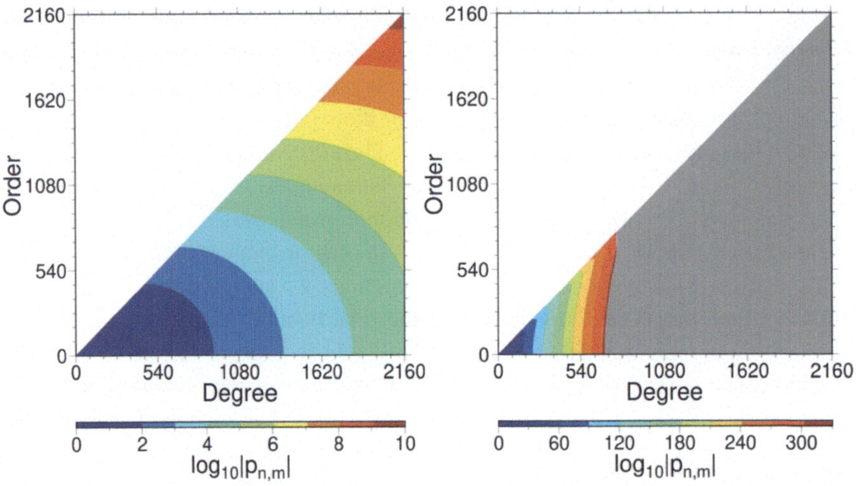

b) Range of value order of magnitude for different values of flattening, namely for $f = 0.01$ (left) and $f = 0.60$ (right).

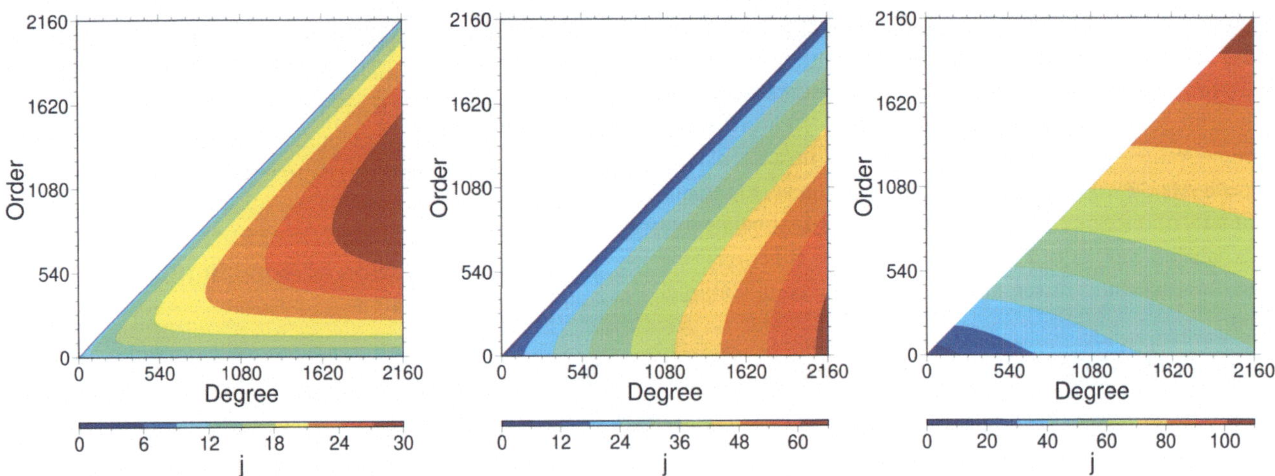

c) Computational efficiency (represented by number of iterations j) of selected types of HG functions described in Tab. 1, namely type 6 (left), 1 (center) and 4 (right).

Fig. 3 Degree and order dependent properties of selected HG functions with respect to their numerical properties and computational efficiency

Table 2 Statistics of the differences between the non-singular and singular expressions over non-polar region. All values are given in $10^{-27}\mathrm{m}^{-1}\mathrm{s}^{-2}$ (signal magnitude is at the order of $10^{-15}\mathrm{m}^{-1}\mathrm{s}^{-2}$)

Derivative	Mean	Std	Min	Max	Derivative	Mean	Std	Min	Max
V_{uuu}	0.01	8.87	−111.40	94.96	$V_{u\lambda\lambda}$	0.00	3.04	−35.00	36.08
$V_{uu\beta}$	−0.03	7.99	−85.15	100.10	$V_{\beta\beta\beta}$	−0.02	7.20	−93.04	70.00
$V_{uu\lambda}$	0.00	3.86	−42.75	45.13	$V_{\beta\beta\lambda}$	0.00	1.91	−25.95	20.25
$V_{u\beta\beta}$	−0.02	7.54	−95.10	111.53	$V_{\beta\lambda\lambda}$	0.00	1.44	−19.06	17.14
$V_{u\beta\lambda}$	0.00	2.38	−36.46	31.51	$V_{\lambda\lambda\lambda}$	0.00	2.79	−66.40	68.83

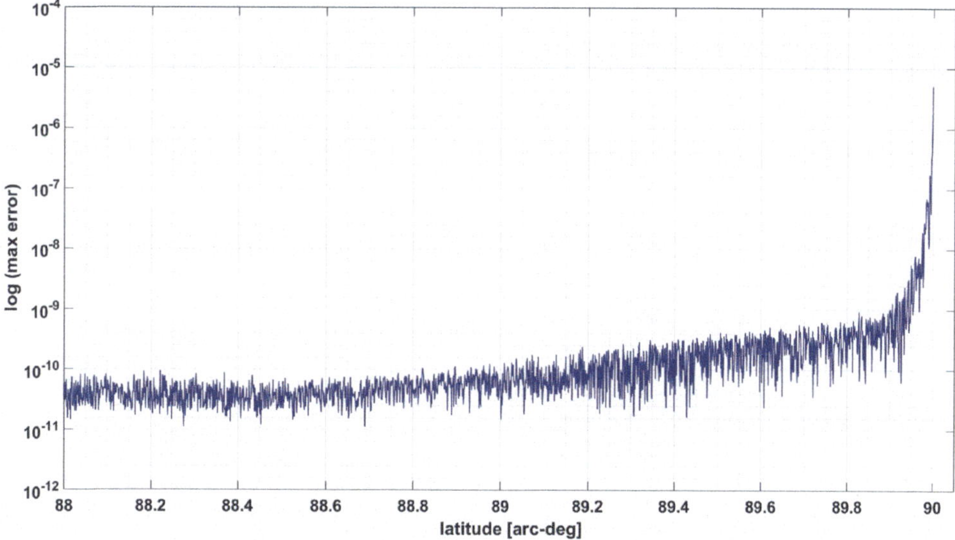

Fig. 4 Differences between the non-singular and singular expressions for $V_{\lambda\lambda\lambda}$ close to the poles. Values are given in $10^{-15}\mathrm{m}^{-1}\mathrm{s}^{-2}$. y axis represents common logarithm of maximal error observed for given latitude

The basic assumption was that comparison between singular and non-singular expressions should indicate very low differences with exception of polar regions, where the singular solution accuracy was expected to decrease rapidly.

Firstly, it was necessary to compare results over non-polar region $-80° \leq \beta \leq 80°$. The gravitational field quantities were synthesized from EGM2008 (Pavlis et al. 2012) to the maximal degree $n = 360$ over the $0.1°$ equiangular grid. The differences were studied for 10 third-order derivatives of the gravitational potential. Resulting statistics are available in Table 2.

Secondly, the rapid increase of differences in regions close to the poles showing innacuracies in singular solution had to be proven. To observe described discrepancies, we have examined differences along meridians (with 1° step between meridians) converging to the North Pole ($88° \leq \beta < 90°$). Example of above-mentioned discrepancies for the derivative $V_{\lambda\lambda\lambda}$ is depicted in Fig. 4. Similar behaviour was observed for the other derivatives with singularities in original expressions.

Results of described experiment confirmed correctness of derived non-singular equations and their software implementation over non-polar regions, where the comparison with singular solution showed only negligible discrepancies. On the other hand, the increase of differences between singular and non-singular expressions over the polar regions complies with assumption about singular solution limitations.

4 Conclusions

In this contribution, we have presented a theory concerning modelling of gravitational field using spheroidal harmonic functions.

We have discussed possible utilization of HG functions with respect to the computation of ALFs and evaluated numerical aspects of this utilization with emphasis on numerical properties and computational efficiency in the first part of numerical experiment. We encountered problems with overflow/underflow of values beyond double precision at points both inside and outside the minimal Brillouin spheroid and for higher values of spheroidal flattening.

Next part of this contribution focused on derivation of non-singular expressions using recursions and thus solving problem with singularities in original equations. After the derivation, expressions were implemented into C language scripts (available on request) and validated in the second part of numerical experiment. Close similarity of values over non-polar region and rapid increase of differences between singular and non-singular solution over the polar areas is in accordance with initial assuption about singular solution limitations.

Regarding future development, implementation of the theory concerning lower part of Eq. (1) and of all analogical equations for other derivatives is planned together with the implementation of non-singular equivalents.

Acknowledgements Authors were supported by the project No. 23-07031S of the Czech Science Foundation. The authors thank the reviewers and editors for their time, constructive suggestions and comments.

References

Freeden W, Michel V (2004) Multiscale potential theory (with applications to geoscience). Birkhäuser Verlag, Boston, USA. https://doi.org/10.1007/978-1-4612-2048-0

Fukushima T (2012) Numerical computation of spherical harmonics of arbitrary degree and order by extending exponent of floating point numbers. J Geodesy 86:271–285. https://doi.org/10.1007/s00190-011-0519-2

Hamáčková E, Šprlák M, Pitoňák M, Novák P (2016) Non-singular expressions for the spherical harmonic synthesis of gravitational curvatures in a local north-oriented reference frame. Comput Geosci 88:152–162. https://doi.org/10.1016/j.cageo.2015.12.011

Heiskanen WA, Moritz H (1967) Physical geodesy. Freeman and Co., San Francisco, USA

Hobson EW (1965) The theory of spherical and ellipsoidal harmonics. Chelsea Publishing Company, New York, USA

Hu X, Jekeli C (2015) A numerical comparison of spherical, spheroidal and ellipsoidal harmonic gravitational field models for small non-spherical bodies: examples for the Martian moons. J Geodesy 89:159–177. https://doi.org/10.1007/s00190-014-0769-x

Marchi S, Raymond CA, Russell CT (2022) Vesta and Ceres: Insights from the dawn mission for the origin of the solar system. Cambridge University Press, Cambridge. https://doi.org/10.1017/9781108856324

Novák P, Šprlák M, Tenzer R, Pitoňák M (2017) Integral formulas for transformation of potential field parameters in geosciences. Earth Sci Rev 164:208–231. http://doi.org/10.1016/j.earscirev.2016.10.007

Novák P, Šprlák M (2018) Spheroidal integral equations for geodetic inversion of geopotential gradients. Surv Geophys 39:245–270. https://doi.org/10.1007/s10712-017-9450-2

Ophaug V, Gerlach C (2017) On the equivalence of spherical splines with least-squares collocation and Stokes's formula for regional geoid computation. J Geodesy 91:1367–1382. https://doi.org/10.1007/s00190-020-01375-7.

Pavlis NK, Holmes SA, Kenyon SC, Factor JK (2012) The development and evaluation of the Earth Gravitational Model 2008 (EGM2008). J Geophys Res (Solid Earth) 117:B04406. https://doi.org/10.1029/2011JB008916

Šprlák M, Han SC (2021) On the use of spherical harmonic series inside the minimum Brillouin sphere: Theoretical review and evaluation by GRAIL and LOLA satellite data. Earth Sci Rev 222:103739. https://doi.org/10.1016/j.earscirev.2021.103739

Gravity Signal Variations Implied by a Dynamic Polyhedral Modelling

Georgia Gavriilidou and Dimitrios Tsoulis

Abstract

The melted ice layer between epochs 2009 and 2016 of a part of the Vernagtferner glacier located in the Austrian Alps is modelled dynamically. Three deterministic and two stochastic approaches are applied to compute the induced gravity signal differences. For the deterministic approach, the applied algorithms include the analytical method of right rectangular prism, the line integral analytical solution of general polyhedron and the numerical solution of fully normalized spherical harmonics series expansion. For the stochastic approach, the glacier is approximated as an uncertain general polyhedron of variable shape and as a summation of smaller variable polyhedral masses, where the melted body is expressed in the form of a covariance matrix. The implemented stochastic algorithm, based on the variance propagation law, uses the partial derivatives of the corresponding gravitational functionals with respect to the polyhedral vertex coordinates to derive gravity signal variations implied by the shape changes as described in the covariance matrix. The numerical tests were performed on a set of 20 points which are in fact observation sites of a local gravity survey network. The relative differences of the estimated variations considering the glacier as one polyhedron from the line integral analytical solution are below 7% for gravitational potential, up to 50% for its first and up to 86% for its second order derivatives. These differences between stochastic approach using individual polyhedral elements and analytical solution of a general polyhedron range up to 0.25%, 40% and 60% respectively. Prismatic representation derives relative differences with respect to the analytical general polyhedron up to 53%, 23% and 50% for gravitational potential, its first and second order derivatives, while the differences of harmonic series from the analytical polyhedral solution are 0.35%, 18% and 82% respectively.

Keywords

Dynamic modelling · Gravity signal variations · Right rectangular prism · Spherical harmonic coefficients · Stochastic general polyhedra

1 Introduction

The dynamic modelling of mass changes is an interdisciplinary topic of geodesy. It comprises the evaluation of their induced gravity signal at various time periods and is of interest for investigations such as global mass balance estimation (Swenson and Wahr 2002), water equivalent definition (Boy et al. 2012), satellite orbit determination (Philipp et al. 2018) and geoid heights computation (Novák et al. 2021).

The typical approximation of a time varying gravity field is performed through its continuous deterministic representation over different time periods. Differences between the obtained solutions allow the quantification of gravitational

G. Gavriilidou (✉) · D. Tsoulis
Department of Geodesy and Surveying, Aristotle University of Thessaloniki, Thessaloniki, Greece
e-mail: georgiaga@topo.auth.gr

J. T. Freymueller, L. Sànchez (eds.), *International Symposium on Gravity, Geoid and Height Systems 2024 (GGHS2024)*, International Association of Geodesy Symposia 158, https://doi.org/10.1007/1345_2025_305

effect generated by mass changes (Sośnica et al. 2015). A different approach is implemented by the stochastic representation of the considered mass distribution. In this case, the shape changes are represented as shape uncertainties and the sought gravity signal differences are calculated by applying the variance propagation law to the corresponding gravitational functional expressions (Brandt 2014). The obtained stochastic gravity field is associated with the variability of the examined distribution.

The term stochastic shape refers to a body whose geometry is varying (Teunissen 2000; Hristopulos 2020). It is characterized by the first two stochastic moments, namely the expected value, expressing here the actual shape and the variance which provides the geometry uncertainties (Meissl 1982; Koch 2007). By applying the variance propagation law on specific gravitational functionals, the probability density function of the shape is transformed to the sought gravitational values. The obtained gravity signal variations represent the gravitational functional differences implied by the modelled shape uncertainties. Some investigations that evaluate the effect of the propagated uncertainties of stochastic bodies on the gravitational field generated by them include Muinonen (1998), Muinonen et al. (1997), Muinonen and Lagerros (1998), Muinonen and Pieniluoma (2011), Petrov and Kiselev (2019), Bercovici and McMahon (2019) and Bercovici et al. (2020).

The main goal of the present research is to compare quantitatively various gravitational simulation methods in the frame of dynamic modelling. Several deterministic and stochastic techniques are evaluated. The motivation is to demonstrate that stochastic modelling is a suitable tool for the representation of mass transport in comparison with other well-established approaches.

The present contribution follows the stochastic analysis algorithm proposed by Panicucci et al. (2020) to compute variations on gravitational potential and its derivatives up to second order implied by specific shape changes. The method is based on the expression of the corresponding functionals by means of spherical harmonics using the normalized harmonic coefficients of Werner (1997). The approach is applied to a real case scenario, dynamically modelling the melted ice layer of a part of the Austrian glacier Vernagtferner between 2009 and 2016. The stochastic results are compared with the deterministic evaluation of gravitational potential and its up to second order derivatives. Three algorithms have been used for the deterministic approach, the spherical harmonic series expression (Jamet and Tsoulis 2020), the line integral analytical approach of general polyhedra (Tsoulis 2012; Tsoulis and Gavriilidou 2021) and the analytical solution of a right rectangular prism (Mader 1951). The numerical implementation is applied to a set of 20 computation points placed 20 cm above the surface and outside the ice layer borderline (Fig. 1), which serve as gravity stations for corresponding gravimetric campaigns (Gerlach et al. 2017; Gavriilidou et al. 2024).

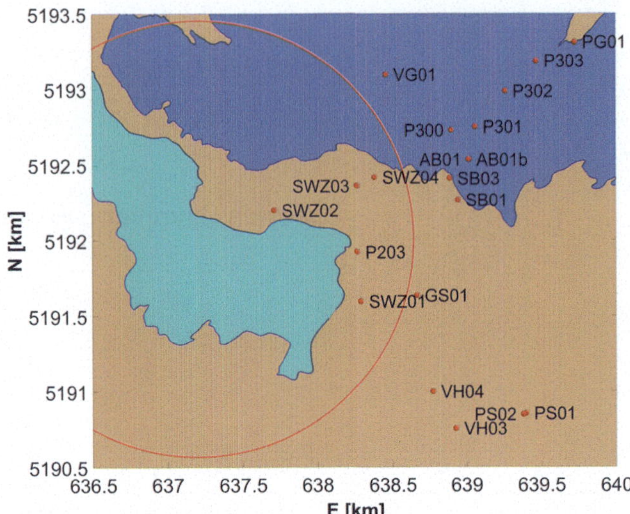

Fig. 1 Computation points with respect to the Vernagtferner glacier boundary (blue line) and Brillouin sphere related to the single polyhedral shape model applied by the stochastic technique (red line). The brown shade represents the rock masses, the cyan the modelled glacier and the dark blue the central part of Vernagtferner glacier, which is not considered in the present contribution

2 Theoretical Background

The gravitational potential induced by a finite homogeneous 3D mass distribution reads (Kellogg 1929)

$$V = G\rho \iiint_U \frac{dU}{\ell}, \tag{1}$$

where ρ is the density, U the total volume, G the Newtonian gravitational constant, dU the infinitesimal volume element and ℓ the distance between computation and integration points.

There are two ways of evaluating the volume integral in Eq. (1). Either by applying a numerical method that replaces it with an infinite series, or by computing it analytically. The expression of Eq. (1) in spherical harmonics reads (Heiskanen and Moritz 1967)

$$\begin{aligned} V(r,\theta,\lambda) &= \tfrac{GM}{r} \textstyle\sum_{n=0}^{\infty} \left(\tfrac{a}{r}\right)^n \sum_{m=0}^{n} \overline{P}_{n,m}(\cos\theta) \\ &\quad \times \left(\overline{C}_{n,m}\cos(m\lambda) + \overline{S}_{n,m}\sin(m\lambda)\right), \end{aligned} \tag{2}$$

where (r,θ,λ) are the spherical coordinates of the computation point, M the total mass of the body, $\overline{P}_{n,m}$ the normalized associated Legendre functions and $\overline{C}_{n,m}, \overline{S}_{n,m}$

the normalized spherical harmonic coefficients. The variable a defines the radius of the Brillouin sphere, a minimum sphere that encloses the whole mass distribution. Numerical convergence for the harmonic series of Eq. (2) is secured only outside the Brillouin sphere (Moritz 1980).

Spherical harmonic expansions are usually applied to model gravity fields of extended bodies, such as the entire Earth. Despite Vernagtferner glacier is small, harmonic series still offer some crucial benefits, as the gravity signal is calculated globally, the needed computational time is reduced while the selection of maximum expansion degree provides tunable precision. Additionally, the harmonic coefficients are also participating in stochastic modelling, thus it is important to assess them independently.

The evaluated scenario of melted ice layer corresponds to a regional signal. However, here it is modelled as an independent three-dimensional shape, like any celestial body. For this mass distribution, potential spherical harmonic coefficients are calculated to represent its global gravity field. Therefore, the same limitations arise as in the case of modelling the entire Earth, e.g., the non-convergence issue inside Brillouin sphere.

3 Deterministic Modelling

The following algorithms provide solutions for the gravity signal by considering the evaluated shape model as deterministic. Thus, the monitoring of a time varying gravity field, especially of mass changes, is performed by computing differences between deterministic solutions.

Numerical and analytical methods are examined for the deterministic evaluations. Spherical harmonic series are implemented to numerically calculate the gravity field induced by the melted ice layer. For the analytical calculation of the same gravity signal, a generally shaped polyhedron and a summation of right rectangular prisms are implemented. The applied analytical methods are related to the fact that the evaluated mass distribution is defined by a DTM.

3.1 Spherical Harmonic Series

Regarding the numerical methods, various algorithms are available for computing coefficients $\overline{C}_{n,m}, \overline{S}_{n,m}$. Here we implement the recursive scheme proposed by Werner (1997) and the non-recursive formulas by Jamet and Tsoulis (2020). The algorithm of Werner (1997) is based on the division of the investigated body into smaller tetrahedra created by exclusively triangular faces. The derived coefficients participate in the stochastic method. Jamet and Tsoulis (2020) divide the distribution into smaller general polyhedra, while this time the computed coefficients are used to calculate the gravity signal induced by the evaluated mass change using spherical harmonic series.

3.2 Line Integral Analytical Solution of a General Polyhedron

For the analytical approach, two different modelling methods are considered. The first one models the 3D distribution as a general polyhedron and evaluates the induced gravity signal by means of the line integral approach (Tsoulis 2012; Tsoulis and Gavriilidou 2021). This shape is formed by smaller general polyhedra, associated with the cells of the applied DTMs. Thus, each polyhedron is defined by four neighboring vertices on the upper and four on the lower boundary of the melted ice layer, set by the two DTMs for the epochs 2009 and 2016 respectively. The eight vertices are connected, creating a total of twelve triangular faces.

3.3 Analytical Solution of a Right Rectangular Prism

The second analytical approach that is evaluated discretizes the distribution into a finite number of right rectangular prisms and computes their gravity signal analytically (Mader 1951; Nagy 1966; Nagy et al. 2000). The upper and lower boundaries of each prismatic element are defined by the elevation values included in the DTMs for 2009 and 2016 respectively. To maintain the horizontal dimensions of the modelled area, a center point is determined by cubic interpolation among four neighboring grid points on the upper and lower surface, that have been previously used to construct the polyhedral elements. These new points represent the elevation of the flat top and bottom rectangular faces of each prismatic element. The horizontal dimensions derive from the resolution of the used DTM.

4 Stochastic Modelling

To involve a geometric uncertainty in the gravity signal quantification, the shape model must be considered as a stochastic quantity (Benedek et al. 2018). In the present contribution stochastic modelling is applied to quantify alterations in the gravity field implied by mass transport providing a different perspective on forward modelling. Shape changes enter the process as an independent parameter in the form of a coordinate covariance matrix. Each element of this matrix contains information for each polyhedral vertex coordinate. Thus, individual mass changes referring to different regions of the evaluated shape model can be represented simultaneously.

Partial derivatives of the examined gravitational functionals remain constant for all investigations related to one shape model and can be used multiple times with different scenarios of covariance matrices to model various mass changes. The modelled coordinate variations are propagated directly into gravity variations without defining any intermediate parameter. This can provide insights into specific locations of the examined mass distribution that highly influence the final results.

For the estimation of gravitational potential variations implied by specific mass changes, the variance propagation law is applied to Eq. (2). The obtained variations are equivalent to the gravity signal differences induced by shape alterations in the investigated mass distribution. Geometry variability is expressed in the form of a covariance matric $\mathbf{P_c}$, which includes polyhedral vertex coordinate variations. The partial derivatives of Eq. (2) with respect to the polyhedral vertex coordinates $\frac{\partial V}{\partial \mathbf{C}}$ are computed as in Panicucci et al. (2020) and Gavriilidou and Tsoulis (2024) based on the formulation of Werner (1997) for the corresponding harmonic coefficients. The gravitational potential variations P_V are finally expressed as (Panicucci et al. 2020)

$$P_V = \left[\frac{\partial V}{\partial \mathbf{C}}\right] \mathbf{P_c} \left[\frac{\partial V}{\partial \mathbf{C}}\right]^{\mathrm{T}}, \tag{3}$$

where $\mathbf{C}$ is the vector that contains all polyhedral vertex coordinates. The algorithmic steps for calculating variations in first and second order derivatives of the gravitational potential are thoroughly presented in Gavriilidou and Tsoulis (2025a, 2025b).

Thereby, the vertex covariances inside $\mathbf{P_c}$ are defined as

$$\sigma_{i,j} = \varrho_{i,j}\sigma_i\sigma_j, \tag{4}$$

where indices i, j refer to the coordinates $i, j = x, y, z$, σ_i is the variation of coordinate i and $\varrho_{i,j}$ the correlation between i and j. It is important to stress that the coordinate variations $\sigma_x = |\Delta x|$, $\sigma_y = |\Delta y|$ and $\sigma_z = |\Delta z|$, are always positive and represent lateral and vertical spatial limits around the corresponding vertex. The coordinate covariance matrix is built by adopting a stochastic model. A detailed analysis on how each stochastic parameter that participate in the covariance matrix definition influence the accuracy of the obtained gravity signal variations is presented in Gavriilidou and Tsoulis (2025c).

In the present contribution, to build the coordinate covariance matrix and especially its non-diagonal elements, the vertices are considered as perfect positively or negatively correlated. Thus, they are characterized by $\varrho_{i,j} = 1$ or $\varrho_{i,j} = -1$. In this way, $\mathbf{P_c}$ represent a uniform vertex transmutation along the inner or outer space of the nominal shape. The correlation $\varrho_{i,j}$ sets a relative position difference between vertices i and j, with the two vertices moving towards the same direction or moving towards different directions for positive or negative $\varrho_{i,j}$ respectively (Brandt 2014; Shevlyakov and Oja 2016).

For the stochastic computation of the induced gravity signal (Panicucci et al. 2020), two approaches are adopted to model the examined ice layer: a single polyhedral body representing the entire ice layer and a division of the ice layer into individual polyhedra defined around four neighboring DTM points.

4.1 Single Polyhedral Representation

To apply the first modelling approach of the stochastic method and represent the ice loss layer as a single polyhedron, the lower boundary of the modelled body is set to a flat surface with an elevation of 2,910 m, while the upper boundary coincides with the actual glacier surface described by the 2016 DTM. A sketch illustrating the geometry of the shape model is shown in Fig. 2. The elevation differences between 2009 and 2016 enter the algorithm as vertex vertical position uncertainties.

The obtained gravity signal variations are expressed in a spherical harmonic series and thus are associated with the definition of the Brillouin sphere, i.e., a minimum sphere enclosing the whole synthetic body. The area modelled as a single polyhedron is illustrated in Fig. 1 with cyan color, while the perimeter of the corresponding Brillouin sphere is outlined with a red line.

4.2 Representation of Individual Polyhedra

The second approach of individual polyhedra consists of polyhedral shape volume elements filling continuously the volume of the glacier and intends to reduce the non-convergence area inside the Brillouin sphere. The upper boundary of each polyhedral element coincides again with one cell of 2016 DTM and the lower boundary is set as the surface created by the 2016 DTM minus 50 m (Fig. 2). Therefore, for each individual polyhedron a new smaller Brillouin sphere is generated. When the entire glacier is considered, this summation of small Brillouin spheres defines a non-convergence area that is close to the actual glacier surface, with respect to the extended Brillouin sphere derived from the single polyhedral representation of the evaluated body. Again, the ice melting between 2009 and 2016 enters the stochastic procedure as differences in z-coordinates for each cell. The computations are performed separately for all polyhedral divisions using the corresponding Brillouin spheres for each one of them. Despite that the melted ice layer is modelled using the same

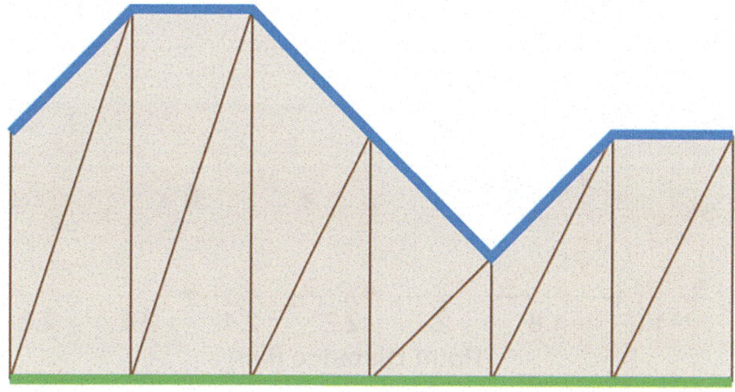

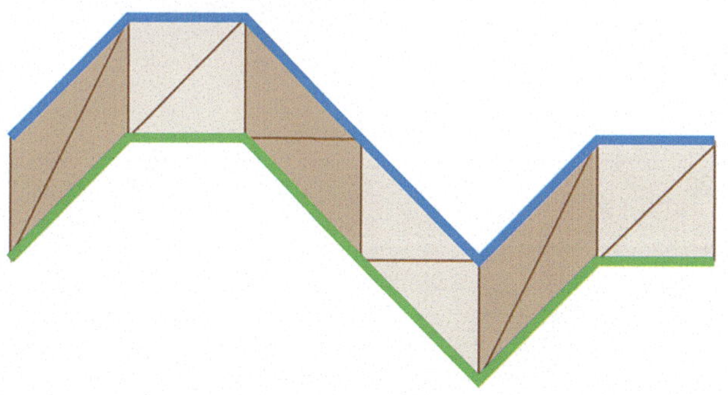

Fig. 2 Sketch that illustrates the geometry of the two shape models implemented by the stochastic approach. The upper model defines the single polyhedron representation, where the upper boundary (blue line) is the DTM surface for the epoch 2016 and the lower boundary (green line) is a surface with 2,910 m elevation. The second shape model defines the representation of individual polyhedra (each one colored with a different shade of brown) where the upper boundary (blue line) remains the same and the lower boundary (green line) coincides with a downward permutation of the DTM for the epoch 2016 by 50 m. The triangular faces participating in both polyhedral representations are outlined with the brown lines

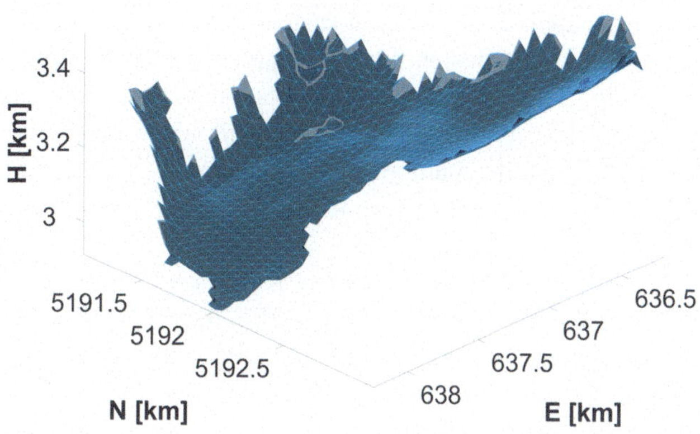

Fig. 3 Ice mass loss in Schwarzwand tongue of the Vernagtferner glacier between 2009 and 2016

polyhedral geometry, the results are affected by the decreased non-convergence area and the estimated variations near the masses are calculated with more precision.

5 Numerical Implementation

Vernagtferner glacier, located in the Austrian Alps, serves as one of the most observed ice bodies worldwide. Its mass changes are observed using a variety of glaciology, hydrometeorology, gravimetry and geodesy techniques (Cuffey and Paterson 2010). One way to quantify the ice mass loss is by comparing DTMs that represent the same area at different epochs. Here, two 6.25 km $\times$ 6.00 km DTMs with 51 m resolution, resampled from respective 1 m resolution DTMs, are used to evaluate ice melting between 2009 and 2016. The maximum elevation difference in this period is 48 m (Fig. 3). Due to the increased CPU time needed by the stochastic technique, computations are limited to a specific part of the glacier, called Schwarzwand tongue, illustrated with cyan color in Fig. 2.

The same test area has been used also in Gavriilidou et al. (2024). There, the entire Vernagtferner glacier, and especially the ice melting between 2009 and 2016, was modelled using general polyhedra and right rectangular prisms followed by an analytical computation of their gravity signal. The scope was to investigate different modelling approximations in the frame of a detailed gravity computation with high accuracy requirements using 1 m resolution DTMs. Here we focus only on the side part of the glacier and evaluate several numerical, analytical and stochastic methods for computing the induced gravity field. Special emphasis is given on different algorithmic processes using the same modelling tools and the benefits gained from each approach. The applied DTMs form processed version of the original DTMs with 1 m resolution. Thus, there is no overlap in the computations and discussion whatsoever, and a direct comparison with the previous results cannot be made.

Variations in gravitational potential and its up to second order derivatives induced by the melted ice between 2009 and 2016 are evaluated analytically, numerically and stochastically, using all of the aforementioned approaches. For the rest of the manuscript the line integral analytical solution of a general polyhedron will be referred to as method A, prismatic as method B and spherical harmonic expansions as method C. For the stochastic approach, the single polyhedral representation will be named as method D and the technique using individual polyhedral elements as method E.

The gravity signal is calculated for a set of 20 computation points, which are in fact actual gravity sites with available gravimetric observations (Fig. 1). The specific configuration of computation points has been adopted as a realistic scenario where the gravity signal of a modelled mass change needs to be evaluated. Their special location, that is very close to the masses, gives insights into the procedure of dynamic modelling. The numerical implementation is separated for points outside and points inside the Brillouin sphere. For points inside the Brillouin sphere only method A, B and E are compared, as the other techniques provide non-convergent results.

Figures 4, 5, 6, and 7 illustrate the relative differences between all implemented methods with respect to method A for gravitational potential and its first order derivatives. The results refer to the 15 points outside the Brillouin sphere (Fig. 1) with increasing distance from the center of the glacier shape model.

Concerning the gravitational potential (Fig. 4), methods E and C indicate same values of relative differences under 0.4%, while method B around 5%. The results of methods D and E provide smaller differences than method B and are highly affected by the location of the computation point. For the horizontal component of the first order derivatives of the gravitational potential V_x, V_y (Figs. 5 and 6) method E provide overall better results in terms of convergence with method A, with the relative differences remaining under 0.5%. The variations derived by applying method D decline most profoundly for V_x, V_y, while as the distance increases their relative differences with respect to method A tend to get smaller approaching the values provided by method B. The vertical component of the first order derivatives indicates a more divergent behavior (Fig. 7), where no approach seems to provide overall better results in terms of convergence with method A. The same plots for V_{xz}, V_{yz} and V_{zz} are shown in Figs. 8, 9, and 10.

For both V_{xz} and V_{yz} components, a clear indication of which method provides the best results is not shown. For V_{xz} the differences between methods A and D range up to 40%, with an exception for one point, while for V_{yz} up to almost 90%. The results for V_{zz} indicate a uniformity with respect to the increasing distance of the computation point. Specifically, the relative differences of method A and C range between 0.1% and 3%, while the differences of A from B between 5.25% and 8.45%. The stochastic variations

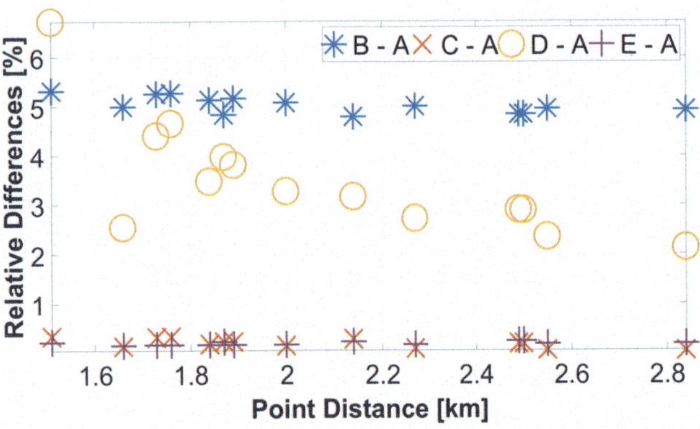

Fig. 4 Relative differences of potential variations with respect to the analytical solution of the line integral analytical approach (method A) due to ice melting between 2009 and 2016 at 15 computation points outside the Brillouin sphere with increasing distance from the center of mass

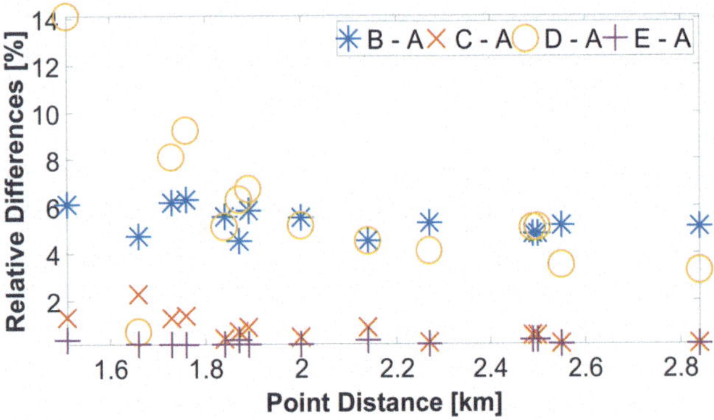

Fig. 5 Same as Fig. 4 for the horizontal component of the first order derivatives of gravitational potential V_x

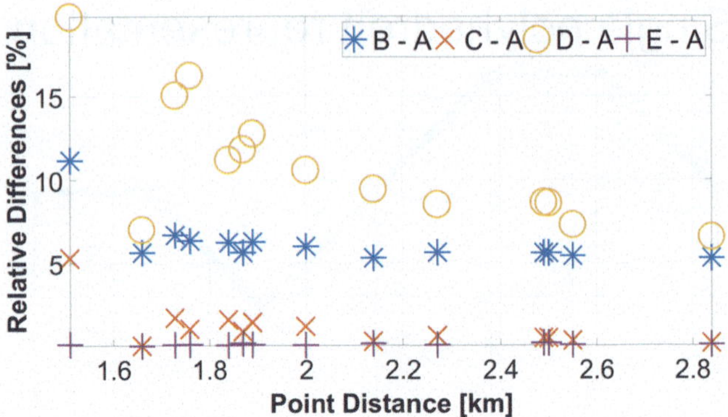

Fig. 6 Same as Fig. 4 for the horizontal component of the first order derivatives of gravitational potential V_y

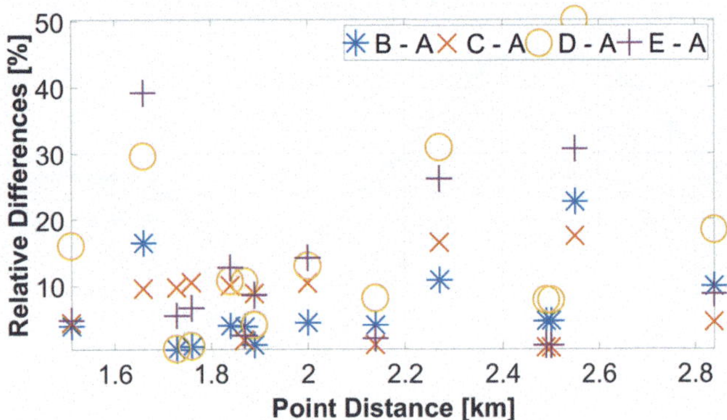

Fig. 7 Same as Fig. 4 for the vertical component of the first order derivatives of gravitational potential V_z

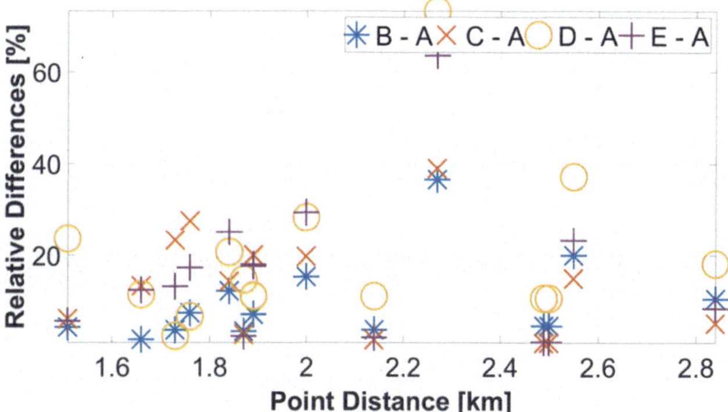

Fig. 8 Same as Fig. 4 for the second order derivatives of gravitational potential V_{xz}

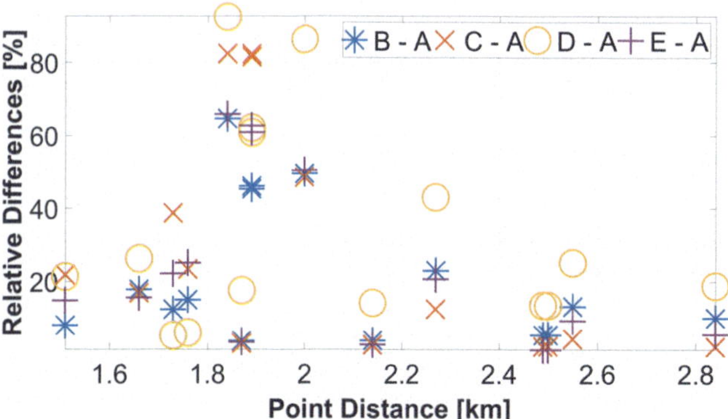

Fig. 9 Same as Fig. 4 for the second order derivatives of gravitational potential V_{yz}

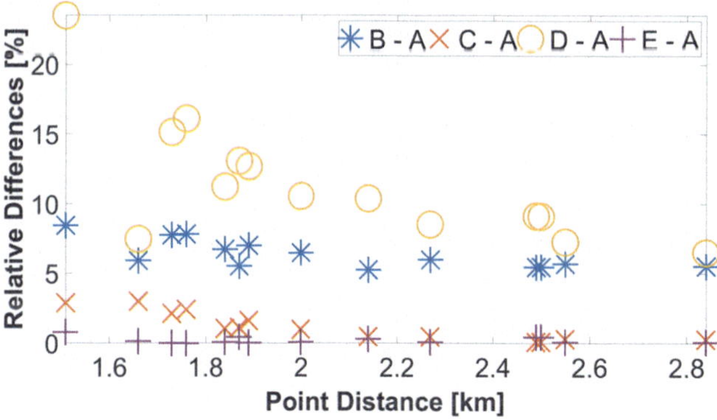

Fig. 10 Same as Fig. 4 for the second order derivatives of gravitational potential V_{zz}

for method D range from 6.5% to 23.5% while for E these differences reduce to 0.01% up to 0.8%.

For points located inside the Brillouin sphere, only results from methods E, B and A are compared. Table 1 shows the results using method A for the melted ice layer, their relative differences from B and E for the potential and its up to first order derivatives, while Table 2 gives the same comparisons for the second order derivatives.

Table 1 Gravity signal differences implied by glacier melting in Vernagtferner between 2009 and 2016 using method A at the first rows of each point in units m^2/s^2 and m/s^2. Second row refers to relative differences of method A from B in %, while the third row differences from E

Name	Method	V	V_x	V_y	V_z
SWZ02	A	1.12E-03	2.31E-07	2.01E-06	2.13E-07
	A – B [%]	5.18	12.59	7.95	5.87
	A – E [%]	0.16	9.69	0.10	13.89
P203	A	9.85E-04	2.44E-06	1.21E-07	4.41E-07
	A – B [%]	8.48	16.62	242.79	1.99
	A – E [%]	0.36	1.12	3.22	12.02
SWZ03	A	6.69E-04	5.51E-07	5.96E-07	4.99E-08
	A – B [%]	5.74	5.20	9.41	6.17
	A – E [%]	0.15	0.14	0.01	11.53
SWZ01	A	8.39E-04	1.39E-06	5.99E-07	2.71E-07
	A – B [%]	5.74	8.89	12.85	5.30
	A – E [%]	0.27	0.79	0.44	8.03
SWZ04	A	5.84E-04	4.48E-07	3.99E-07	2.16E-08
	A – B [%]	5.58	5.91	8.60	12.43
	A – E [%]	0.14	0.07	0.05	18.65

Table 2 Same as Table 1 for potential second order derivatives in s^{-2}

Name	Method	V_{xz}	V_{yz}	V_{zz}
SWZ02	A	1.84E-09	4.52E-10	1.06E-08
	A – B [%]	5.31	3.90	13.00
	A – E [%]	19.30	8.23	1.53
P203	A	4.43E-09	4.46E-10	1.33E-08
	A – B [%]	6.28	292.96	43.42
	A – E [%]	15.90	41.92	3.16
SWZ03	A	1.08E-10	1.35E-11	1.41E-09
	A – B [%]	5.25	96.69	10.90
	A – E [%]	22.52	105.84	0.29
SWZ01	A	2.19E-09	3.91E-10	4.26E-09
	A – B [%]	14.53	28.75	20.76
	A – E [%]	8.92	25.23	2.91
SWZ04	A	2.66E-11	2.88E-11	8.12E-10
	A – B [%]	33.16	76.75	9.68
	A – E [%]	55.69	62.82	0.41

The behavior of the relative differences follows the same patterns as for the points outside the Brillouin sphere. For the gravitational potential, method E provides results under 0.4% while the differences for B range between 5.2 – 8.5% against A. For first order derivatives V_x and V_y, method E derives smaller relative differences. This changes for V_z, where B is more accurate. Nevertheless, the differences from both approaches are at the same order of magnitude. For V_{xz} and V_{yz} method E provides the highest differences. The V_{zz} component indicates the smallest differences for E with respect to V_{xz} and V_{yz}, probably because the corresponding signal values are higher than the rest.

6 Discussion

The stochastic approach results are based on the expression of gravitational potential and its up to second order derivatives into spherical harmonic series. Two modelling techniques are implemented to represent the ice layer as a single polyhedron or as a summation of individual polyhedra. Overall, a better fit between the stochastic results and the analytical polyhedral solution has been obtained for the gravitational potential V and its second order vertical derivatives V_{zz}. Especially the individual polyhedra approach provide results of same or better accuracy than the analytical right rectangular prism. It must be stressed that the large differences between the different methods are expected, as the computation points are located close to the topographic surface and the gravitational masses.

More test cases need to be evaluated to establish the exact influence of the relative position of the computation points with respect to the examined shape model on the obtained gravity signal variations. In addition, more shape models of different volumes need to be modelled in order to find a threshold of shape changes where the stochastic representation can derive more accurate results.

Acknowledgments The first author (GG) was supported by the Hellenic Foundation for Research and Innovation (HFRI) under the fourth Call for HFRI PhD Fellowships (Fellowship Number: 10834). We would like to thank Prof. Gábor Papp and an anonymous reviewer for their detailed comments, that helped us improve the clarity of the presentation. The toolbox SHBUNDLE was used for the computation of the associated Legendre functions. Its latest version is provided via download from the Institute of Geodesy (GIS), University of Stuttgart: https://www.gis.uni-stuttgart.de/en/research/downloads/shbundle/

References

Benedek J, Papp G, Kalmár J (2018) Generalization techniques to reduce the number of volume elements for terrain effect calculations in fully analytical gravitational modelling. J Geodesy 92:361–381. https://doi.org/10.1007/s00190-017-1067-1

Bercovici B, McMahon JW (2019) Inertia parameter statistics of an uncertain small body shape. Icarus 328:32–44. https://doi.org/10.1016/j.icarus.2019.02.016

Bercovici B, Panicucci P, McMahon J (2020) Analytical shape uncertainties in the polyhedron gravity model. Celest Mechan Dyn Astron 132(5):29. https://doi.org/10.1007/s10569-020-09967-3

Boy J-P, Hinderer J, de Linage C (2012) Retrieval of large-scale hydrological signals in Africa from GRACE time-variable gravity fields. Pure Appl Geophys 169(8):1373–1390. https://doi.org/10.1007/s00024-011-0416-x

Brandt S (2014) Data analysis. Springer International Publishing. https://doi.org/10.1007/978-3-319-03762-2

Cuffey KM, Paterson WSB (2010) The physics of glaciers, 4th edn. Elsevier, Oxford

Gavriilidou G, Tsoulis D (2024) Dynamical evaluation of gravity spherical harmonic coefficients due to generally shaped polyhedra. In: International Association of Geodesy Symposia. Springer. https://doi.org/10.1007/1345_2024_256

Gavriilidou G, Tsoulis D (2025a) Stochastic modelling of polyhedral gravity signal variations. Part I: First-order derivatives of gravitational potential. J Geodesy 99(3):22. https://doi.org/10.1007/s00190-025-01937-7

Gavriilidou G, Tsoulis D (2025b) Stochastic modelling of polyhedral gravity signal variations. Part II: Second-order derivatives of gravitational potential. J Geodesy 99(2):16. https://doi.org/10.1007/s00190-025-01938-6

Gavriilidou G, Tsoulis D (2025c) Implementation of different stochastic models in the frame of a dynamic polyhedral gravitational approach. GEM Int J Geomathem 16(1):15. https://doi.org/10.1007/s13137-025-00272-5

Gavriilidou G, Gerlach C, Tsoulis D (2024) Analytical computation of local gravitational effects of mountain glacier mass change from polyhedral and prismatic modeling – test case Vernagtferner, Austrian Alps. Global Planet Change 234:104378. https://doi.org/10.1016/j.gloplacha.2024.104378

Gerlach C, Ackermann C, Falk R, Lothhammer A, Reinhold A (2017) Gravimetric investigations at Vernagtferner. In: Vergos G, Pail R, Barzaghi R (eds) International Symposium on Gravity, Geoid and Height Systems 2016. International Association of Geodesy Symposia, vol 148. Springer, Cham. https://doi.org/10.1007/1345_2017_2

Heiskanen WA, Moritz H (1967) Physical Geodesy. San Francisco, CA, W. H. Freeman and Company

Hristopulos DT (2020) Random fields based on local interactions. Springer, Dordrecht. https://doi.org/10.1007/978-94-024-1918-4_7

Jamet O, Tsoulis D (2020) A line integral approach for the computation of the potential harmonic coefficients of a constant density polyhedron. J Geodesy 94(3):30. https://doi.org/10.1007/s00190-020-01358-8

Kellogg OD (1929) Foundations of potential theory. Verlag von Julius Springer

Koch KR (2007) Introduction to bayesian statistics. Springer, Berlin, Heidelberg. https://doi.org/10.1007/978-3-540-72726-2

Mader K (1951) Das Newtonsche Raumpotential prismatischer Körper und seine Ableitungen bis zur dritten Ordnung. Sonderheft 11 der Österreichischen Zeitschrift für Vermessungswesen. Österreichischer Verein für Vermessungswesen, Wien

Meissl P (1982) Least squares adjustment: a modern approach. Technischen Universität Graz

Moritz H (1980) Advanced physical geodesy. Abacus Press

Muinonen K (1998) Introducing the Gaussian shape hypothesis for asteroids and comets. Astron Astrophys 332:1087–1098

Muinonen K, Lagerros JSV (1998) Inversion of shape statistics for small solar system bodies. Astron Astrophys 333:753–761

Muinonen K, Pieniluoma T (2011) Light scattering by Gaussian random ellipsoid particles: first results with discrete-dipole approximation. J Quant Spectrosc Radiat Transfer 112(11):1747–1752. https://doi.org/10.1016/j.jqsrt.2011.02.013

Muinonen K, Lamberg L, Fast P, Lumme K (1997) Ray optics regime for Gaussian random spheres. J Quant Spectrosc Radiat Transfer 57(2):197–205. https://doi.org/10.1016/S0022-4073(96)00127-6

Nagy D (1966) The gravitational attraction of a right rectangular prism. Geophysics 31(2):362–371. https://doi.org/10.1190/1.1439779

Nagy D, Papp G, Benedek J (2000) The gravitational potential and its derivatives for the prism. J Geodesy 74(7–8):552–560. https://doi.org/10.1007/s001900000116

Novák P, Šprlák M, Pitoňák M (2021) On determination of the geoid from measured gradients of the Earth's gravity field potential. Earth-Sci Rev 221:103773. https://doi.org/10.1016/j.earscirev.2021.103773

Panicucci P, Bercovici B, Zenou E, McMahon J, Delpech M, Lebreton J, Kanani K (2020) Uncertainties in the gravity spherical harmon-

ics coefficients arising from a stochastic polyhedral shape. Celest Mechan Dyn Astron 132(4):23. https://doi.org/10.1007/s10569-020-09962-8
Petrov DV, Kiselev NN (2019) Conjugated gaussian random particle model and its applications for interpreting cometary polarimetric observations. Solar Syst Res 53(4):294–305. https://doi.org/10.1134/S0038094619040075
Philipp D, Woeske F, Biskupek L, Hackmann E, Mai E, List M, Lämmerzahl C, Rievers B (2018) Modeling approaches for precise relativistic orbits: analytical, Lie-series, and pN approximation. Adv Space Res 62(4):921–934. https://doi.org/10.1016/j.asr.2018.05.020
Shevlyakov GL, Oja H (2016) Robust correlation. Wiley. https://doi.org/10.1002/9781119264507
Sośnica K, Jäggi A, Meyer U, Thaller D, Beutler G, Arnold D, Dach R (2015) Time variable Earth's gravity field from SLR satellites. J Geodesy 89(10):945–960. https://doi.org/10.1007/s00190-015-0825-1
Swenson S, Wahr J (2002) Estimated effects of the vertical structure of atmospheric mass on the time-variable geoid. J Geophys Res 107(B9):ETG 4-1–ETG 4-11. https://doi.org/10.1029/2000JB000024
Teunissen PJG (2000) Adjustment theory: an introduction. Delft, The Netherlands, VSSD
Tsoulis D (2012) Analytical computation of the full gravity tensor of a homogeneous arbitrarily shaped polyhedral source using line integrals. Geophysics 77(2):F1–F11. https://doi.org/10.1190/geo2010-0334.1
Tsoulis D, Gavriilidou G (2021) A computational review of the line integral analytical formulation of the polyhedral gravity signal. Geophys Prospect 69(8–9):1745–1760. https://doi.org/10.1111/1365-2478.13134
Werner RA (1997) Spherical harmonic coefficients for the potential of a constant-density polyhedron. Comput Geosci 23(10):1071–1077. https://doi.org/10.1016/S0098-3004(97)00110-6

Height Anomaly Determination Using a Tile-Based LSC Approach: Program Development and Case Study

Neda Darbeheshti and Jack McCubbine

Abstract

Height anomaly computation from gravity anomalies follows a *remove-compute-restore* approach, where global gravity and terrain effects are removed, the height anomaly is predicted, and these effects are restored to obtain the final model. This study presents a methodology for height anomaly determination across Australia, employing least squares collocation (LSC) for prediction.

The LSC process involves two main steps: first, generating a gravity anomaly grid for the region, and second, using these gridded anomalies to compute height anomaly heights. Gravity data sources in Australia include terrestrial datasets, satellite altimetry, and airborne gravimetry and gradiometry. This study focuses on LSC computations for New South Wales, integrating these diverse data types to produce a gridded height anomaly using the *remove-compute-restore* technique.

Our results highlight the effectiveness of LSC in height anomaly modeling, demonstrating its capability to merge multiple datasets into a precise and regionally tailored height anomaly solution.

Keywords

Airborne gravimetry and gradiometry · Height anomaly · Least squares collocation

1 Introduction

The height anomaly, denoted by ζ, represents the difference between the ellipsoidal height (h) of a point and its normal height H. It is defined by the following relation:

$$\zeta = h - H$$

This project introduces a MATLAB-based approach for regional height anomaly determination using least squares collocation (LSC). Leveraging MATLAB's capabilities, we incorporate additional covariance models, particularly for gravity gradients, enhancing height anomaly modeling accuracy.

Our work focuses on two key aspects:

- **Topographic Corrections:** We refine Forsberg's terrain correction method (Forsberg and Tscherning 2014) using Nagy's formula (Nagy et al. 2000), introducing a modification to reduce aliasing effects in high-frequency components.
- **Least Squares Collocation:** LSC enables height anomaly computation from gravity anomalies and gradients via covariance functions and matrix operations, offering a robust geodetic framework.

This paper details our approach, building on Tscherning's work and studies from Ohio State University, with the broader aim of advancing international height anomaly accuracy to the one-centimeter level.

N. Darbeheshti (✉) · J. McCubbine
National Geodesy, Geoscience Australia, Canberra, ACT, Australia
e-mail: neda.darbeheshti@ga.gov.au; jack.mccubbine@ga.gov.au

J. T. Freymueller, L. Sànchez (eds.), *International Symposium on Gravity, Geoid and Height Systems 2024 (GGHS2024)*, International Association of Geodesy Symposia 158, https://doi.org/10.1007/1345_2025_293

2 Method

In summary, computing a height anomaly from gravity data (anomalies and gradients) can be conceptualized as a sequence of "remove-compute-restore" operations. First, a global gravity model (GGM) and terrain effects are removed, a height anomaly is predicted, and then the effects are restored to obtain the final height anomaly model. This is achieved through the following steps:

- **Remove:** Subtract the GGM and terrain effects from the observed gravity.
- **Compute:** Convert gravity data to the height anomaly using least squares collocation.
- **Restore:** Add back the terrain and GGM effects to obtain the final height anomaly.

2.1 Topographic Effects

Topographic effects are divided into two components based on the DEMs used: long and full wavelengths. Initially, the visible topographic correction is computed using the full digital elevation model (DEM). The smoothed long-wavelength DEM topography is then calculated using a filter and subtracted:

$$\delta g_{TE} = \delta g_{TE}^{full} - \delta g_{TE}^{long} \tag{1}$$

To compute the vertical component of the gravitational potential due to a rectangular prism at a point outside the prism, this calculation follows equation (8) in Nagy et al. (2000), which yields the full topographic effect for gravity:

$$\delta g_{TE}^{full} = |||x \log(y + r) + y \log(x + r) - z \arctan\left(\frac{xy}{zr}\right)|_{x_1}^{x_2}|_{y_1}^{y_2}|_{z_1}^{z_2} \tag{2}$$

Similarly, equation (24) in Nagy et al. (2000) provides the topographic effect for gravity gradient:

$$\delta dg_{TE} = ||| - \arctan\left(\frac{xy}{zr}\right)|_{x_1}^{x_2}|_{y_1}^{y_2}|_{z_1}^{z_2} \tag{3}$$

where $r = \sqrt{x^2 + y^2 + z^2}$, except for a negative sign.

A filtered DEM is created by Fourier transform and low-pass filter mask. This filtered DEM is used to calculate long wavelength topographic effect, which can be approximated using a Bouguer slab when using a very long wavelength GGM:

$$\delta g_{TE}^{long} = 0.0419\rho(DEM)^{long} \tag{4}$$

where $(\mathrm{DEM})^{\mathrm{long}}$ is the digital elevation model filtered at long wavelengths, and ρ is the density, typically taken as the standard value of 2670 kg/m^3. We refer to this term as the long-wavelength Bouguer slab.

2.2 Least Squares Collocation

Covariance calculation is a fundamental component of Least Squares Collocation (LSC). The following section outlines the four steps implemented in the software to construct covariance matrices for LSC.

2.2.1 Building Covariance Matrices

Compute Spherical Empirical Covariance

The first step is to compute an empirical covariance function for scalar quantities on a sphere by taking the mean of the product sums of sample scalar values (here Δg) (Darbeheshti and Featherstone 2009):

$$cov(\psi_j) = \frac{1}{N_j}\sum_{k,i}^{N_j} \Delta g(\phi_k, \lambda_k)\Delta g(\phi_i, \lambda_i) \tag{5}$$

where ψ_j is the Haversine distance and N_j is the number of pairs for each interval. The Haversine formula is used to calculate the great-circle distance between two points on a sphere given their longitudes and latitudes. The formula is as follows:

$$\psi = 2r \arcsin\left(\sqrt{\sin^2\left(\frac{\phi_2 - \phi_1}{2}\right) + \cos(\phi_1)\cos(\phi_2)\sin^2\left(\frac{\lambda_2 - \lambda_1}{2}\right)}\right) \tag{6}$$

where:

- ψ is the distance between the two points along a great circle of the sphere.
- r is the radius of the sphere.
- ϕ_1 and ϕ_2 are the latitude of point 1 and latitude of point 2.
- λ_1 and λ_2 are the longitude of point 1 and longitude of point 2.

The formula assumes a perfect sphere for Earth, which is not entirely accurate but provides a close approximation of distances. Here, r is assumed to be 1.

Fit Empirical Covariance

The second step involves fitting the spherical empirical covariances of gravity anomalies to the covariance function of gravity anomalies. The anomaly degree variance model from equation (68) in Tscherning and Rapp (1974) is used:

$$\sigma_n(\Delta g, \Delta g) = \frac{A(n-1)}{(n-2)(n+B)} \quad (7)$$

For the covariance function of gravity anomalies:

$$cov\left(\Delta g_P, \Delta g_Q\right) = \sum_{n=2}^{n_{\max}} \sigma_n(\Delta g, \Delta g) s^{(n+2)} P_n(\cos\psi) \quad (8)$$

where $s = \frac{R_B^2}{r_P r_Q}$, ψ is the spherical distance in radians, and P_n is the Legendre polynomial. r_P and r_Q are the geocentric radii to the points P and Q, which are separated by ψ. R_B is the radius of the Bjerhammar sphere. The mathematical formulation of the radius of the Bjerhammar sphere is as follows:

$$R_B = \frac{ab}{\sqrt{(a\sin\varphi)^2 + (b\cos\varphi)^2}} \quad (9)$$

where a is the major axis of the reference ellipsoid, b is the minor axis of the reference ellipsoid, and φ is the latitude in radians. The coefficients A and B are estimated using values of $s = 1$ and $n_{\max} = 250$, which were found suitable for our dataset. Note that these parameters may vary depending on the data.

Precompute Covariance Function

This step calculates the spherical covariance of six functionals of the gravity field. The autocovariance of gravity anomaly and potential, as well as the cross-covariance between gravity anomaly and potential, are computed using equations from Tscherning and Rapp (1974):

$$cov\left(\Delta g_P, \Delta g_Q\right) = \sum_{n=2}^{n_{\max}} \sigma_n(\Delta g, \Delta g) s^{(n+2)} P_n(\cos\psi) \quad (10)$$

$$cov\left(T_P, T_Q\right) = R_B^2 \sum_{n=2}^{n_{\max}} \frac{\sigma_n(\Delta g, \Delta g)}{(n-1)^2} s^{(n+1)} P_n(\cos\psi) \quad (11)$$

$$cov\left(\Delta g_P, T_Q\right) = r_Q \sum_{n=2}^{n_{\max}} \frac{\sigma_n(\Delta g, \Delta g)}{(n-1)} s^{(n+2)} P_n(\cos\psi) \quad (12)$$

Auto-covariance of the vertical gradient of gravity anomaly (gravity gradient) is calculated using equation (2.69) of Jekeli (1978):

$$cov\left(dg_P, dg_Q\right) = \frac{\partial}{\partial r_P}\left(\frac{\partial}{\partial r_Q} cov\left(\Delta g_P, \Delta g_Q\right)\right) = \frac{1}{R_B^2} \sum_{n=2}^{n_{\max}} \sigma_n(\Delta g, \Delta g)(n+2)^2 s^{(n+3)} P_n(\cos\psi) \quad (13)$$

The cross covariance of the gravity gradient and gravity anomaly is calculated using equation (3.66) of Zhu (2007):

$$cov\left(dg_P, \Delta g_Q\right) = \frac{\partial}{\partial r_P}\left(\frac{\partial}{\partial r_P} cov\left(T_P, \Delta g_Q\right)\right) = \frac{1}{r_P} \sum_{n=2}^{n_{\max}} \frac{\sigma_n(\Delta g, \Delta g)(n+1)(n+2)}{(n-1)} s^{(n+2)} P_n(\cos\psi) \quad (14)$$

The cross-covariance of the gravity gradient and potential is calculated using:

$$cov\left(dg_P, T_Q\right) = \frac{\partial}{\partial r_P}\left(\frac{\partial}{\partial r_P} cov\left(T_P, T_Q\right)\right) = \frac{R_B^2}{{r_P}^2} \sum_{n=2}^{n_{\max}} \frac{\sigma_n(\Delta g, \Delta g)(n+1)^2}{(n-1)^2} s^{(n+1)} P_n(\cos\psi) \quad (15)$$

Interpolate Covariance Function

Finally, we construct the auto- and cross-covariance matrices required for Least Squares Collocation (LSC) between the observation points (Q) and prediction points (P) using MATLAB's `griddedInterpolant`. The observation points correspond to gravity measurement locations, while the prediction points align with the DEM grid. The goal is to interpolate a regular grid of height anomalies from the LSC results. All necessary covariance matrices for performing LSC are shown in the following section.

2.2.2 Solving LSC

LSC is performed twice using gravity and gravity gradient data. In the following equations, we reserve the superscript for data points, meaning "obs" at the observation point and DEM meaning DEM points, and the subscript for the type of the data, for example, "res" meaning residuals.

$$\begin{aligned} \Delta g_{\text{res}}^{\text{obs}} &= \Delta g^{\text{obs}} - \Delta g_{\text{GGM}} - \delta g_{\text{TE}} \\ dg_{\text{res}}^{\text{obs}} &= dg^{\text{obs}} - dg_{\text{GGM}} - \delta dg_{\text{TE}} \end{aligned} \quad (16)$$

Then, in combination with Eq. (1), we can present the extended form for gravity anomalies:

$$\Delta g_{\text{res}}^{\text{obs}} = \Delta g^{\text{obs}} - \Delta g_{\text{GGM}} - \delta g_{\text{TE}}^{\text{full}} + \delta g_{\text{TE}}^{\text{long}} \quad (17)$$

First, we predict residual gravity anomalies at DEM points using residual gravity anomalies and residual gravity gradients at observation points.

$$\Delta g_{res}^{DEM} = \begin{bmatrix} C_{\Delta g,dg}^{DEM,obs} & C_{\Delta g,\Delta g}^{DEM,obs} \end{bmatrix} \begin{bmatrix} C_{dg,dg}^{obs,obs} + E_{dg,dg}^{obs,obs} & C_{\Delta g,dg}^{obs,obs} \\ C_{dg,\Delta g}^{obs,obs} & C_{\Delta g,\Delta g}^{obs,obs} + E_{\Delta g,\Delta g}^{obs,obs} \end{bmatrix}^{-1} \begin{bmatrix} dg_{res}^{obs} \\ \Delta g_{res}^{obs} \end{bmatrix} \quad (18)$$

$E_{\Delta g,\Delta g}^{obs,obs}$ and $E_{dg,dg}^{obs,obs}$ come from formal error of preprocessing of gravity anomaly and gravity gradient data; and to compute error variance covariance matrix for predicted residual gravity anomalies:

$$\Sigma_{\Delta g,\Delta g}^{DEM,DEM} = C_{\Delta g,\Delta g}^{DEM,DEM} - \begin{bmatrix} C_{\Delta g,dg}^{DEM,obs} & C_{\Delta g,\Delta g}^{DEM,obs} \end{bmatrix} \begin{bmatrix} C_{dg,dg}^{obs,obs} + E_{dg,dg}^{obs,obs} & C_{\Delta g,dg}^{obs,obs} \\ C_{dg,\Delta g}^{obs,obs} & C_{\Delta g,\Delta g}^{obs,obs} + E_{\Delta g,\Delta g}^{obs,obs} \end{bmatrix}^{-1} \begin{bmatrix} C_{\Delta g,dg}^{obs,DEM} \\ C_{\Delta g,\Delta g}^{obs,DEM} \end{bmatrix} \tag{19}$$

Then

$$\Delta g_{Bouguer}^{DEM} = \Delta g_{res}^{DEM} - \delta g_{TE}^{long} \tag{20}$$

We first grid the Bouguer anomaly in a remove compute restore scheme. A long wavelength Bouguer anomaly (composed of the satellite GGM free air corrected for long wavelength terrain) is subtracted from pointwise Bouguer anomaly data. These residuals are interpolated onto the topographic surface using LSC, and the long wavelength Bouguer anomaly (satellite GGM free air corrected for long wavelength topography) is restored. Then Faye anomalies are computed by the equation (2.14) from McCubbine (2016).

$$\Delta g_{Faye}^{DEM} = \Delta g_{Bouguer}^{DEM} + 0.0419\rho DEM \tag{21}$$

Note that the second term in Eq. (21) uses the full-resolution DEM, in contrast to Eq. (4). This process ultimately returns gridded Faye anomalies at the DEM points at full resolution and avoids any topography aliasing which might be present in using the pointwise free air anomalies directly for geoid computations.

Finally, to predict the height anomaly at DEM points:

$$\zeta_{res}^{DEM} = C_{T,\Delta g_{Faye}}^{DEM,DEM} \left(C_{\Delta g_{Faye},\Delta g_{Faye}}^{DEM,DEM} \right)^{-1} \Delta g_{Faye}^{DEM} / \gamma \tag{22}$$

where γ is the normal gravity; and to compute error variance covariance matrix for predicted height anomaly:

$$\Sigma_{\zeta,\zeta}^{DEM,DEM} = C_{T,T}^{DEM,DEM} - \begin{bmatrix} C_{T,dg}^{DEM,obs} & C_{T,\Delta g}^{DEM,obs} \end{bmatrix} \begin{bmatrix} C_{dg,dg}^{obs,obs} + E_{dg,dg}^{obs,obs} & C_{\Delta g,dg}^{obs,obs} \\ C_{dg,\Delta g}^{obs,obs} & C_{\Delta g,\Delta g}^{obs,obs} + E_{\Delta g,\Delta g}^{obs,obs} \end{bmatrix}^{-1} \begin{bmatrix} C_{T,dg}^{obs,DEM} \\ C_{T,\Delta g}^{obs,DEM} \end{bmatrix} \tag{23}$$

A key numerical challenge in solving Least Squares Collocation (LSC) is matrix inversion, as shown in (18). To avoid inverting large matrices, stepwise LSC is commonly used. In this study, however, we employ a tilewise LSC approach. This method partitions the computation area into tiles, each with a size of $1° \times 1°$. Height anomaly calculations are performed independently within each tile, followed by the application of a Gaussian filter to blend the tiles into a continuous height anomaly grid.

The Gaussian filter, applied iteratively, functions as a moving average filter that smooths the input grid data. Repeated applications help reduce discontinuities at tile boundaries and ensure a seamless transition between adjacent tiles. A circular Gaussian-like filter kernel is constructed using a specified filter size and radius, with values empirically set to $15'$ and $10'$, respectively. These values were determined through experimentation for tiles of size $1° \times 1°$. The kernel is applied iteratively using MATLAB's two-dimensional convolution function (`conv2`) a total of six times to progressively smooth the input grid and suppress edge effects. Intermediate multiplications with the original input grid reinforce the spatial structure during filtering. The resulting weighted filter attenuates contributions near tile edges and enhances continuity across boundaries, ensuring seamless integration of individual tile solutions into a continuous grid surface.

To further minimise edge effects, a buffer zone is introduced between tiles. In our calculations, we set the buffer width to $0.5°$, which we found to be optimal for computing gravity anomalies and height anomalies at a spatial resolution of $1' \times 1'$. It is worth noting that our input DEM also has a spatial resolution of $1' \times 1'$.

3 Numerical Result

We computed the gravimetric height anomaly for New South Wales, Australia, using the methodology outlined earlier and input data in Table 1. The study incorporated four types of gravity data (Fig. 1): Terrestrial gravity data, gravity observations from satellite altimetry, airborne gravimetry and airborne gradiometry.

Table 1 Gravity data sources

Data type	Source
Terrestrial gravity	Geoscience Australia's national gravity database
Airborne gravity	Gravity anomaly and gravity gradient
Satellite altimetry gravity	One-minute arc DNSC08 (Andersen et al. 2010)
Digital elevation model	One-minute arc (Gallant et al. 2011)
Global gravity model	GOCE (GO_CONS_GCF_2_DIR_R6 (Bruinsma et al. 2014))

Topographic effects were computed using the prism method (Fig. 2), accounting for terrain-induced variations in gravity and gradients.

Figure 3 illustrates the least squares collocation (LSC) process. Faye residual gravity anomalies were computed for each tile, empirical covariances were estimated, and a covariance function was fitted before performing LSC to grid the anomalies and derive the residual height anomaly.

A total of 660 tiles were computed, each taking approximately 3 minutes on a high-performance laptop, with matrix sizes ranging from 3000×3000 to $16{,}000 \times 16{,}000$, leading to a total computation time of up to 29 hours. The final height anomaly was obtained by merging the tiles, and height anomaly sigma errors were evaluated with and without airborne gravity data to assess its impact on uncertainty (Fig. 4). Table 2 shows that the mean height anomaly error reduced by 33.33% when using airborne data.

Table 2 shows the formal LSC estimation error, which is typically optimistic. The plots below compare the difference between the gravimetric height anomaly and the geometric height anomaly from 2017 GPS/levelling data, with and without the use of airborne gravity data (Fig. 5). This is an initial comparison; further analysis is underway, as recent campaigns aim to assess the impact of airborne gravimetry. Notably, the differences are most evident along the coast, where terrestrial gravity data was sparse (see white gaps in Fig. 1).

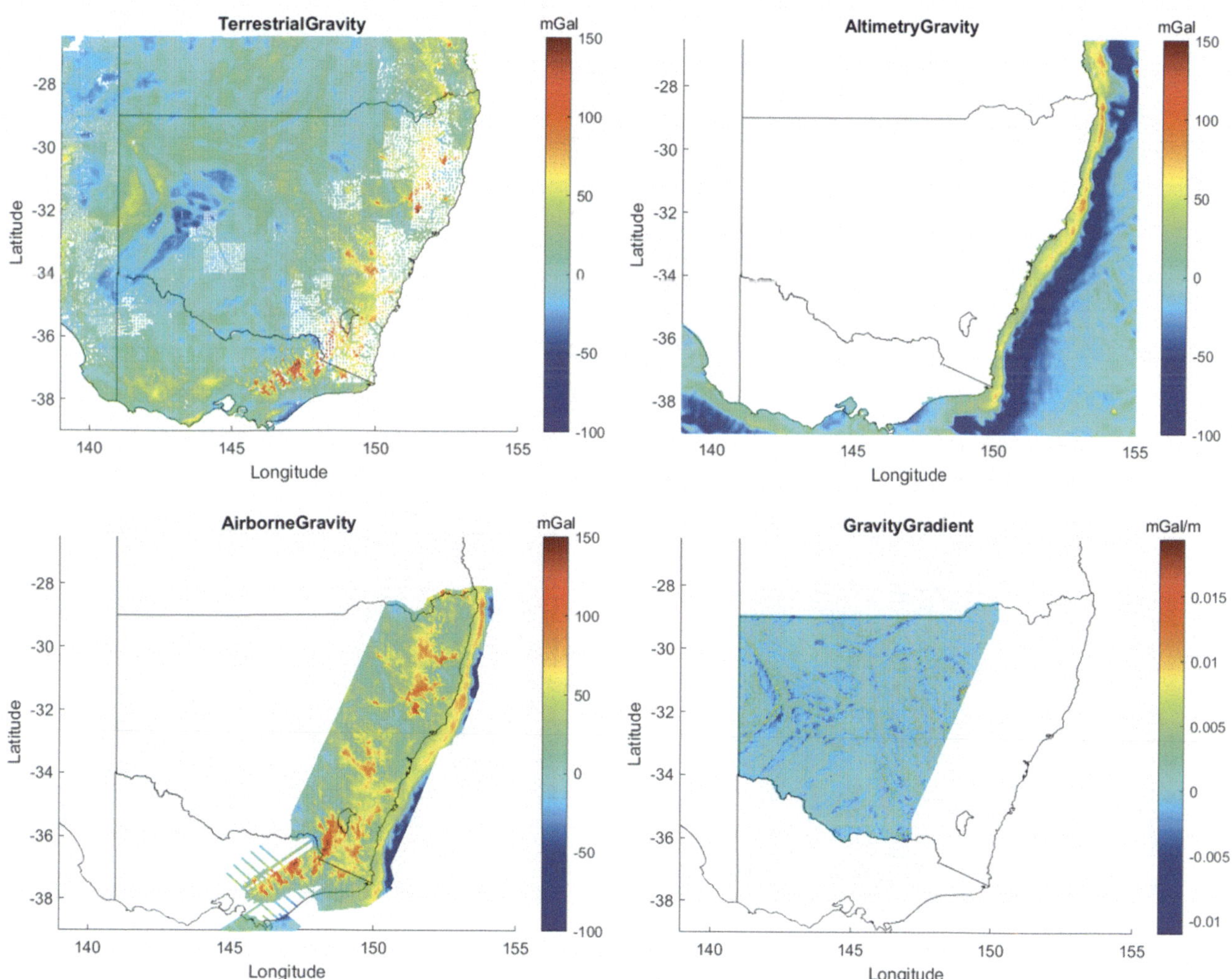

Fig. 1 Gravity data sources: (**a**) terrestrial, (**b**) satellite altimetry, (**c**) airborne gravimetry, (**d**) airborne gradiometry

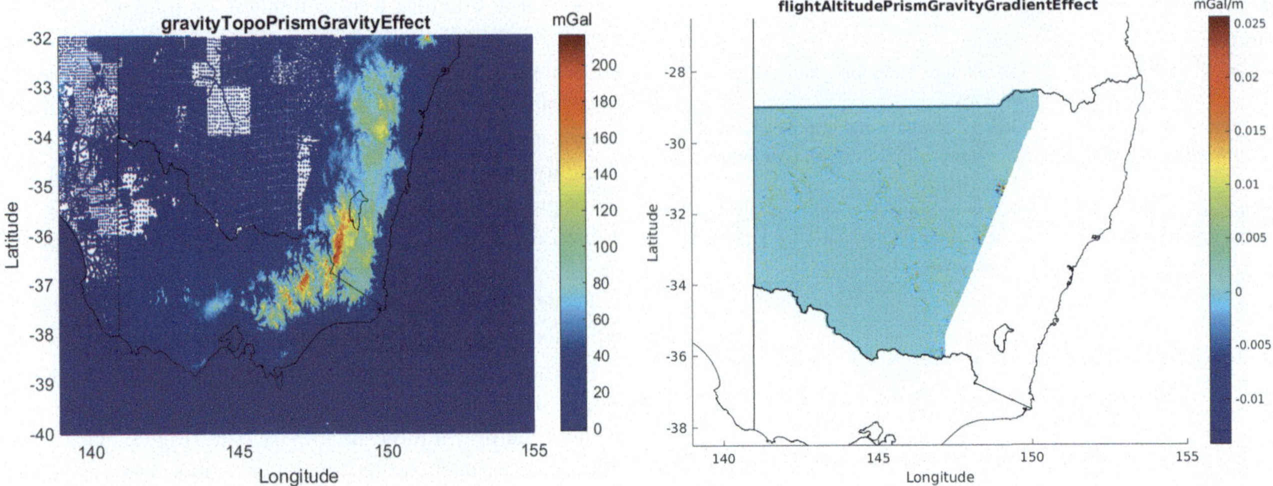

Fig. 2 Topographic effects on gravity and gravity gradient data

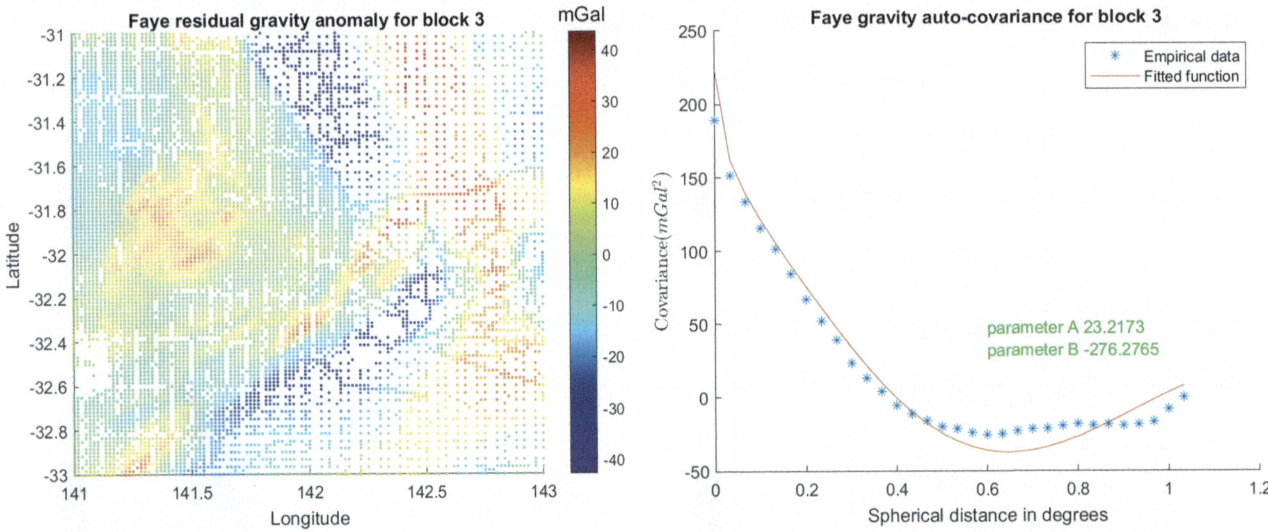

Fig. 3 LSC procedure: Faye residuals and covariance functions. The fitted function is multiplied by parameter A

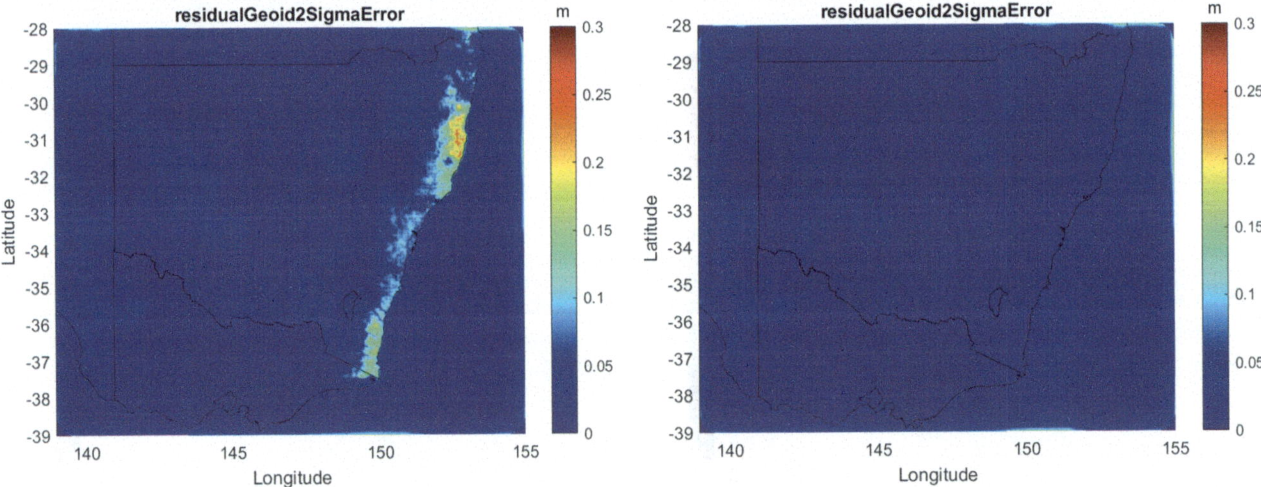

Fig. 4 Height anomaly 2-sigma error: (left) without and (right) with airborne gravity data

Table 2 Height anomaly sigma error with and without airborne data

Height anomaly Sigma Error (m)	Min	Max	Mean	Std
Without airborne data	0.0026	0.1148	0.0114	0.0106
With airborne data	0.0009	0.0864	0.0076	0.0057

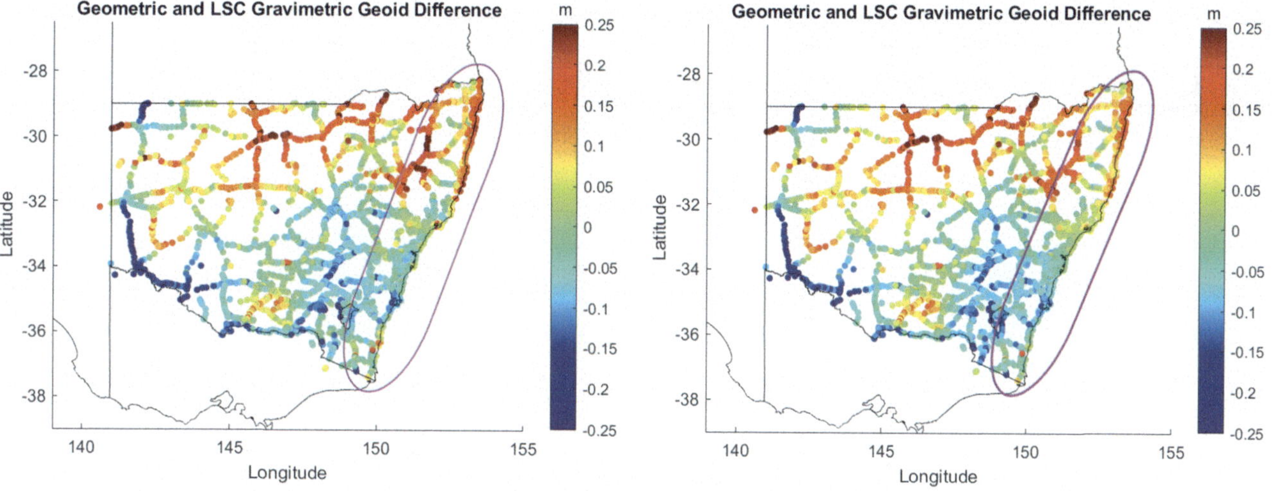

Fig. 5 Difference between gravimetric and geometric height anomalies (left) without airborne gravity data; (right) with airborne gravity data

4 Conclusion

We developed an open-source software for gravity data analysis and height anomaly computation, available for download https://github.com/GeoscienceAustralia/analysis-ready-gravity-data-workflow. This software uses Least Squares Collocation (LSC) for height anomaly calculation, incorporating various gravity functionals such as gravity anomalies, gravity gradients, and airborne gravity data.

We applied this software in New South Wales to compute the height anomaly. Our tests, using airborne gravity data–including airborne gravimetry and airborne gradiometry–showed that the LSC formal uncertainty improved when incorporating airborne data.

Acknowledgements This work was supported by AuScope and the Australian Government through the National Collaborative Research Infrastructure Strategy. We also acknowledge the New South Wales State Government Spatial Services for funding the airborne data collection. This paper is published with the permission of the CEO, Geoscience Australia.

References

Andersen OB, Knudsen P, Berry PAM (2010) The DNSC08GRA global marine gravity field from double retracked satellite altimetry. J Geodesy 84(3):191–199

Bruinsma S, Förste C, Abrikosov O, Lemoine J, Marty J, Mulet S, Rio M, Bonvalot S (2014) ESA's satellite-only gravity field model via the direct approach based on all GOCE data. Geophys Res Lett 41(21):7508–7514

Darbeheshti N, Featherstone WE (2009) Non-stationary covariance function modelling in 2D least-squares collocation. J Geodesy 83:495–508

Featherstone WE, Kirby JF (2000) The reduction of aliasing in gravity anomalies and geoid heights using digital terrain data. Geophys J Int 141(1):204–212

Forsberg R, Tscherning CC (2014) An overview manual for the GRAVSOFT geodetic gravity field modelling programs, 3rd edn. Denmark.

Gallant JC, Dowling TI, Read AM, Wilson N, Tickle P, Inskeep C (2011) 1 second SRTM derived digital elevation models user guide. Geoscience Australia

Jekeli C (1978) An investigation of two models for the degree variances of global covariance functions. Ohio State University, Research Foundation, Report No. 275

McCubbine J (2016) Airborne gravity across New Zealand - For an improved vertical datum. Open Access Te Herenga Waka-Victoria University of Wellington, Thesis

Nagy D, Papp G, Benedek J (2000) The gravitational potential and its derivatives for the prism. J Geodesy 74:552–560

Tscherning CC, Rapp RH (1974) Closed covariance expressions for gravity anomalies, geoid undulations, and deflections of the vertical implied by anomaly degree variance models, Report 208, Department of Geodetic Science, The Ohio State University, Columbus, USA

Zhu L (2007) Gravity gradient modeling with gravity and DEM, Report 483, Geodetic science and surveying, The Ohio State University, Columbus

Reviewers

Axel Rülke
Carlo De Gaetani
Christian Gerlach
Claudia N. Tocho
Daniela Carrion
Derek van Westrum
Dimitrios Piretzidis
Dimitrios Tsoulis
Felix Johann
Gabor Papp
Gabriel do Nascimento Guimaraes
George Vergos
Hussein Abd-Elmotaal
Ismael Foroughi
Jianliang Huang
Jonas ågren
Kevin Ahlgren
Laura Sánchez
Lorenzo Rossi
Ludger Timmen
Manuel Schilling
Mirko Reguzzoni
Öykü Koç
Pavel Novàk
Petr Holota
Przemyslaw Dykowski
Riccardo Barzaghi
Srinivas Bettadpur
Thomas Gruber
Thorben Döhne
Tim Enzlberger Jensen
Vadim Vyazmin
Vassilios Andritsa
Vassilios Andritsanos
Vassilios Grigoriadis
Walyeldeen Godah

J. T. Freymueller, L. Sànchez (eds.), *International Symposium on Gravity, Geoid and Height Systems 2024 (GGHS2024)*, International Association of Geodesy Symposia 158, https://doi.org/10.1007/978-3-032-22865-9

Author Index

J. T. Freymueller, L. Sànchez (eds.), *International Symposium on Gravity, Geoid and Height Systems 2024 (GGHS2024)*,
International Association of Geodesy Symposia 158, https://doi.org/10.1007/978-3-032-22865-9

Subject Index

J. T. Freymueller, L. Sànchez (eds.), *International Symposium on Gravity, Geoid and Height Systems 2024 (GGHS2024)*, International Association of Geodesy Symposia 158, https://doi.org/10.1007/978-3-032-22865-9

The manufacturer's authorised representative in the EU is Springer Nature Customer Service Centre GmbH, Europaplatz 3, 69115 Heidelberg, Germany. If you have any concerns regarding our products, please contact ProductSafety@springernature.com

Printed and bound by CPI Group (UK) Ltd, Croydon, CR0 4YY
07/07/2026
02160935-0001